AA003215

Advanced Fossil Energy Utilization 2011

Topical Conference at the 2011 AIChE Spring Meeting and 7th Global Congress on Process Safety

Chicago, Illinois, USA
13 – 17 March 2011

ISBN: 978-1-61782-890-4

Printed from e-media with permission by:

Curran Associates, Inc.
57 Morehouse Lane
Red Hook, NY 12571

Some format issues inherent in the e-media version may also appear in this print version.

Printed by Curran Associates, Inc. (2011)

For permission requests, please contact AIChE
at the address below.

AIChE
3 Park Avenue
New York, NY 10016-5991

Phone: (203) 702-7660
Fax: (203) 775-5177

www.aiche.org

Additional copies of this publication are available from:

Curran Associates, Inc.
57 Morehouse Lane
Red Hook, NY 12571 USA
Phone: 845-758-0400
Fax: 845-758-2634
Email: curran@proceedings.com
Web: www.proceedings.com

TABLE OF CONTENTS

Reactivation of CaO-based Sorbents for CO_2 Capture: Some Aspects of the Carbonation of $Ca(OH)_2$

*John Blamey[1], Scott Champagne[2], Dennis Lu[2], Paul Fennell[1], E.J. Anthony[2]**

[1]Department of Chemical Engineering, Imperial College, South Kensington, London, SW7 2AZ, UK
[2]CanmetENERGY, 1 Haanel Dr., Ottawa, ON, K1A 1M1, Canada
*Corresponding author. Tel: +1 613 996-2868; fax: +1 613 992-9335; e-mail: banthony@nrcan.gc.ca

Abstract

Calcium-looping is an emerging technology for CO_2 capture, which uses a regenerable CaO-based sorbent. Here, a novel hydration reactivation strategy proposed by Industrial Research Limited of New Zealand is investigated: they have shown that the temperature of dehydration can be increased in a CO_2 atmosphere, resulting in higher conversions to $CaCO_3$ and a more mechanically stable sorbent. The mechanism proposed by Industrial Research Limited is of free water being formed within $Ca(OH)_2$ prior to dehydration before interacting with CO_2 leading to proton injection into the lattice further delaying dehydration. In contrast, we propose a pore-occlusion mechanism.

To investigate this, $Ca(OH)_2$ powder and pellets have been heated in various different atmospheres in a TGA with a mass spectrometer performing online gas analysis of the off-gas. Under a nitrogen atmosphere, both powder and pellets decompose directly to CaO, with decomposition starting at around 400 °C and resulting in a single peak of H_2O being observed in the off-gas. Under a carbon dioxide atmosphere, carbonation of both pellets and powder appears to occur slightly earlier at around 350 °C and at the same time steam is observed in the off-gas. However: powder proceeded to ~ 90% carbonation by ~ 600 °C with a single peak of H_2O being observed on the mass spectrometer; whereas, pellets approached to a stable mass at ~ 600 °C before a second release of H_2O, and an increase in mass to ~ 90% carbonation. The carbonation behaviour of hydrated calcined dolomite ($CaCO_3.MgCO_3$) has also been investigated. In this case, a single peak of steam associated with dehydration was observed upon heating in a carbon dioxide atmosphere, whereas two peaks associated with dehydration were observed in a nitrogen atmosphere.

These findings are consistent with an occluded pore model, in which an impermeable carbonate layer prevents H_2O diffusion until occlusions rupture (the molar volumes of $CaCO_3$ and $Ca(OH)_2$ are 36.9 and 33.7 cm^3/mol respectively). The carbonate layer has a critical thickness that is not reached in the case of the $Ca(OH)_2$ powder, but is in the case of the $Ca(OH)_2$ pellets. The network of MgO in the dolomitic particles results in CaO grains that are not large enough to accommodate the impermeable carbonate layer.

1. Introduction

Calcium-looping is an emerging technology applicable to both flue and fuel gas streams for CO_2 capture [1]. A typical proposed process is described as follows: CaO-based sorbent is repeatedly cycled between two fluidised beds; in the lower temperature carbonator CaO-based sorbent strips CO_2 from a flue or fuel gas via the exothermic formation of $CaCO_3$ (forwards of reaction (I)); in the calciner, operating at a higher temperature, $CaCO_3$ decomposes to provide a pure stream of CO_2 (reverse of reaction (I)), suitable for storage, and regenerates CaO. There are several advantages to this system including the ability to reclaim high-grade heat from the carbonation reaction and the use of cheap non-toxic sorbents, which have the potential for application in the cement industry after use [2].

$$CaO_{(s)} + CO_{2(g)} = CaCO_{3(s)} \quad (I)$$

$$Ca(OH)_{2(s)} + CO_{2(g)} = CaCO_{3(s)} + H_2O_{(g)} \quad (II)$$

Unfortunately, the sorbent reduces in reactivity towards CO_2 over repeated cycles of reaction. This can occur for a number of reasons including loss of reactive surface area and porosity through sintering, competing reactions with sulphurous molecules and mass loss through attrition [1]. One method of resolving the reactivity loss is by periodic hydration of the sorbent, which has the effect of increasing the reactivity of the sorbent for CO_2 by increasing its porosity.

A reactivation strategy using a hydration method has been proposed by Industrial Research Limited of New Zealand [3]. They have shown that the temperature of dehydration can be increased in a CO_2 atmosphere to above the temperature at which the $Ca(OH)_2$ would typically decompose at a reasonable rate, resulting in higher conversions to $CaCO_3$ and a more mechanically stable sorbent. In lab-scale fluidised bed tests, the best performing reactivation procedure involved: (i) hydrating calcined sorbent in 15 % CO_2, balance air, starting at 270 °C, increasing to 350 °C because of the exothermic reaction; (ii) switching to a pure CO_2 environment; (iii) heating to 520 °C and annealing (under CO_2 still); and (iv) heated to 620 °C and holding for 20 min. Under this regime, an impressive average activity of 60 % was obtained over 27 cycles with reactivation every third cycle, while only creating 3.25 % of fines.

The proposed mechanism to explain the improvement in mechanical stability is annealing while the sorbent is held at 520 °C. The authors have also suggested that the ability of $Ca(OH)_2$ to exist above its decomposition temperature, labelled as "Superheated Dehydration", is a result of $Ca(OH)_2$ becoming more stable in the presence of CO_2, which prevents significant conversion to $CaCO_3$ through reaction (II). The mechanism proposed for this is that the superheated state may result from the absorption of H^+ ions generated by water and CO_2.

Here, we have investigated this process in a TGA with a mass spectrometer downstream. The mass spectrometer detects any H_2O emitted during the process allowing investigation of the emission of steam during the carbonation – and "superheated dehydration" – of $Ca(OH)_2$. Our particular interest is to understand the role of direct carbonation (reaction (II)), which has been shown to occur at reasonable rates at the temperatures of interest [4].

2. Experimental

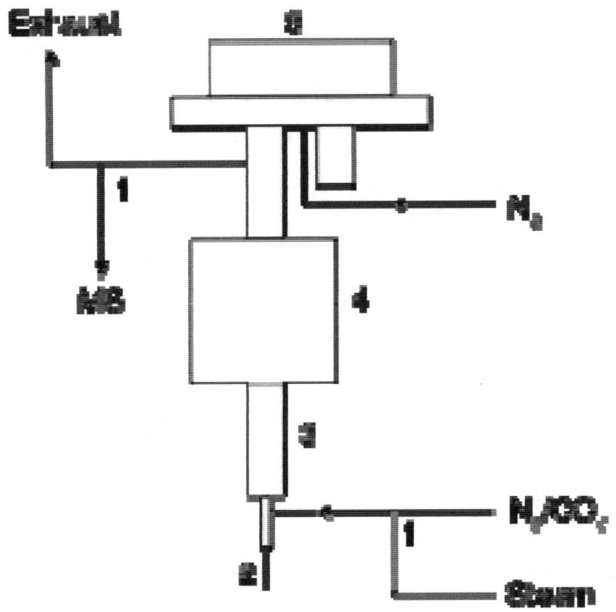

Experiments were performed in a Cahn 1100 Pressurized Thermogravimetric Analyzer (PTGA, Figure 1). The PTGA operates at temperatures up to 950 °C with pressures up to 2.0 MPa, though this work was performed at atmospheric pressure. A platinum pan, which contains the sample, hangs centrally in an inconel 600 alloy reactor tube (inner diameter of 25 mm), which is externally heated by a tube furnace. The temperature was controlled using a K-type thermocouple within the furnace and the sample temperature was monitored by a K-type thermocouple below the sample pan. Digital mass flow controllers regulate the gas flow-rate into the PTGA, both to the sample and to the balance, while a syringe pump fed a known flow-rate of water into a steam generation system to the PTGA when necessary. A heated sample line (20 mL/min) from the outlet of the PTGA fed to a Gastrace-E Portable Mass Spectometer (MS, Aston Analytical), which was used to measure the concentrations of CO_2, N_2 and H_2O, calibrated using standard calibration gases and by reference to the known flow-rate of H_2O into the system.

Six different types of sample were used in this work: (a) $Ca(OH)_2$ powder (Graymont, 92-99 % purity), < 30 µm; (b) $Ca(OH)_2$ pellets [5] made with a Glatt Ait TMG granulator using $Ca(OH)_2$ powder, of size fraction 400-800 µm (c) $Ca(OH)_2$ pellets of size fraction 1000-1410 µm; (d) hydrated calcined BC (British Columbia) dolomite, produced *ex situ* (batch calcination of 2 g in a muffle furnace at 900 °C under air for 15 mins, before hydration in a pressure bomb (Parr 4522M Pressure Reactor) at 15 bar for 15 mins), 400-800 µm; (e) hydrated calcined Havelock limestone, produced *in situ* to allow different calcination environments to be studied, size fraction 500-710 µm; (f) hydrated calcined Havelock, produced *in situ*, size fraction 1000-1410 µm. The purities of the samples, as determined by x-ray fluorescence (Bruker AXS S4 Explorer), are given in Table 1.

Figure 1. Simplified schematic of the pressurised TGA: (1) trace-heated lines; (2) thermocouple; (3) reaction tube; (4) furnace; (5) balance

Table 1. Purities of carbonate equivalents of Ca and Mg for original samples

	Powder	Havelock	Dolomite
F_{CaCO3}	0.99	0.96	0.55
F_{MgCO3}	0.00	0.02	0.41

Table 2. Experimental Matrix

	(1) N_2	(2) CO_2	(3) 15% CO_2
(a) $Ca(OH)_2$ powder	POW-N2	POW-CO2	n/a
(b) $Ca(OH)_2$ pellets	PEL-N2	PEL-CO2	PEL-15CO2
(c) $Ca(OH)_2$ pellets, large	n/a	PEL-CO2-LG	n/a
(d) Hydrated calcined dolomite	DOL-N2	DOL-CO2	n/a
(e) Hydrated calcined Havelock	n/a	-LS-CO2 -LS-CO2	n/a
(f) Hydrated calcined Havelock, large	n/a	-LS-CO2-LG	n/a

Note: denotes calcination under N_2, denotes calcination under CO_2

A set of simple experiments were performed whereby the sample was heated at full power (mean 35-40 °C/min) from room temperature to 900 °C under different atmospheres: (1) under 100 % N_2; (2) under 100% CO_2; and (3) under 15% CO_2, balance N_2. The only change to this procedure was in the cases of the hydrated calcined Havelock samples, which were subjected to prior calcination and hydration, whereby heating started at 250 °C under the experimental conditions rather than at room temperature (because of the long cooling time of the system). The calcination conditions for the Havelock samples were: ( ) 15 minutes at 850 °C under 100 % N_2; ( ) 15 minutes at 915 °C under 100 % CO_2. The hydration conditions for the calcined Havelock samples were 10 minutes at 250 °C under 100% H_2O. For each experiment, the total flow-rate to the sample in the PTGA was 100 mL/min under standard conditions and the pressure within the PTGA was atmospheric; the total flow-rate of nitrogen to the PTGA balance was also 100 mL/min. For each experiment, approximately 40 mg of sample was used. An experimental matrix is shown in Table 2.

3. Results

Figures 1(a-d) show plots of two sets from each experiment: (i) the mass change in g per g of *calcined* sorbent, μ, as calculated from PTGA data using Eq. (1); and (ii) the steam concentration in the outlet gas, X_{H2O}, as calculated from the currents obtained in the MS using Eq (2); both plotted as a function of sample temperature.

$$\Delta\mu = \frac{m_t}{m_{calcine}} - 1 \qquad \text{Eq. (1)}$$

where m_t is the mass at time *t* and $m_{calcine}$ is the calcined mass of the sample. μ gives an indication as to the initial extent of hydration of the sample and its reactivity towards CO_2. For example, one mole of pure $Ca(OH)_2$ would give have an initial $\mu = (74.09/56.08) - 1 = 0.32$ (note that this value would be increased if the sample was slightly carbonated or contained moisture) and if this produced one mole of pure $CaCO_3$, this would give a final $\mu = (100.09/56.08) - 1 = 0.78$.

$$X_{H_2O} = \frac{I_{18}/F_{H_2O}}{I_{18}/F_{H_2O} + I_{14}/F_{N_2} + I_{44}/F_{CO_2}} \quad \text{Eq. (2)}$$

where X_i is the mole fraction of species i in the outlet gas, I_n is the current detected in the MS for mass number n and F_i is the calibration factor for species i. Mass numbers of 18 and 44 were chosen for H_2O and CO_2 because they correspond to their largest peaks; the mass number of 14 was chosen for N_2 because the secondary peak of CO_2 (i.e., the peak relating to CO) interferes with its largest peak at 28. There was some background noise, likely a result of condensation in the lines and reactor that evaporated as the equipment was heated, that was corrected for by a third order polynomial through data before and after the main H_2O emission peak(s). Figure 1 presents the data for the steam concentration post-treatment. As a result of the background noise, mass balances during experiments frequently did not close, however the trends presented in this work are believed to be valid and there scope for future work to resolve this problem. The steam peaks appear to have occurred at slightly higher temperatures than the mass changes; this is because of the time taken for steam to pass from the sample to the mass spectrometer.

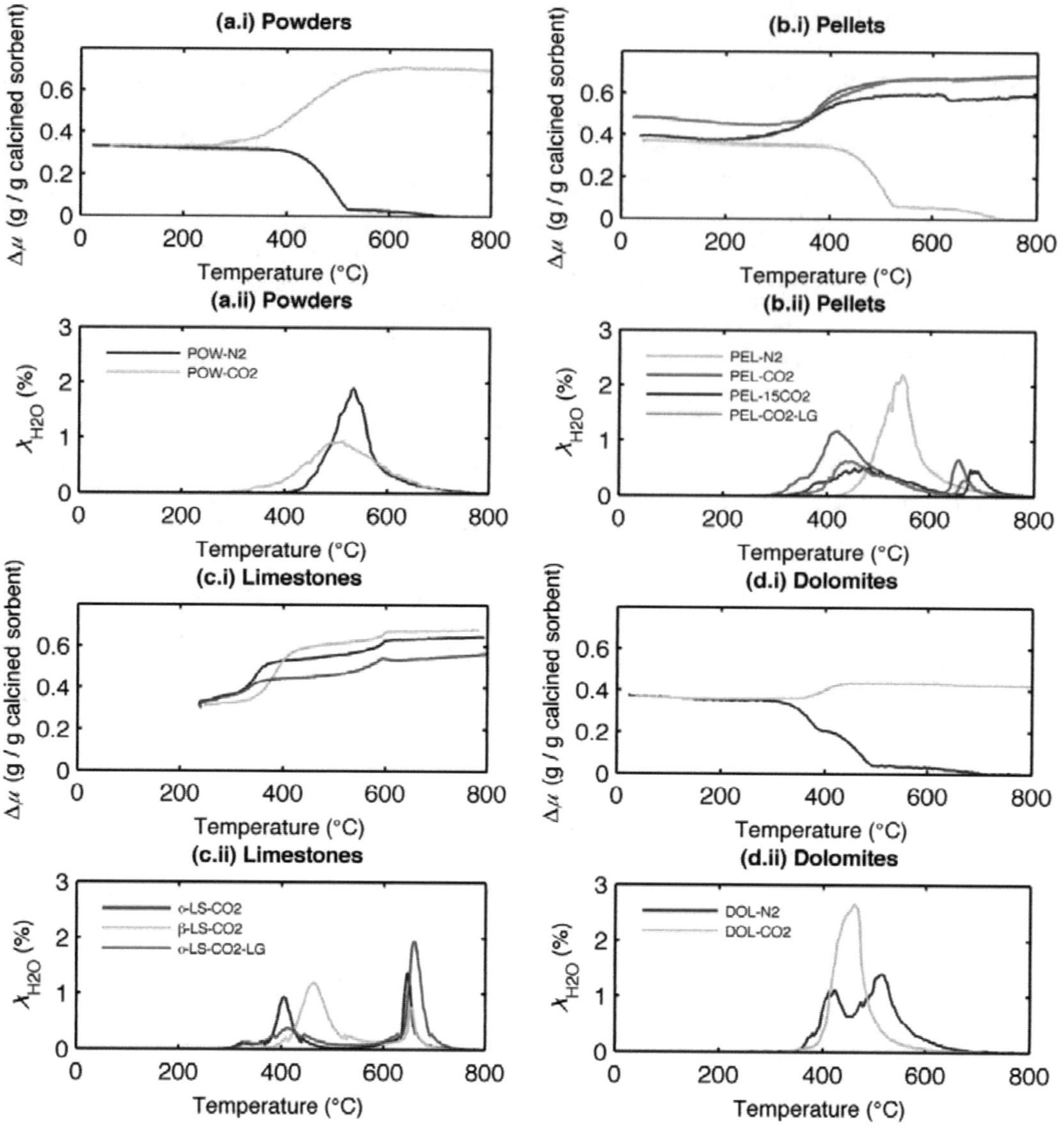

Figure 1. Experimental data obtained for the matrix shown in Table 2, with plots of (i) the mass change in g per g of *calcined* sorbent, μ, and (ii) the steam concentration in the outlet gas, X_{H2O}, plotted as a function of bed temperature.

In Figure 1, there are four pairs of plots: (a.i) and (a.ii) relating to the $Ca(OH)_2$ powder experiments; (b.i) and (b.ii) relating to the $Ca(OH)_2$ pellet experiments; (c.i) and (c.ii) relating to the hydrated calcined limestone experiments; and (d.i) and (d.ii) relating to the hydrated calcined dolomite. All samples show a characteristic change in mass upon heating, with samples heated under N_2 reducing in mass and the samples heated under CO_2 increasing in mass. These mass changes were typically coincident with a release of H_2O, dehydration. The samples heated under N_2 typically observed one peak of H_2O release (apart from the dolomite samples) and the samples heated under CO_2 observed either one (in the case of $Ca(OH)_2$ powder and dolomite samples) or two (in the case of $Ca(OH)_2$ pellets and limestone samples) peaks associated with H_2O release.

Table 3. Original conversions of available oxide sites to hydroxide and carbonate as determined by heating up in N_2

	POW-N2	PEL-N2	DOL-N2
$X_{hydroxide}$	0.970	0.980	0.996
$X_{carbonate}$	0.037	0.078	0.088

From the data obtained from experiments performed under N_2 in the TGA, the original conversions of available oxide sites to hydroxide and carbonate have been determined (see Table 3). A small degree of carbonation was found for the pellet and powder samples, with the extent of carbonation appearing to increase during the pelletisation process. The hydrated calcined dolomite sample also appeared to have some slight carbonate conversion from either the hydration process or incomplete calcination. The conversion to hydroxide will also include any condensed water on the surface of particles, which results in apparent conversions of greater than unity. Carbonate conversions for the samples heated under CO_2 were calculated as mol CO_2/mol CaO at the maximum conversion conversion (see Table 4).

Table 4. Final conversions of available oxide sites to carbonate following heating up in CO_2

	POW-CO2	PEL-CO2	PEL-15CO2	PEL-CO2-LG	DOL-CO2	-LS-CO2	-LS-CO2	-LS-CO2-LG
$X_{carbonate}$	0.910	0.906	0.765	0.897	0.979	0.872	0.915	0.782

4.1 $Ca(OH)_2$ Powders. The first thing to note with the $Ca(OH)_2$ powder experiments is that for the experimental data obtained with pure N_2 the dehydration is found to occur fairly rapidly in the range of 400-500 °C. The mass does not return to zero on the scale because of mild carbonation of the powder during manufacture/storage/sample preparation. When the powder is heated under CO_2 however, the dehydration appears to start at a lower temperature (approx. 350 °C) and steam is released over a larger temperature range. The mass increase of the sample corresponded with a peak of steam release, suggesting that direct carbonation – reaction (II) – was occuring.

4.2 $Ca(OH)_2$ Pellets. The $Ca(OH)_2$ pellets behaved in a similar manner to the powder under N_2, in that one peak of steam evolution was observed, with the corresponding mass change somewhere in the range 400-500 °C. Again, there is evidence of mild carbonation of the original sample (to a slightly greater extent than the original $Ca(OH)_2$ powder). However, when the $Ca(OH)_2$ pellets were heated under CO_2, two peaks of steam evolution were observed. One occurred at a similar temperature (approx. 350 °C) to that obtained for the powders under CO_2 and one at an elevated temperature (600-650 °C). The former is associated with a large increase in mass and is likely to be associated with direct carbonation and the latter is associated with a small 'blip' in mass and is the "superheated dehydration" effect described by Materic et al. [3] It is interesting to note that the sample carbonated at 15 % CO_2 (a more realistic concentration of CO_2 for the carbonator) reaches a much lower carbonation extent than in 100 % CO_2.

4.3 Hydrated Calcined Limestone. The hydrated calcined limestone that exhibited the highest degree of conversion was prepared with a calcination environment of 900 °C under CO_2, followed by the small particles calcined under N_2, followed by the large particles calcined under N_2. It seems counter-intuitive that the sample calcined under a CO_2 atmosphere should be the one that was most reactive, because this will be the most highly sintered/least porous sample [1]. A potential reason for this could be the fragmentation of the sorbent upon hydration, a phenomonon that has been found to be more pronounced for more highly sintered sorbents [6]. The sample with the largest limestone particle size had the lowest reactivity, which likely to be on account of its lower surface-to-area ratio. It is also interesting that the more reactive the samples are, the higher the mass at the end of the direct carbonation stage (at 500 °C) and the larger the first steam peak (and the smaller the second steam peak); this suggests that much of the improvement in the reactivity is associated with the reactivity of the sorbent during the direct carbonation stage, rather than the capicity of the sorbent to increase its dehydration temperature.

4.4 Hydrated Calcined Dolomite. Hydrated calcined dolomite heated under N_2 decomposed in two stages, one associated with $Mg(OH)_2$ and one with $Ca(OH)_2$. When heated under CO_2, the two peaks converged into one peak. This could be a result of a similar effect to that described for the $Ca(OH)_2$ samples, ie., that steam release from direct carbonation occurred at lower temperatures to that from dehydration under N_2.

4. Discussion

The results described above show that the "superheated dehydration" effect was not observed for all samples tested: it was observed for hydrated calcined limestones and $Ca(OH)_2$ pellets, but not for $Ca(OH)_2$ powder or hydrated calcined dolomite. These results suggest that the effect of "superheated dehydration" is not an intrinsic property of $Ca(OH)_2$ itself, but is a kinetic effect, which is more likely to occur with increasing particle size. This is apparent because the effect is not observed for the powder (< 30 μm), but is for the larger $Ca(OH)_2$ pellets and hydrated calcined limestone. We hypothesize that an occluded pore mechanism is more likely, whereby the larger $CaCO_3$ (36.9 cm^3/mol) forms on the surface of $Ca(OH)_2$ (33.6 cm^3/mol) [7] (the $Ca(OH)_2$ may be expected to contract further if it goes through a CaO (16.9 cm^3/mol) intermediate), closing off pores in the process. This then forms an impenetrable shell of $CaCO_3$ around the $Ca(OH)_2$, providing a barrier to mass transfer of H_2O from the centre of the particle to the surface. Formation of a protective carbonate layer in the carbonation of $CaCO_3$ has been discussed with relation to carbonation of $Ca(OH)_2$ at lower temperatures, when dehydration is not expected to occur at appreciable rates [8-10]. $Ca(OH)_2$ has a low melting point (835 °C at 1000 bar [11]) and it is likely that the core of the particle will be prone to sintering, enhancing the annealing process and potentially enhancing shell formation by reduction of surface area. In this model, a gas pressure is allowed to build up within the locked pores in the particles and the particles are found to rupture when this pressure exceeds rupture pressures of the weak spots formed over the pores; note that at 625 °C, the steam pressure over $Ca(OH)_2$ is 6 bar [12]. This, simultaneously, opens up new surface area to carbonation, however it would appear from Figure 1(c.i) that carbonation in the first – direct carbonation – stage is the largest contributor to the final conversion. The reason that the effect is not observed for the hydrated calcined dolomite, where CaO exists in small grains formed amongst MgO grains upon calcination of dolomite [13], is likely to be a result of porosity maintained by the network of unreacted MgO.

5. Conclusion

We have investigated "superheated dehydration" as a reactivation strategy for CaO-based sorbent for CO_2. Steam release at elevated temperatures under CO_2 atmospheres has been observed for $Ca(OH)_2$ pellets and hydrated calcined limestone, but not for $Ca(OH)_2$ powder or hydrated calcined dolomite. We propose that this is a result of occluded pores formed upon direct carbonation of $Ca(OH)_2$, which prevent mass transfer of H_2O from the centre of the particle to the surface.

Acknowledgements

Acknowledgements are given to Y. Wu, V. Manovic and I. He of CanmetENERGY for providing the $Ca(OH)_2$ pellets.

References

1. Blamey, J., Anthony, E.J., Wang, J., and Fennell, P.S., *The calcium looping cycle for large-scale CO_2 capture.* Progress in Energy and Combustion Science, 2010. 36(2): p. 260-279.
2. Dean, C.C., Blamey, J., Florin, N.H., Al-Jeboori, M.J., and Fennell, P.S., *The calcium looping cycle for CO_2 capture from power generation, cement manufacture and hydrogen production.* Chemical Engineering Research and Design. In Press, Corrected Proof.
3. Materic, V., Edwards, S., Smedley, S.I., and Holt, R., *$Ca(OH)_2$ Superheating as a Low-Attrition Steam Reactivation Method for CaO in Calcium Looping Applications.* Industrial & Engineering Chemistry Research, 2010. 49(24): p. 12429-12434.
4. Criado, J.M., Macias, M., and Macias-Machìn, A., *Analysis of the system CaO---CO_2---H_2O for storage of solar thermal energy.* Solar Energy, 1992. 49(2): p. 83-86.
5. Wu, Y., Manovic, V., He, I., and Anthony, E.J. *High-performance CO_2 capture with modified CaO-based pellet sorbents: TGA results.* AIChE Spring Meeting, 2011. Chicago, Il, USA.
6. Blamey, J., Paterson, N.P.M., Dugwell, D.R., and Fennell, P.S., *Mechanism of Particle Breakage during Reactivation of CaO-Based Sorbents for CO_2 Capture.* Energy & Fuels, 2010. 24: p. 4605-4616.
7. Perry, R.H. and Green, D.W., *Perry's Chemical Engineers' Handbook, seventh edition,* 1997. McGraw-Hill, New York, USA.
8. Montes-Hernandez, G., Pommerol, A., Renard, F., Beck, P., Quirico, E., and Brissaud, O., *In situ kinetic measurements of gas-solid carbonation of $Ca(OH)_2$ by using an infrared microscope coupled to a reaction cell.* Chemical Engineering Journal, 2010. 161(1-2): p. 250-256.
9. Gillott, J.E., *Carbonatation of $Ca(OH)_2$ investigated by thermal and x-ray diffraction methods of analysis.* Journal of Applied Chemistry, 1967. 17(7): p. 185-189.
10. Beruto, D.T. and Botter, R., *Liquid-like H_2O adsorption layers to catalyze the $Ca(OH)_2/CO_2$ solid-gas reaction and to form a non-protective solid product layer at 20 °C.* Journal of the European Ceramic Society, 2000. 20(4): p. 497-503.
11. Wyllie, P.J. and Tuttle, O.F., *The system CaO-CO2-H2O and the origin of carbonatites.* J. Petrology, 1960. 1(1): p. 1-46.
12. McBride, B.J., Zehe, M.J., and Gordon, S., *NASA Glenn Coefficients for Calculating Thermodynamic Properties of Individual Species.* 2002, National Aeronautics and Space Administration: Clevelan, Ohio, US.
13. Wilsdorf, H.G.F. and Haul, R.A.W., *X-ray study of the thermal decomposition of dolomite.* Nature, 1951. 167(4258): p. 945-6.

(2b) Chemical Looping Gasification - A Study on Calcination/Carbonation Looping Cycle

Bishnu Acharya[1], Animesh Dutta[2] and Prabir Basu[1], (1)Mechanical Engineering, Dalhousie University, Halifax, NS, Canada, (2)School of Engineering, University of Guelph, Guelph, ON, Canada

At present there are no two minds on the fact that we are facing a great environmental challenge in the form of global warming and climate change caused by higher concentration of carbon dioxide in atmosphere which in turn is produced from combustion of fossil fuels. But the present economy is all carbon based, so issues right now is how easily we can shift from this carbon based economy to renewable based without compromising on the energy need. For that do we have the technology to capture carbon dioxide as well as provide us with the energy we wanted? To overcome this challenge author's has come up with the concept of Chemical Looping Gasification. It produces hydrogen rich gas from the biomass using steam gasification technology which can be used for fuel cell technology, as well captures carbon dioxide during the process and regenerates it producing the pure stream of carbon dioxide for direct sequestration.

The system consists of two reactors: one is a bubbling fluidized bed gasifier and another is a circulating fluidized bed regenerator. The fuel is fed into the gasifier that receives calcium oxide from the regenerator and steam from an external source. During gasification, the carbon dioxide produced will be captured by the calcium oxide that makes up the bubbling fluidized bed. Immediate removal of the reaction products, CO_2 from the system will increase the rate of water-gas shift reaction, and therefore yield higher hydrogen. The calcium carbonate formed in the gasifier will move into a circulating/ Transport regenerator where it decomposes (calcines) into calcium oxide and carbon dioxide. The calcium oxide is returned to the gasifier, and the carbon dioxide is taken for sequestration. A pilot plant of 5 kW capacities has been built and the tests are underway.

However the limitation is imposed by the loss in the activity of the sorbent CaO for capturing carbon dioxide which may be due to sintering of the sorbent, deposition of tar and char around the sorbents. To identify the loss in activity of the sorbent, an experiment is being conducted by taking the sample going into and out of the gasifier. Sample going into will shows the capacity of regeneration while coming out will give the absorbing capacity. SEM technique will be used to analyse the samples and identify the sintering rate. The results from this experiment will be presented in this paper.

See more of this Session: Calcium Looping and Chemical Looping Processes for CCS (I)
See more of this Group/Topical: Topical 2: Advanced Fossil Energy Utilization

(2c) Development and Performance of Cobalt Oxide as Oxygen Carrier In Dimethyl Ether (DME) Fueled Chemical-Looping Combustion

Tao Han[1], **Hui Hong**[2], Fengjuan He[1] and Hongguang Jin[2], (1)Graduate University of Chinese Academy of Sciences, Beijing, China, (2)Institute of Engineering Thermophysics, Chinese Academy of Sciences, Beijing, China

With ever growth concerns on energy security and environmental pollution, dimethyl ether (DME) is considered as a promising alternative fuel in the future. It is expected as a candidate for use in a diesel engine or gas turbine to produce power generation. Although it is a relatively clean fuel, direct combustion of DME still causes CO_2 diluted by N_2, which complicates the separation of CO_2. Recovering CO_2 from a DME fueled power cycle will cause the thermal efficiency decreased by about 7.7 percentage points. Thus, we have to face a serious issue how to more efficiently and compatibly utilize DME fuel in the power generation systems with CO_2 capture of super-lower energy penalty.

In our previous study, a thermal power cycle combined with DME fueled chemical-looping combustion (CLC) was proposed, and no extra energy penalty is required for CO_2 separation of this system. Besides, compared with natural gas, DME has high oxygen content. Its C-O bond energy is smaller than that of the C-H and the distortion of C-O bonds in the DME molecule weakens the boning strength. As a result, the C-O bond breaks easier than the C-H bond and the reaction between DME and looping material may be feasible at relatively low temperatures. Thus the lower grade heat such as waste heat or solar thermal energy can be supplied to the endothermic reaction in the fuel reactor, which would further increase the amount of heat produced in the high-temperature air reactor, improving the thermal efficiency of the CLC power generation system.

To identify suitable looping material candidates for DME fueled chemical-looping combustion, we have conducted experiments by using a thermogravimetrical analyzer. Seven oxygen carriers, with Fe_2O_3, NiO and CoO as solid reactants and Al_2O_3, $MgAl_2O_4$ and YSZ as binders, respectively, have been prepared by dissolution method. The reductions at 400°C showed that, compared with Fe and Ni based oxygen carrier, the Co based one have higher reactivity and may be an appropriate looping material for chemical-looping combustion fueled by C-O bonded fuels like DME. Kinetic behaviors of CoO/$CoAl_2O_4$ and CoO/YSZ were studied at the reduction temperature of 400°C, 450°C and 500°C, and the results suggested that suitable reduction temperature is around 450°C. Scanning electron microscopy (SEM) was used to characterize the morphological features of the fresh, reduced and oxidized oxygen carriers, and it is observed that porous structure were formed after a reaction cycle, which facilitated the diffusion of gaseous reactant into the core of oxygen carrier. The energy-dispersive X-ray (EDX) identified that serious carbon deposition occurred during the reduction process, and its rate became lower with the growth of reduction temperature. To restrain carbon deposition, DME was saturated with a mole ratio of H_2O/DME as 1/1, 2/1, 3/1, and 4/1, respectively. It is found that when the ratio of H_2O/DME is above 3/1, carbon deposition could be completely avoided. The findings of this paper may broad the CLC field in the utilization of alternative fuel for CO_2 capture.

See more of this Session: Calcium Looping and Chemical Looping Processes for CCS (I)
See more of this Group/Topical: Topical 2: Advanced Fossil Energy Utilization

ELSEVIER

Available online at www.sciencedirect.com

ScienceDirect

Energy Procedia 00 (2010) 000–000

Energy Procedia

www.elsevier.com/locate/XXX

Post-combustion CO_2 capture with CaO. Status of the technology and next steps towards large-scale demonstration

A. Sánchez-Biezma[1], J.C. Ballesteros[1], L. Diaz[2], E. de Zárraga[2], F.J. Álvarez[3], J. López[3], B. Arias[4*], J.C. Abanades[4**], E.J. Anthony[5]

1 Endesa Generación, Riberal del Loira 60, 28042 Madrid (Spain)
2. Hunosa, Avenida de Galicia 44, 33005 Oviedo (Spain)
3.Foster Wheeler Energía S.L.U, Gabriel García Márquez, 2, 28232 Madrid (Spain)
4.Spanish Research Council,CSIC-INCAR (Oviedo-Spain), CSIC-ICB** (Zaragoza-Spain)*
5.Department of Chemical and Biological Engineering, University of Ottawa (Canada)

Abstract

Post-combustion CO_2 capture using CaO requires a large-scale circulating fluidized bed (CFB) reactor as CO_2 absorber, operating between 600-700°C. In addition, a large-scale oxy-fired CFBC must be interconnected to this reactor to allow for the decomposition of $CaCO_3$ formed in the carbonator. This allows for continuous regeneration of the CaO sorbent and the production of a CO_2-rich stream suitable for final purification and compression. Despite the known limitations of this technology (mainly sorbent deactivation, solid attrition, and high energy requirements in the calciner), clear operating windows have been identified at which the process could be implemented in practice using CFBC technology. Theoretical studies on thermal integration of a carbonation/calcination loop in new and existing power plants have shown that the technology has the potential to achieve a substantial reduction (around 30%) in capture cost and energy penalties with respect to stand-alone oxy-fired systems. Since the solid materials and operating conditions in the CFB units are similar to those present in existing large-scale CFBCs, prospects for rapid scale-up of the technology are very promising. Rapid development is currently taking place by demonstrating the key concepts in laboratory-scale test rigs of 10s of kW. However, it is essential to move on to the next phase of pilot testing and validate the results in conditions fully comparable with those expected in large-scale units. We describe here the design of a 1.7MW_{th} pilot plant to capture 70-95% of the CO_2 contained in the flue gas from a 1/150 side stream emitted by an existing 50MW_e CFB power plant. The pilot is made up of two interconnected CFB reactors of 15 m height. Construction of the pilot has been initiated and is expected to enter into full operation in the first half of 2011, providing the necessary experimental results to decide on the launching of an aggressive programme that aims to demonstrate the technology at large scale well before 2020.

Keywords: post-combustion CO_2 capture, Ca-looping cycle

1. Introduction

Fossil fuels account for more than 80% of the primary energy consumption worldwide and this scenario is likely to continue during the next few decades. Combustion of fossil fuels increases the concentration of greenhouse gases (GHG) in the atmosphere, contributing to climate change. For this reason, abatement of GHG derived from fossil fuel utilisation is now an urgent challenge. Between the different alternatives to stabilize GHG concentrations in the atmosphere, CO_2 capture and storage (CCS) technologies are a key option in the portfolio of solutions because they could contribute 15-55% to the total mitigation effort during this century (Metz, 2005). The main technical drawback of existing CO_2 capture technologies for coal-based power plants is the penalty associated with the capture process. CO_2 capture technologies can be classified in two main groups due to their state of development:

near-commercial and emerging technologies. The first group includes technologies already developed at commercial scale for other processes, but that must be adapted to capture the high CO_2 flows from fossil fuel power plants and all other pollutants. The second group includes less developed technologies, but with a real potential for reducing the energy penalty and CO_2 capture costs. In the emerging technologies, two categories can be distinguished: processes with no analogous reactor of sufficiently large scale in operation and that require scale-up from laboratory size and theoretical studies; and new processes that use new materials in reactors already commercially established at large scale. Calcium looping is a promising CO_2 capture technology that fits in this second sub-category as it relies on the utilization of CFB reactors operating with circulating materials (mainly CaO particles, their derived products and coal ashes) and key operating conditions close to those present in large-scale CFB boilers.

Calcium looping uses the reversible reaction of CO_2 with CaO to separate it from a flue gas stream in a cyclic process. Low-priced natural limestones and dolomites are in principle the best CaO precursor for large-scale CO_2 capture in power plants. The CO_2 absorption capacity of natural CaO sorbents is known to decay rapidly with the number of carbonation/calcination cycles (Curran, 1967) and a large body of literature is now available trying to better understand this process and to design methods to overcome it (see most recent review by Blamey et al., 2010). However, it is now well known (Grasa, 2006) that there is a residual activity of the CaO particles that allows for the operation of the capture loop by using sufficiently large solid circulation rates. For those conditions where this is not possible (because of attrition losses, deactivation of CaO by SO_2, excessive heat requirements in the calciner, etc.), a large make-up flow of fresh limestone is required. In these cases, the operation with high consumption of limestone may be allowed by the low cost of natural limestone in many locations close to power plants and the potential synergies with the cement industry (making use of the purge rate of deactivated CaO and coal ashes).

In this paper we briefly review the status of development of the technology from the process perspective, in order to present a design of a 1.7MW_{th} pilot plant aimed at the capture of 70-95% of the CO_2 contained in a side stream of the flue gases from an existing CFB power plant. The design of the pilot combines the knowledge acquired in recent years when testing the Ca-looping concept at laboratory scale of 10s of kW, and the industrial experience designing and operating large-scale CFBCs.

2. Process description and status

Figure 1 shows a concept of post-combustion capture where CO_2 is captured from the combustion flue gas of a new or an existing power plant using CaO as sorbent in a CFB carbonator operating between 600-700ºC. The stream of partially carbonated solids leaving the carbonator is directed to a second CFB where solids are calcined to regenerate the sorbent (CaO) and release the CO_2. In order to calcine the $CaCO_3$ formed in the carbonator and to produce a highly concentrated stream of CO_2 suitable for purification and compression in the calciner, coal is burned under oxy-fuel conditions at temperatures above 900ºC. Therefore, exhaust gases from the calciner contain the CO_2 captured from the flue gases of the power plant and that resulting from the oxy-fired combustion of coal in the calciner. Other advanced concepts where heat is transferred to the calciner using a hot stream of solids circulating from an air-fired combustor have been proposed, to avoid the penalty associated with the air separation unit (Abanades, 2005). However, these processes will need a longer period of time to be developed. The option of calcining the sorbent using an oxy-fired CFB, first proposed by Shimizu et al. (1999), is one of the strengths of this process and may speed up the development of calcium looping technology. This is due to the fact that the oxy-combustion CFB technology for CO_2 capture is being developed as an independent route and it is already a more mature technology in a near-commercial stage.

A full programme to demonstrate and validate OxyCFB technology is already on-going in a collaboration among Foster Wheeler, Fundación Ciudad de la Energía (CIUDEN) and ENDESA. This programme that counts with the support of the European Energy Program for Recovery (EEPR), includes R&D testing campaigns at the 0.8MW_{th} oxy-fuel CFB pilot plant (CanmetENERGY) and at CIUDEN´s 30MW_{th} OxyCFB boiler currently under design and construction in El Bierzo, Spain. The goal of this ambitious project will be a 300MW_e commercial demonstration plant planned to start at the end of 2015 in ENDESA´s Compostilla power plant, if the project is feasible. In addition, and due to the higher amount of limestone to calcine in this process compared with an OxyCFB boiler, the endothermic calcination reaction can make the operation of the oxy-fired calciner easier as it may facilitate temperature control and reduce CO_2 recycling requirements.

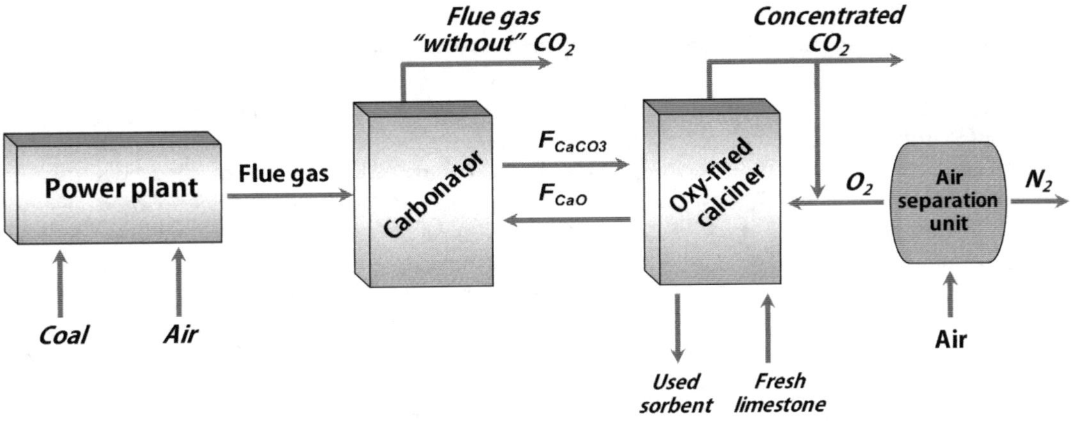

Figure 1. Schematic of a calcium looping system to capture CO_2 from an existing power plant.

Calcium looping technology shows a low energy penalty compared with other CO_2 capture technologies (Romeo, 2008; Ströhle, 2009). A main feature of this process is the possibility of producing additional power from different high-temperature sources involved in the system. In fact, the calciner of the process (Figure 1) can be viewed as a new oxy-fired FB which increases the total installed power of the global process (power plant+capture process).

Some experimental work for testing calcium looping in small FB pilot plants with different configurations (10s of kW) has been published in recent years. By operating a CFB pilot rig at CANMET as a bubbling FB carbonator, Abanades et al. (2004) demonstrated the effective capture of CO_2 from flue gases in a FB reactor with sufficient quantity of active CaO (the part of the CaO particles reacting in the fast reaction regime). Lu et al. (2008) reported stable capture conditions in a similar semi-continuous 75 kW pilot plant composed of an oxy-fired CFB calciner and a bubbling FB carbonator. They showed that regeneration of CaO under calcination conditions does not significantly reduce the activity of the sorbent. Charitos et al. (2010) performed continuous experimental tests on a dual $10kW_{th}$ pilot plant composed of a FB carbonator and a CFB calciner. One of the most important features of this experimental facility is the precise control of the solid circulation between reactors by means of a high-temperature cone valve. They reported a parametric study of the main operation variables in the bubbling FB carbonator. To achieve operation conditions closer to those expected in CFB reactors, Alonso et al. (2010) carried out experimental work in a $30kW_{th}$ pilot plant composed of a CFB carbonator coupled with a CFB calciner. They reported CO_2 capture efficiencies of 70-97 % under realistic operation conditions in the CFB carbonator reactor. A more detailed comparison of the results of these testing facilities can be found elsewhere (Rodriguez, 2010a).

Experimental data available at small pilot -scale have served to establish the key variables in the carbonation process and to develop models for the carbonation reactor (Alonso, 2009; Hawthorne, 2008). A recent analysis of the results obtained in a CFB carbonator by Rodriguez et al. (2010b) indicates that the overall reaction rate is limited mainly by the carbonation reaction characteristic of highly cycled solids. One of the most important variables in the capture process is the inventory of active solids in the carbonator bed, composed of a mixture of $CaCO_3$, CaO, $CaSO_4$ (from the reaction of SO_2 and CaO), and ash from the coal used in the calciner. CO_2 capture efficiency increases with the inventory of active sorbent in the carbonator bed. This is in turn dependent on the average residence time of particles in the reactor, established by the ratio between the bed inventory and the solid circulation flow between reactors.

Quantitative information and modelling work available from laboratory studies on the carbonator reactor have been used to design a pilot plant, following the scheme of Figure 1. However, the importance of the oxy-fuel CFB calciner, as well as the need to obtain in the pilot plant certain parameters (bed inventories, and solid circulation rates) at specific conditions, has required the application of all necessary know-how on CFB technology achieved by

Foster Wheeler during decades of experience in the design and construction of CFB boilers. We summarize in the following paragraphs the main design criteria, the main features of the pilot and the expected work plan.

3. Pilot plant design to test calcium looping technology for post-combustion CO_2 capture

As has been discussed above, in order to advance industrial application of Ca-looping technology, the next steps towards larger-scale demonstration must be carried out as soon as possible. For this objective, an agreement between ENDESA (a major European utility), Foster Wheeler (a leading world manufacturer of fluidized bed combustion technology), HUNOSA (the largest coal mining company in Spain) and CSIC (Spanish Research Council) was signed in 2009, and an ambitious roadmap was proposed to develop this technology to a commercial scale before 2020. This consortium was substantially reinforced by leading R&D partners in the field in Europe (IFK of the U. Stuttgart, Lappeenranta University, Imperial College) and in Canada (University of Ottawa), through the project "CaOling" (Development of postcombustion CO_2 capture with CaO in a large testing facility, www.CaOling.eu), funded by the European Union 7th Framework Programme-FP7 (Dec 2009-Nov 2012). The main project targets are the detailed design and construction of a pilot plant of 1.7MW$_{th}$, to perform experimental testing at this scale and to demonstrate the performance of calcium looping as post-combustion CO_2 capture technology.

The pilot plant will be built and integrated with "La Pereda" power plant to treat a small fraction of the flue gas produced. This coal power plant consists of a CFB boiler with an installed capacity of 50MW$_e$. The plant, owned by HUNOSA, is located in Asturias (northern Spain). Figure 2 shows an aerial picture of "La Pereda" and the site where the calcium looping pilot plant will be located. A bituminous coal with high ash content (ash, 36.0 %; volatile matter, 16.6 %; S, 0.7 %; HHV, 4516 kcal/kg) is used in the power plant. Typical composition of flue gas produced at "La Pereda" is 5.5%$_{vol}$ O_2, 12.6%$_{vol}$ CO_2, 7.0%$_{vol}$ H_2O and 700 ppm SO_2.

One of the most important aims of the pilot plant is to work under operation conditions representative of those encountered in industrial CFB boilers. Following this objective, a minimum facility size was established according to standard design criteria and the pilot plant was dimensioned to treat the flue gas corresponding to the generation of 1.7MW$_{th}$ in "La Pereda" power plant. This flow corresponds to 1/150 of the total flue gas produced. The pilot plant was designed to work in an operating window where process variables are kept under reasonable values that may be expected in the industrial application of calcium looping for post-combustion CO_2 capture. The variables chosen to define the operation window were fresh limestone flow, heat demand in the calciner and solids circulation. Having defined the gas flow to be treated in the pilot plant and the desired capture efficiency in the carbonator, the remaining variables involved in the process can be calculated by solving the respective mass and energy balances, using "La Pereda" coal as reference, and taking into account that fresh limestone flow and heat demand in the calciner must be minimized in a calcium looping process. The presence of SO_2 in the carbonator and ashes and sulphur in the coal fed in the calciner affect the requirements of fresh limestone. Flow of limestone was fixed to a maximum value of 1 kg$CaCO_3$/kg coal fed in the calciner as the scenario of high purge of solids can be acceptable if it is used as feedstock in the cement industry. On the other hand, the minimum make-up flow was determined by fixing the maximum heat demand in the calciner to a value of 1.7MW. This value results in a scenario with high energy consumption in the calciner which cannot be economically feasible. However, it allows the possibility of increasing the capacity of the carbonator to values higher than 1.7MW$_{th}$ and testing carbonation efficiency under higher gas velocities. Regarding the dimensions of the reactors, the diameter was fixed to achieve gas velocities similar to those encountered in CFB boilers (3-5 m/s). To ensure enough gas residence time and inventory of solids, the risers were designed with a height of 15 m.

Inside this operation window, solid circulation between reactors was calculated assuming a conservative value of sorbent conversion in the carbonator (5% lower than the maximum average conversion). Circulation rates expected inside the operation window can vary between 1-5 kg/m^{2}s, which are in the typical range of CFB reactors. Another important aspect of the design was the heat dissipation in the carbonator, which depends on the amount of reacted CO_2 and solid circulation from the calciner. The heat released by the exothermic reaction is fixed by the amount of CO_2 captured in the carbonator. This second term is approximately constant once the carbonation efficiency and the flow of CO_2 to the carbonator are fixed. However, the heat transported with the solids coming from the calciner can vary greatly depending on the activity of the solids in the system. In the most extreme case, it was calculated that the heat dissipation in the carbonator can change between 0.1 and 0.7MW_{th}.

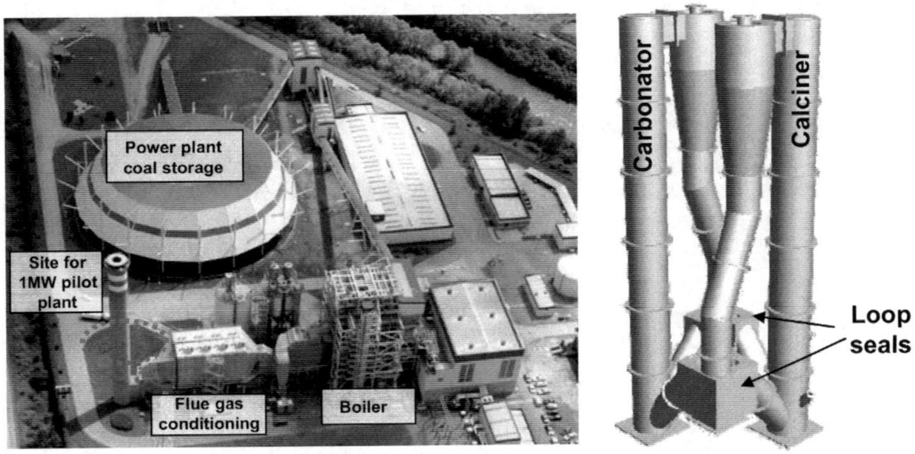

Figure 2. "La Pereda" power plant and diagram of the carbonator and calciner reactors.

The core of the CO_2 capture pilot plant consists of two interconnected CFBs (see Figure 2). Typical operation temperatures will be 650ºC in the carbonator and 950ºC in the calciner. Each reactor is equipped with a high-efficiency cyclone (cut size 5 μm) and a loop seal, which can divert the stream of solids coming from the stand pipe to the same reactor (internal circulation) or to the opposite reactor. The design of the loop seal allows adjusting the solid flow between reactors. A dedicated high-pressure fan is used to fluidize the loop seals and to control solid circulation between reactors. Table 1 shows the variation range of some characteristic parameters of the pilot plant.

Table 1. Main inputs to the 1.7MW_{th} pilot plant.

Flue gas flow to carbonator (kg/h)	680-2300
Maximum coal flow to calciner (kg/h)	325
Maximum fresh limestone flow (kg/h)	300
Oxygen flow to calciner (kg/h)	300-600
CO_2 flow to calciner (kg/h)	700-2250
Air flow to calciner (kg/h)	600-2500

Figure 3 shows a layout of the integration of the pilot plant with "La Pereda" power plant. A stream of flue gas from "La Pereda" will be taken after the electrostatic precipitator and will be sent to the CO_2 capture pilot plant. A fan with an average gas flow of 1400 kg/h will be used to increase the pressure of the flue gas before entering the carbonator. The maximum flow of this fan can increase the capacity of the carbonator to a value up to 1.6MW_{th}. The gas leaving the carbonator at high temperature (approximately 650ºC) will be cooled down and returned to the main "La Pereda" flue gas stream before the electrostatic precipitator.

Typical temperature of flue gas from the calciner will be around 950ºC, and will be mixed with that leaving the carbonator before being cooled down. The coal feed system was dimensioned to introduce a maximum flow rate of 325 kg/h. The system can use different types of coal or pretreated solid fuels. The fuel is discharged in an 8m^3 feed hopper using a bigbag handling system. From the feed hopper, coal is pneumatically transported to an intermediate bin which has a capacity of 1.1 m^3. From this bin, coal is discharged to a common hopper where limestone and coal is mixed. A rotary feeder isolates mixture feeding from furnace overpressure. Finally a screw drives the solids into the calciner bed. As indicated above, coal can be burned in the calciner using an oxy-fired or air-fired mode. It was decided to supply O_2 and CO_2 from tanks of liquefied gases as this is the more cost effective and the more flexible solution. The liquid O_2 and CO_2 pass through several atmospheric vaporizers and an electrical conditioner. O_2 and CO_2 flows are controlled and blended in a mixer skid. This facility enables the mix of a variable O_2 and CO_2

percentage. The temperature of this flow is increased in a gas heater. An additional steam line joins to the flow previous to the calciner inlet. This steam flow that comes from “La Pereda” can be controlled to simulate different compositions of the gas, increasing the operational flexibility of the plant. In order to work under air-firing mode, a fan will be installed to supply the air needed for the combustion of coal.

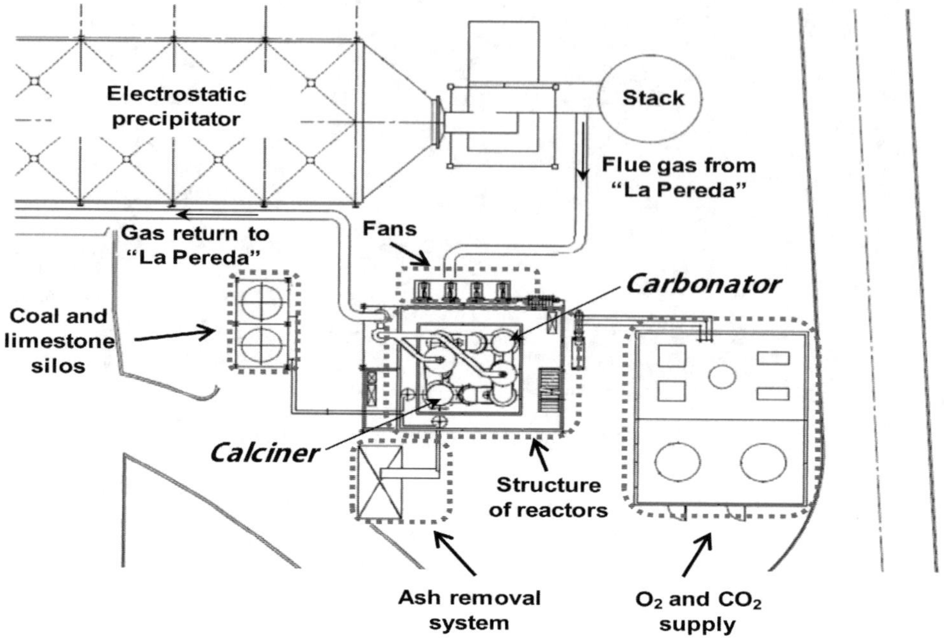

Figure 3. Layout of the integration of the 1.7MW$_{th}$ pilot plant with “La Pereda” power plant.

To work inside the limits of the operating window, the limestone feeding system was dimensioned to give a maximum mass rate of 300 kg/h. Limestone is discharged directly from trucks to a feed silo which is close to 50 m^3. From this silo, limestone is pneumatically transported to an intermediate bin with a capacity of 1.1 m^3 before being mixed with coal. To remove the solids from the system two pneumatic slide gates are located under the reactors. The calciner has an automatic purge removal system which consists of a water cooled screw conveyor 6m long. This screw conveyor discharges the cold ashes to a chain conveyor that leads the material to a 12 t container. The carbonator material purge can only be made manually and after plant operation is stopped.

A water cooling system using double-pipe cooling tubes will be installed in the carbonator to control reactor temperature. This system was designed to dissipate the heat of the stream of solids from the calciner and the heat produced during the carbonation reaction. The system consists of four retractable vertical bayonet tubes of 13 m to control reactor temperature. Cooling water circulates in the bayonets and extracts the heat released by the system. A retractable system permits moving the bayonet tube to modify the heat exchange surface, regulating the total heat extracted. This system gives great flexibility and permits operation of the plant under a wide range of conditions.

During normal operation of the pilot plant, the heat will be supplied by feeding coal. However, for start-up sequences, each reactor is equipped with a start-up burner fired with propane. These burners will be used to preheat the refractory of the reactors, increasing their temperature up to solid fuel ignition. A liquid propane tank with all the associated equipment to gasify the fuel is provided. A full skid of valves control the feeding process to the burners

The pilot plant instrumentation was defined attending to control requirements and to obtain information to analyze the experimental results (inventory of solids, solid samples, gas composition, temperature, etc.). Different ports along the risers, stand pipes and loop seals will measure different parameters inside the reactors and collect solid samples. Gas composition will be analyzed at the exit of each reactor and along the risers. A matrix of experiments will be carried out in order to find the operation conditions to optimize capture efficiency. Additionally, design data gained from testing of the 1.7MW_{th} pilot plant will be the basis for the detailed design of large-scale units. Due to the nature of calcium looping which involves a stream of sorbent circulating between two interconnected CFB reactors, the controllability/stability of the process is a key aspect that has to be analyzed and evaluated in this experimental facility during the testing program. Moreover, the matrix of experiments will be focussed on two important aspects of this process: the need of make-up flow of fresh limestone to achieve a specific activity of the solids in the system; and the energy consumption in the calciner.

The experimental campaign will start in mid-2011, and will run for 12 months. From the information and experience obtained during the operation of the 1.7MW_{th} pilot plant, a conceptual design of a pre-industrial plant of 20-30MW_{th} will be done. Construction of this mid-sized plant will be the next step in the scale-up route, if results obtained in the 1.7MW_{th} pilot plant are successful and the validity of calcium looping under realistic CFB conditions is demonstrated.

4. Conclusions

Calcium looping is a promising technology for post-combustion CO_2 capture. In order to scale up the technology, a 1.7MW_{th} pilot plant has been designed and is being constructed. This facility is expected to enter into full operation in the first half of 2011. The aim of this plant will be to advance the experimental validation of this technology and speed up the deployment of calcium looping at commercial scale. It has been possible to design this pilot using design information, equipment and material that are standard in existing large-scale CFBC power plants. This highlights the possibility for rapid scale-up of calcium looping technology if the testing program on this pilot confirms the good results obtained so far in small lab-scale units.

5. References

1. Abanades J.C., Anthony E.J., Wang J., Oakey J.E. Fluidized bed combustion systems integrating CO_2 capture with CaO. Environ. Sci. Technol. 2005; 39: 2861-2866.
2. Abanades J.C., Rubin E.S., Anthony E.J. Sorbent cost and performance in CO_2 capture systems. Ind. Eng. Chem. Res. 2004; 43 (13): 3462-3466.
3. Alonso M., Rodríguez N., González B., Grasa G., Murillo R., Abanades J.C. Carbon dioxide capture from combustion flue gases with a calcium oxide chemical looping. Experimental results and process development. Int. Journal of Greenhouse Gas Control 2010; 4: 167-176.
4. Alonso M., Rodríguez N., Grasa G., Abanades J.C. Modelling of a fludized bed carbonator reactor to capture CO_2 from a combustion flue gas. Chemical Engineering Science 2009; 64: 883-891.
5. Blamey J., Anthony E.J., Wang J., Fennel P.S. The calcium looping cycle for large-scale CO_2 capture. Progress in Energy and Combustion Science 2010; 36: 260-279.
6. Charitos A., Hawthorne C., Bidwe A.R., Sivalingam S., Schuster A., Spliethoff H., Scheffknecht G. Parametric investigation of the calcium looping process for CO_2 capture in a 10kWth dual fluidized bed. International Journal of Greenhouse Gas Control 2010, doi: 10.1016/j.ijggc.2010.04.009.
7. Curran G.P., Fink C.E., Gorin E. CO_2 acceptor gasification process. Studies of acceptor properties. Adv. Chem. Ser. 1967; 69: 141-161.
8. Grasa G.S, Abanades J.C. CO_2 capture capacity of CaO in long series of carbonation/calcination cycles. Ind. Eng. Chem. Res. 2006; 45: 8846-8851.
9. Hawthorne C., Charitos A., Perez-Pulido C.A., Bing Z., Scheffkenecht G. Design of a dual fluidized bed system for the post-combustion removal of CO_2 using CaO. Part I. CFB carbonator model. Proceedings of the 9th International Conference on Circulating fluidized Beds, Hamburg, Germany 2008, 759-764.
10. Lu D.Y., Hughes R.W., Anthony E.J. Ca-based sorbent looping combustion for CO_2 capture in pilot-scale dual fluidized beds. Fuel Processing Technology 2008; 89: 1386-1395.
11. Metz B., Davidson O., Coninck H.D., Loos M., Meyer L. Special report on carbon dioxide capture and storage. Intergovernmental Panel on Climate Change (IPCC), New York, NY, USA, 2005.
12. Rodríguez N., Alonso M., Abanades J.C., Charitos A., Hawthorne C., Schuster A., Lu D., Anthony E.J. Comparison of experimental results from three fluidized bed test facilities capturing CO_2 with CaO. International Conference on Greenhouse Gas Control Technologies 2010a, submitted.
13. Rodríguez N., Alonso M., Abanades J.C. Experimental investigation of a circulating fluidized bed reactor to capture CO_2 with CaO. AIChE Journal 2010b, doi 10.1002/aic.12337.
14. Romeo L.M., Abanades J.C., Escosa J.M., Paño J., Jiménez A., Sánchez-Biezma A., Ballesteros J.C. Oxyfuel carbonation/calcination cycle for low cost CO_2 capture in existing power plants. Energy Conversion Management 2008; 49: 2809-2814.
15. Shimizu T., Hirama T., Hosoda H., Kitani K., Inagaki M., Tejima K. A twin-bed reactor for removal of CO_2 from combustion processes. Trans. IChemE. 1999; 77 (part A): 62-68.
16. Ströle J., Galloy A., Epple B. Feasibility study on carbonate looping process for post-combustion combustion CO_2 capture from coal-fired power plants. Energy Procedia 2009; 1; 1313-1320.

(14a) The Energy Challenge: Providing Sustainable Supplies to Meet Growing Demand

F. Emil Jacobs, ExxonMobil Research and Engineering Company, Annandale, NJ

As economies grow and living standards improve, the world faces a dual energy challenge: meeting the growing demand for energy while mitigating emissions and their impact on the environment. ExxonMobil believes that addressing this dual challenge requires an integrated set of solutions which includes increasing the supply of all commercially viable energy sources, reducing emissions, and improving energy efficiency in all sectors of the economy. Technology will be a key enabler of continued progress in all of these areas.

Many experts agree that the immense scale of global energy demand necessitates a continuing role for oil and gas as a substantial portion of the energy supply mix for at least the next several decades. These experts also forecast that the overall energy mix will shift toward a relatively higher level of sustainability through a combination of increased efficiency across the economy, growth in lower-carbon sources of energy, and development and deployment of technologies that help mitigate emissions.

ExxonMobil has developed a number of technologies that contribute to increased supplies of energy, reduced emissions, and higher efficiency in its operations and in consumer use of its products. The presentation will provide an overview of ongoing research and development activity in a broad range of technology areas including algae-based biofuels. It will also highlight some of company's current technologies including lithium-ion battery separator films, lightweight plastics that improve automotive efficiency, and processes that could advance carbon capture and storage.

See more of this Session: Spring Meeting Keynote Address - Sponsored by the AIChE Fuels & Petrochemicals Division
See more of this Group/Topical: Liaison Functions

(24a) An Assessment of Chemical Looping Combustion (CLC) for CO_2 Capture In a Coal-Based IGCC Power Plant

Hari C. Mantripragada and Edward S. Rubin, Engineering and Public Policy, Carnegie Mellon University, Pittsburgh, PA

Chemical looping combustion (CLC) is an indirect combustion process in which fuel is combusted without direct contact with air. Transfer of oxygen between air and fuel takes place with the aid of an oxygen-carrier (OC). The oxygen-carrier extracts O_2 from air in one reactor and then transfers it to fuel in a subsequent reactor. Since the fuel does not come in direct contact with air, the products of combustion contain only carbon dioxide (CO_2) and water (H_2O). A CO_2 stream of very high purity can then be obtained by condensing the water vapor. Gaseous fuels such as natural gas or syngas (CO and H_2) can be used as fuels in CLC, for inherent CO_2 capture. The focus of this paper is the use of CLC for CO_2 capture in a coal-based integrated combined cycle power plant (IGCC), in lieu of the conventional approach to capture.

CO_2 capture in conventional IGCC designs involves a water gas shift reaction to convert CO in the syngas from the gasifier to CO_2, followed by subsequent capture of CO_2 using a physical absorption solvent (Selexol or Rectisol). Alternatively, CLC can be used for CO_2 capture in a coal-based IGCC plant. The products from the CLC reactor are separate streams of oxygen-depleted air and combustion products (CO_2 and H_2O), both at high temperature and pressure. These streams can be expanded in separate turbines to generate electricity. The hot exhaust from these turbines can be used to generate steam for further electricity production. Water can be condensed from the CO_2-rich stream in order to obtain high purity CO_2 for further compression and storage.

The use of CLC in place of a conventional CO_2 capture process in an IGCC system thus involves both addition and removal of certain processes and equipment. It is important to study the advantages and disadvantages of CLC in terms of its effects on the performance of the whole system, compared to a conventional IGCC. In this paper a performance model is developed to perform a detailed thermodynamic analysis of the IGCC plant using CLC for CO_2 capture. Chemical equilibrium models are used to simulate the CLC reactions using NiO/Ni as the oxygen carrier. The effect of varying the syngas composition and process operating conditions (.e.g., temperature and pressure) also are studied.

The performance results are compared to a conventional IGCC power plant with CO_2 capture in order to understand the relative advantages and disadvantages of CLC when applied to IGCC. The needs for further model development also are elaborated.

This work is supported under a contract from the U.S. Department of Energy's National Energy Technology Laboratory (DOE/NETL)

See more of this Session: Calcium Looping and Chemical Looping Processes for CCS (II)
See more of this Group/Topical: Topical 2: Advanced Fossil Energy Utilization

(24b) A Parametric Study of a Fluidized Bed Carbonator to Remove CO_2

Rongrong Zhai and Yongping Yang, North China Electric Power University, Beijing, China

Post-combustion method has been viewed as the most suitable way for exiting power plants' CO_2 mitigation. This paper mainly focuses on the utilization of reversible carbonation reaction of CaO to capture CO_2 from the flue gas and the effects of superficial gas velocity, the particle diameter and the calcination/carbonation cycle number on the carbonation process. A model has been built based on the KL fluidized bed model, and the results of the model have been compared with other researchers' experimental data. It is concluded that the model can well present the experimental data; with the increase of superficial gas velocity, the carbonation process first turns worse then better; with the decrease of particle size, the absorption of CO_2 turns better, but there is still some limit to avoid the sintering; with the increase of cycles, the activity of the absorbent decays and the absorption turns worse. Therefore, in order to optimize the carbonation process the superficial gas velocity can increase gradually during the reaction and smaller-sized absorbent should be chosen. However, the limits of superficial gas velocity and absorbent size need to be taken into consideration as well to avoid sintering.

See more of this Session: Calcium Looping and Chemical Looping Processes for CCS (II)
See more of this Group/Topical: Topical 2: Advanced Fossil Energy Utilization

Packed-bed Chemical-Looping Combustion: Theoretical investigation and experimental validation

M. van Sint Annaland*, S. Noorman, F. Gallucci, J.A.M. Kuipers

Chemical Process Intensification, Multiphase Reactors, Department of Chemical Engineering and Chemistry, Eindhoven University of Technology, Eindhoven, The Netherlands

Chemical looping combustion in dynamically operated packed bed reactors has been investigated both theoretically and experimentally. The effects of operating temperature, feed oxygen concentration, steam addition and type of oxygen carrier material on the process performance have been evaluated, with detailed models and with dedicated experiments.

The prevailing phenomena during reduction and oxidation cycles that the oxygen carrier particles undergo in Chemical Looping Combustion processes have been studied in great depth using an advanced numerical model in which the effects of reaction kinetics and internal and external mass and heat transfer processes were incorporated in detail. The particle model was validated through TGA experiments carried out on different oxygen carriers at different operating conditions. With the validated model, improved understanding of the effect of mass and heat transfer processes on the rate at which oxygen carriers are reduced and oxidized has been obtained, in particular the effect of drift fluxes, which is essential for an accurate prediction of the performance of packed-bed CLC processes at larger scales.

Subsequently, a combined reactor and particle model was developed with which the effect of mass and heat transfer limitations inside the oxygen carrier particles on the axial concentration and temperature profiles in packed-bed CLC can be properly accounted for. The model was validated via experiments carried out in a lab-scale dynamically operated packed bed for CLC of methane. Conversion and selectivities during both oxidation and reduction cycles along with the axial temperature profiles in the bed have been evaluated and compared with modeling results.

With these models, more reliable predictions of the axial concentration and temperature profiles in packed-bed CLC can be given, which are necessary to arrive at a proper design and a quantitative assessment of the techno-economical viability of packed-bed CLC.

Keywords: Chemical looping combustion, packed bed, dynamic operation, oxygen carrier

* Corresponding Author: Martin van Sint Annaland, e-mail: M.v.SintAnnaland@tue.nl tel: +31 40 247 2241 fax: F +31 40 247 5833

11 AIChE
Spring Meeting, Chicago, IL

2011 AIChE Spring Meeting
T2 - Topical Conference: Advanced Fossil Energy Utilization
Calcium Looping and Chemical Looping Processes for Carbon Capture and Sequestration (CCS)

Extended Abstract
Paper # 211278

High-temperature CO_2 capture with modified CaO-based pellet sorbents: TGA results

Yinghai Wu, Vasilije Manovic, Ian He and Edward J. Anthony*
CanmetENERGY, Natural Resources Canada
1 Haanel Dr. Ottawa, ON K1A1M1

Introduction

Carbon dioxide is one of the major greenhouse gases that contribute to anthropogenic global climate change. One possible mitigation proposal is the capture of CO_2 from large stationary sources, such as fossil fuel (especially coal) combustion power plants, for sequestration in geological formations. However, the CO_2 concentration produced in normal power plants is diluted. The goal of carbon capture technology is the production of a concentrated stream of CO_2 that can be readily liquefied, piped and sequestered. Solid looping cycles offer an important new technique for CO_2 separation. These cycles employ a solid carrier to either bring oxygen to the fuel or remove CO_2 from combustion or gasification gases to be released as a pure CO_2 stream, with subsequent regeneration and re-use of the carrier [1]. CaO-based calcium looping cycles, which utilize reversible carbonation and calcination reactions (Eq. 1), offer a promising technology to separate CO_2 from flue or syngas while producing a high-purity CO_2 stream suitable for sequestration and, in the case of syngas application, for sorption-enhanced H_2 production.

$$CaO(s) + CO_2(g) = CaCO_3(s)\ \Delta H < 0 \qquad (1)$$

Preliminary economic analyses suggest that such processes are economically attractive, and an important advantage of using CaO-based sorbent is that limestone ($CaCO_3$) is widely and inexpensively available. It has been proposed that CO_2 separation from flue gas is possible in a multi-cycle process in a dual FBC reactor [2]. This involves reaction of CaO with CO_2 from flue gas/syngas in a carbonator and regeneration of sorbent and obtaining of a concentrated CO_2 stream in a calciner.

* Corresponding author. Tel.: +1-613-943-7773; fax: +1-613-992-9335.
E-mail address: ywu@nrcan.gc.ca

Naturally occurring limestones (mostly calcitic) have so far been the subject of intensive research [3]. However, natural sorbents lose their CO_2 carrying capacity quickly with increasing numbers of reaction cycles. Hence, hydration of the spent sorbents has been proposed as a method to recover the lost carrying capacity; unfortunately, the resulting reactivated sorbents tend to be fragile and are unsuitable for fluidized bed conditions. To overcome this drawback, a novel method using pelletization of the sorbents is proposed here. In this study, the sorbents were pelletized, with or without calcium aluminate cements as the binder for pelletization and the performance of the pelletized sorbents was investigated in a thermogravimetric analyzer (TGA) reactor with simulated cyclic conditions.

Experimental

Powders of quick lime (CaO) and hydrated lime ($Ca(OH)_2$) (<75 μm), obtained from Graymont Limestone Inc., were used as starting material for pelletization. The pellets were made by a mechanical pelletizer in batch mode, using calcium aluminate cements (CA-14M) as the binder. The powdered sorbent and binder (~300 g) were mixed in the pelletization vessel (1 L column container) with the weight ratio of sorbent to cement being 9:1. The type of binder and the sorbent/binder weight ratio were selected based on previous experience of binder screening [4, 5]. The CaO and $Ca(OH)_2$ pellets without addition of binder were also made for blank tests. A certain amount of water was sprayed intermittently during pelletization with a nozzle, which can produce the micron-sized water droplets (~5 μm) required for the pelletization process. The amount of water varied depending on the starting material used. The water droplet size and the total amount of water added appear to be the most important factors which affect the size of the pellets. The pellet size was also controlled by the speed of a pair of rotor blades attached to the vessel, i.e., one agitator located on the bottom and one chopper on the side. Typically one batch of pellets can be produced in 20-30 min. After pelletization, the pellets were sieved to desired size ranges and dried in open air for days before storage. In this work, most tests were carried out with pellets in the size range of 425-1000 μm, although other size ranges, such as 1-1.4 mm, were also investigated for some conditions.

The pellet sorbents were examined using nitrogen physisorption analysis for pore surface area and pore size distribution. Multi-cycle CO_2 carrying capacities for pellets were determined with a Mettler Toledo TGA using ~30 mg samples in an alumina crucible. The gas flow rate, controlled by a flowmeter, was 0.04 dm^3/min, and the temperature and gas used were controlled by the TGA software. Data on sample mass during the experiments were monitored continuously and it is assumed that mass change occurs only because of the formation/decomposition of $CaCO_3$. Both carbonation and calcination were performed isothermally at 800 °C unless specified otherwise: carbonation in 20% CO_2 (N_2 balance) for 30 min and calcination in 100% N_2 for 15 min. Typically 30 cycles of carbonation and calcination were conducted.

Results and discussion

The surface property analysis of $Ca(OH)_2$-precursor samples are shown in Table 1. The addition of cement as the binder resulted in produced pellets with increased surface area, pore diameter and pore volume compared with the pellets made from $Ca(OH)_2$ powder only.

Figure 1 shows the typical CO_2 carrying capacity of pellet sorbents under repeated carbonation and calcination cycles in the TGA.

Table 1. Surface analysis of pellet sorbents with $Ca(OH)_2$ as the starting material

Sample	Pellet size range, µm	BET, m^2/g	Average pore dia., nm	Total pore volume, cm^3/g
$Ca(OH)_2$ powder	425-1000	7.5—11.4	7.1—8.5	0.014—0.024
	<425	13.4	9.1	0.03
	850-1400[2]	9.6	7.5	0.018
$Ca(OH)_2$ powder + cement	425-1000	10.3—12.5	10.5—10.9	0.027—0.034
	<425[2]	13.1	12	0.039
$Ca(OH)_2$ pellets + cement	425-1000	13.2	9.1	0.03
	1000-1400	10.6	8.9	0.024
	1400-2000	11.0	9.4	0.026

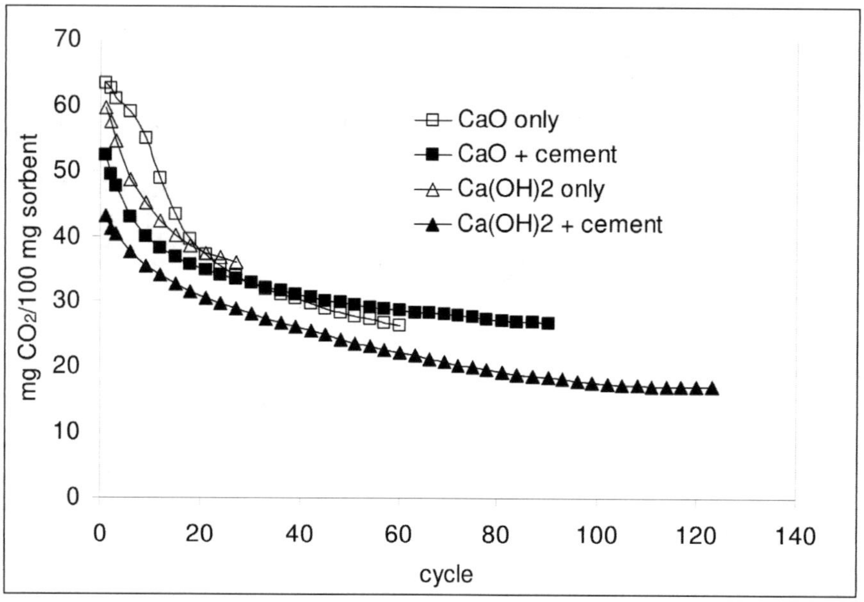

Figure 1. CO_2 carrying capacity over multiple reaction cycles

The carrying capacity of CaO samples without cement addition was higher than that of the samples with cement addition for the first 35 cycles. After that samples with cement addition maintained a higher carrying capacity. In contrast, the $Ca(OH)_2$ samples without addition of cement always had a higher carrying capacity than samples with addition of cement. However, in the longer term (i.e., > 80-100 cycles), the carrying capacity of pellet samples with the addition of cement appeared to be more stable or even slightly increased. The increase in carrying capacity over longer-term cycles was not investigated for the natural limestone samples in this study because, based on extensive prior work, it is expected to be lower. It is also noted in the TGA tests that the particle size of the pellets had limited impact on the carrying capacity.

Conclusions

This work has produced novel CO_2 sorbents using a mechanical pelletizer for calcium looping. The performance of these pellet sorbents was evaluated in the TGA environment under simulated cyclic conditions for carbonation and calcination. The samples with addition of cement as the binder for pelletization showed a stable carrying capacity over long-term multi-cycle tests. Future work will study the attrition properties of these pellet sorbents, using a small-scale bubbling fluidized bed and the CO_2

carrying capacity will also be investigated under more realistic conditions using CanmetENERGY's pilot plant facility.

References

1. Anthony, E.J., *Solid Looping Cycles: A New Technology for Coal Conversion.* Industrial & Engineering Chemistry Research, 2008. **47**(6): p. 1747-1754.
2. Shimizu, T., et al., *A twin fluid-bed reactor for removal of CO2 from combustion processes.* Trans IChemE, 1999. **77, Part A**: p. 62-68.
3. Blamey, J., et al., *The calcium looping cycle for large-scale CO_2 capture.* Progress in Energy and Combustion Science, 2010. **36**(2): p. 260-279.
4. Manovic, V. and E.J. Anthony, *Screening of Binders for Pelletization of CaO-Based Sorbents for CO_2 Capture.* Energy & Fuels, 2009. **23**(10): p. 4797-4804.
5. Manovic, V. and E.J. Anthony, *Long-Term Behavior of CaO-Based Pellets Supported by Calcium Aluminate Cements in a Long Series of CO_2 Capture Cycles.* Industrial & Engineering Chemistry Research, 2009. **48**(19): p. 8906-8912.

Scale-up of fixed-bed chemical looping combustion

E. Kimball, P. Geerdink, E.L.V. Goetheer
TNO Separation Technology
Delft, the Netherlands

Summary

Fixed-bed chemical looping combustion is currently being studied experimentally with the optimization of systems up to 1 kW. One of the key advantages for using the fixed-bed configuration over fluidized-beds is the flexibility in operating parameters that allows for application of CLC not only to power production, but also to niche applications, such as the decoupled production of heat and CO_2 for the greenhouse industry. Another advantage is the modularity of the systems that greatly simplifies scale-up once the optimal single reactor design is realized. This is currently being studied by optimizing the production of pure CO_2 while minimizing the pressure drop over the reactor. With results for three different sizes of reactors, optimal packing and operating conditions can be predicted for increasingly larger systems while being tailored to the specific application.

Introduction

Chemical Looping Combustion (CLC) is a technology in which the combustion of a fuel is accomplished in two steps. First, the oxygen is taken up from air through the oxidation of a bed of metal particles, forming metal oxides and producing a stream of hot, depleted nitrogen. Second, the fuel is combusted as it flows through the metal oxide bed, reducing the bed back to metal and producing a stream of CO_2 and water. In this way, the metal particles, the *oxygen carriers*, are looped between oxidized and reduced states. The key advantage in relation to CCS is the elimination of the need to separate the CO_2 from the nitrogen at the outlet of a conventional combustion reactor. CO_2 is usually present in the nitrogen in low concentrations and must be separated by energy intensive absorption processes in order to achieve the required purity needed for storage or re-use, now considered to be at least 95%. With CLC, the separation is inherent. Simple condensation of water from the combustion product stream yields essentially pure CO_2 the can be readily compressed for transport and storage.

The conventional configuration for chemical looping systems is with dual circulating fluidized beds—a smaller reactor where the fuel is fed to reduce the metal oxides and a larger reactor where high volumes of air oxidize the metal. The size of the particles and the gas flow rates must be chosen correctly so that the bed fluidizes, the particles have the right residence time in the reactor, and the fuel conversion is as high as possible. Between the air and fuel reactors, the particles must be separated from the high flow and hot depleted air. This requires a large cyclone and seals that make the system more costly and increase the mechanical degradation of the particles. These factors become much more difficult to manage and introduce new technical difficulties with each new step up in scale.

As an alternative to the circulating fluidized bed configuration, TNO has developed a fixed-bed CLC process that allows for much more flexibility in operating parameters and has no need for expensive cyclones and loop seals. It has been noted that in a fixed-bed configuration, large valves will be needed and will be costly, but initial designs indicate that this will not be a showstopper. The concept, shown schematically in Figure 1, is to switch the feed gas to the reactor between air (for the oxidation step) and fuel (for the reduction step). For continuous flows, such as the production of a hot stream of nitrogen to be sent to a gas turbine, at least two reactors would be operated simultaneously, one always being oxidized and another always being reduced. Scale-up is accomplished by first finding the optimal size for an individual reactor based on costs and operability, then simply adding more reactors running in parallel. The design is modular—no new technical hurdles are introduced with each consecutive step in scale and the time from development to demonstration is expected to be much shorter than with fluidized beds.

Currently, the optimal parameters and operating conditions for an individual reactor are still being determined as the focus of several projects at TNO. The fixed-bed concept was originally applied to the niche application of CO_2 and heat production for the greenhouse industry in Holland and is still an active area of research. The performance of the system is measured by the ability to avoid slippage of fuel at the outlet of the reactor and maintaining a low pressure drop (< 1 bar) through the reactor. The ability to operate a fixed-bed system under a range of conditions has the added advantage of allowing for comparison between potential oxygen carrier (OC) materials and possible degradation testing. Furthermore, the knowledge gained from the experimental results has also been translated into process models of CLC integrated into an IGCC power plant to study the benefits of CLC on an industrial scale.

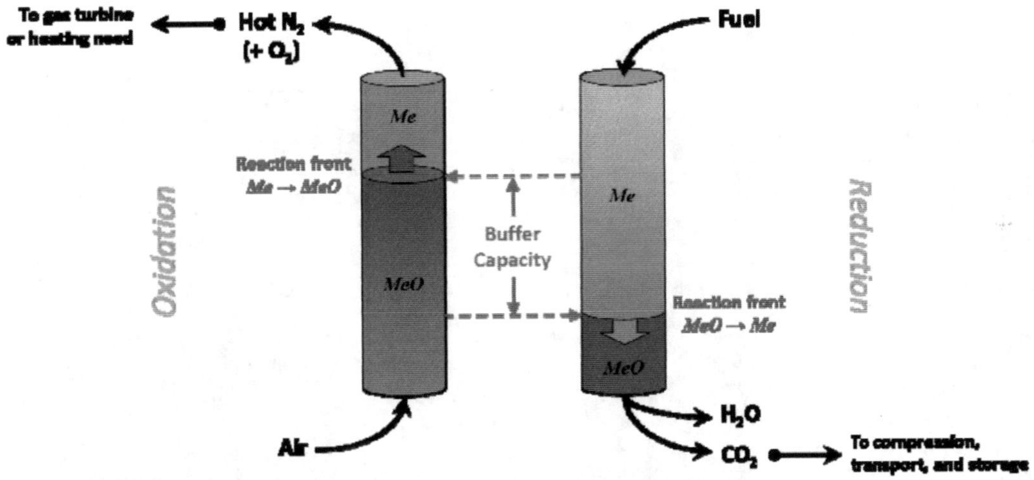

Figure 1 Schematic of fixed-bed CLC showing both the oxidation and reduction cycles. A single reactor can be used for sequential heat and CO_2 production or for buffering while multiple reactors operating in parallel at different steps would provide continuous flows.

Development through niche application

Fixed-bed CLC technology was originally developed at TNO for the Dutch greenhouse industry. In the Netherlands alone, greenhouses occupy about 106 million m^2 of land area and would greatly benefit from the development of a system for uncoupled CO_2 and heat production. During the day, the concentration of CO_2 inside the greenhouse must be increased to accelerate the growth of the produce. For a normal sized greenhouse of 50,000 m^2, this is a requirement of about 1 kton CO_2/yr. During the night, CO_2 is no longer needed, but instead, heat must be produced as the outside temperatures drop below those which the plants can withstand. Traditionally, the required heat is produced with a local CHP plant, consuming about 800,000 m^3 of natural gas per year for the same sized greenhouse. In some cases, the CO_2 produced from the combustion of natural gas to produce heat can be used, but this is problematic for two main reasons. The first is that the CO_2 is not needed during the night when it is produced and so must be stored until it can be used during the day. The second is that several pollutants are present in the exhaust of the CHP system and must be separated before the CO_2 is used, especially ethylene which can be extremely harmful to plants, fruits, and vegetables. Often, greenhouse owners must resort to purchasing CO_2 separately from what they produce themselves from the CHP system.

By uncoupling the CO_2 and heat production while at the same time producing much cleaner CO_2, a CLC system is uniquely suited to fulfill the requirements of the greenhouse industry. Fixed-bed CLC in particular allows for the flow rates of the feed gases to be chosen so that CO_2 and heat are both supplied continuously for the individual amount of time needed. For example, CO_2 may be needed for 16 hours during the day, but heat may only be needed for 8 hours when the temperature decreases at night. Without the need for fluidization of the bed particles, adjustments are easily accomplished and the flows can be tailored to match the seasonal and regional dependences of the area where the particular CLC system operates. The performance of the process is measured by the production of pure CO_2 (before non-combusted fuel begins to break through at the outlet of the reactor) relative to the maximum amount of CO_2 able to be produced based on the amount of oxygen carrier material in the bed. The pressure drop is a further indicator to guide the system design.

Experimental results from scale-up

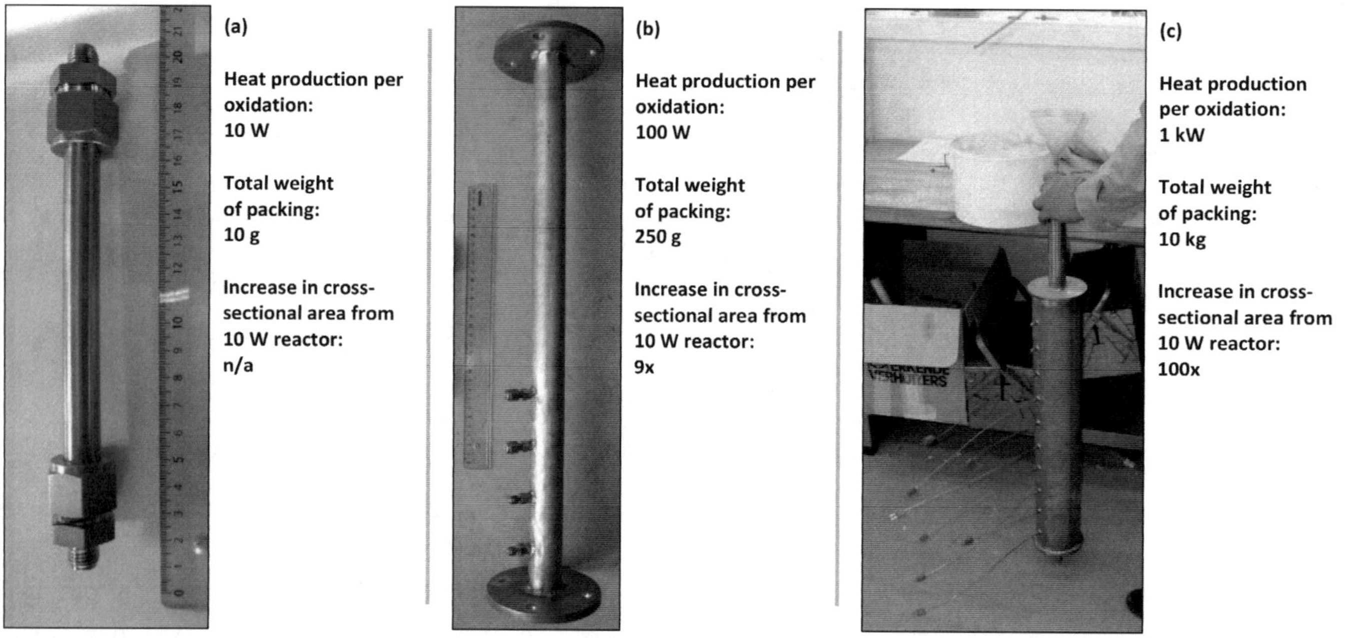

Figure 2 Photos of the fixed-bed CLC reactors at TNO for each successive step in scale during development.

The development of fixed-bed CLC has mostly been accomplished through experimental work with successively larger reactors, as shown in Figure 2. Experiments were first conducted with a small 1 cm D x 15 cm L tubular reactor placed inside an oven for temperature control. The estimated heat production was 10 W_{th} with a total packing of 10 g. The method of packing and operational parameters were chosen to minimize the pressure drop while maximizing the production of pure CO_2. Following successful results, the next larger scale reactor with a cross sectional area and heat production both about 10 times greater was constructed. Operation was also done inside the oven for temperature control. It should be noted that changing the length of the reactor changes the total amount of packing and thus the cycle times for oxidation and reduction, but does not change the optimal operating parameters to optimize the pressure drop and CO_2 production.

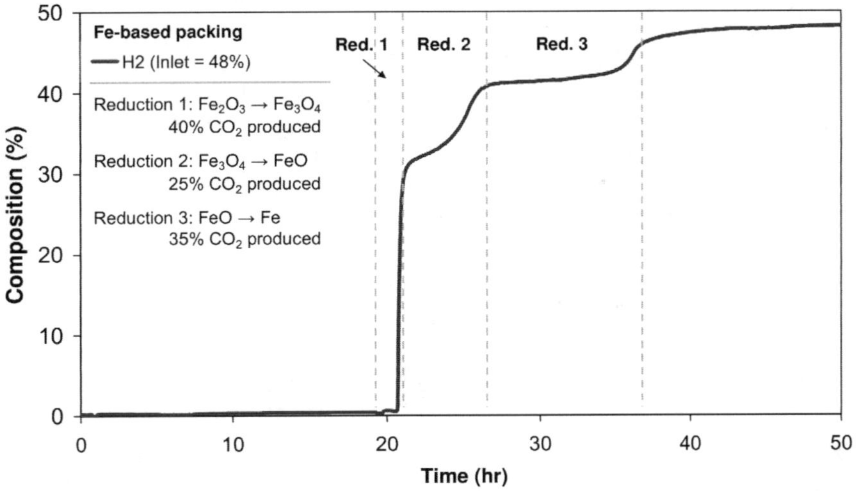

Figure 3 Percentage of H_2 at the outlet of the fixed-bed CLC reactor with an iron-based packing and syngas as the fuel. The different oxidation states of iron are clearly seen and lead to incomplete combustion of the fuel and contamination of the CO_2 stream.

The parameters that are changed easily between experiments include the choice of OC material, reactant flow rates, and feed composition. For large scale energy production, the required amount of material is on the order of hundreds of tons and must be cheap for the system to be economically feasible. Iron ores are especially attractive in this regard, but have been shown to be problematic. Due to the different oxidation states of iron, the reduction does not complete and non-combusted fuel slips out of the reactor with the CO_2. Figure 3 shows this very clearly with the results for the reduction of an iron-based bed with a syngas mixture. During the first stage of the reduction, hematite (Fe_2O_3) is converted to magnetite (Fe_3O_4), which reacts with the syngas completely to produce only CO_2 and water at the outlet. This is indicated by 0% H_2 measured at the outlet. When magnetite begins to be converted into wüstite (FeO), the syngas slips and both H_2 and poisonous CO are present in the CO_2 stream. Finally, when wüstite is converted into metallic iron (Fe), the reaction is even less complete and still more reactants slip. A fixed-bed system greatly aids in seeing these effects of the oxidation states of the OC and can further be used to study similar phenomena from the use of novel OC materials. It is worth noting that it is possible to simply stop the reduction before the second reduction stage starts and avoid the slip of reactants, but this reduces the efficiency of the catalyst utilization in the bed. An alternative approach was found by TNO that still allows for the use of large amounts of iron for the process, but also ensures much better control over the reactant slip while maintaining a high catalyst utilization.

Figure 4 shows the experimental results for a modified iron-based packing in the 100 W reactor (Figure 2(b)), still utilizing syngas as the fuel. A reduction and an oxidation cycle are shown as the composition of the gases at the outlet of the reactor, both resulting in sharp breakthrough curves of the reactant gases—CO and H_2 for the reduction and O_2 for the oxidation. The pressure drop over the reactor is also shown with only a very small rise of 0.015 bar or less during each cycle. The temperature of the oven was set at 700 °C in this case, but experiments have also been run successfully up to 900 °C. With correctly chosen flow rates and conditions, it was possible to produce 92% of the maximum amount of CO_2 (based on the amount of oxygen taken up by the OC material in the bed) before the fuel breakthrough.

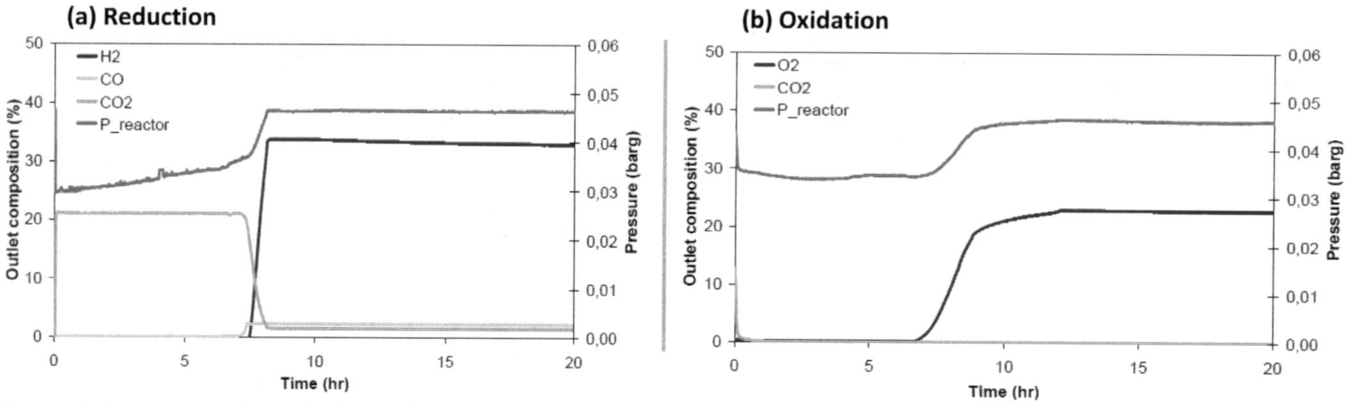

Figure 4 Experimental results for both a reduction and an oxidation cycle with a fixed-bed CLC reactor utilizing a modified iron-based bed and syngas as the feed. Sharp breakthrough curves are observed as the cycles complete and the bed is fully reduced (part a) or oxidized (part b).

The production of CO_2, relative to the theoretical maximum amount, was affected by several factors. The most important was the choice of flow rates. For the greenhouse application, where CO_2 at high purity was one of the desired products, low flow rates were desired to maximize the amount of pure CO_2 produced, but also advantageous for supplying CO_2 to the greenhouse steadily over several hours. For power production, the requirements are very different—large volumes of gases and short cycle times—but a high fuel conversion is still important. Experiments were carried out in the 100 W reactor to investigate the trade-off between high fuel conversion (also leading to high CO_2 production) and fast cycle times (achieved with higher flow rates). The results for a specific set of conditions, given in Figure 5, show a relationship between flow rate and CO_2 production (part a) that reaches a maximum of about 54% at lower flow rates with no benefit when reducing the flow rate from 400 to 200 ml/min. However, as shown in part (b), decreasing the flow from 400 to 200 does cause a significant increase in the cycle time, which may be undesirable, especially for power production. It is also evident that increasing the flow rates much beyond 1000 ml/min will not decrease the cycle time much below 4 hours for the particular fixed-bed used in these experiments. Instead, with higher flow rates, more reactants will be present at the outlet and the fuel conversion efficiency will be lowered.

Following successful operation at the 100 W scale, a 1 kW reactor was constructed, as shown in Figure 2(c). The operation is still in the process of being optimized, but results have been promising with controllable breakthrough curves and small pressure drops. The scale-up of the fixed-bed CLC system is currently following a relatively fast timeline with only 6 months of development required between each order of magnitude of scale, and the trend is expected to continue. Figure 6 shows the results for one oxidation-reduction cycle utilizing a copper-based packing and syngas as the fuel. The experiments were conducted inside an oven set at 700 °C, but future experiments will be conducted without external heat control while monitoring the temperature profile within the reactor during operation.

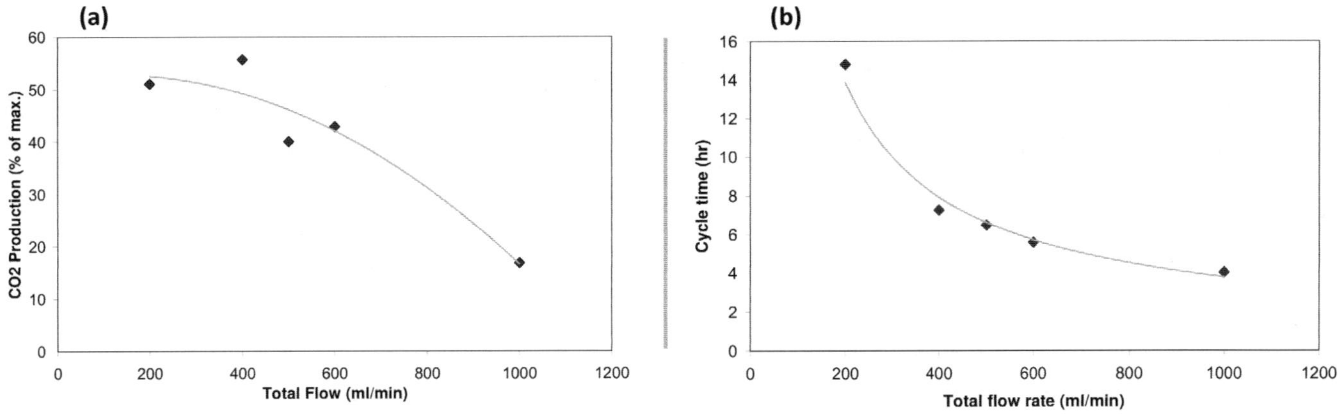

Figure 5 Total production of pure CO_2 for one reduction cycle versus total flow rate of a syngas mixture to the fixed-bed reactor with a modified iron-based packing. The data follow power law and second-order polynomial relationships for parts (a) and (b), respectively. The light trendlines are given only as a guide and will require several more experimental results to be validated.

The curves given in Figure 6 indicate the composition of the gas at the outlet of the reactor. During the oxidation step, only air is fed to the reactor so that the percentage of oxygen at the outlet indicates the degree of oxidation of the bed—the bed is fully oxidized when the composition reaches 21% O_2. For about 5 hours, the oxygen is completely consumed, but then gradually increases in concentration at the outlet to 21% over the next 25 hours. This is an extremely long cycle time and shows a need for higher flow rates or a modification in the bed packing. The reduction step, on the other hand, shows much sharper breakthrough curves, especially for hydrogen, and a shorter cycle time of about 15 hours. Given the importance of fuel efficiency and CO_2 purity, this is the more significant result. It is also important to note the pressure drop over the reactor. The pressure increases from 0.8 bar to only about 1.1 bar during the oxidation, but then decreases back to about 0.95 during the reduction. A small fluctuation of about 0.2 bar is observed as the reduction completes, but then the pressure restabilizes before the next oxidation is started.

Given the results from the three different reactor sizes, relationships for the dependence of the operational parameters on the scale of the system can be inferred. As the reactor becomes larger and the gas flows become higher, there is more chance for mass transfer resistances and reaction kinetics to become important and limit the operation. Furthermore, since all of the experiments conducted so far have been inside an oven, temperature changes due to the heats of reaction will become necessary to consider for larger reactors, where ratio of the heat transferred through the walls of the reactor to the heat generated or consumed by the reaction is decreased.

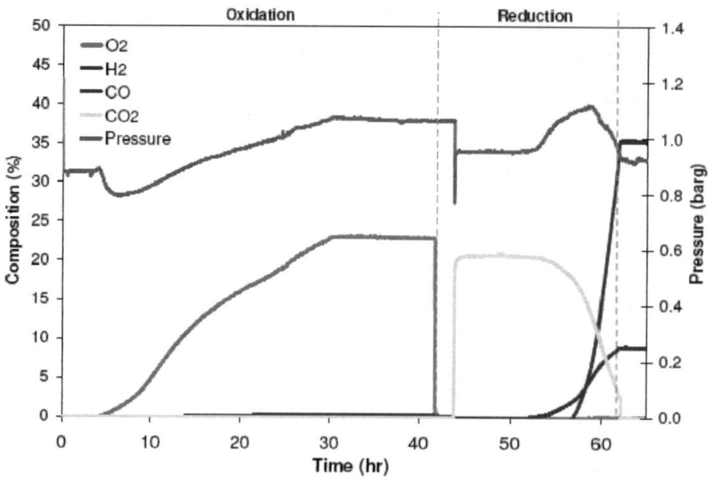

Figure 6 Operational results for the 1 kW scale reactor utilizing a copper-based bed and syngas as the fuel. Both an oxidation and a reduction step are shown by the composition of the gas at the outlet of the reactor and the pressure drop over reactor.

These factors will effect the operation of the CLC reactor in terms of the fuel efficiency. As shown in Figure 7, the efficiency can be expressed as the fuel or air requirement per amount of OC material to complete a full reduction or oxidation cycle, respectively. The fuel requirement for the reduction step appears to follow a logarithmic relationship (noting the log scale on for the x-axis). This is encouraging for scale-up and indicates that with each step in scale, the relative increase in amount of fuel required is decreased. Improving the packing of the bed or adjusting the flow rates should allow for this relationship to become less and less dependent on the reactor size. The relationship for the air requirement to complete one oxidation cycle is shows a higher dependence on the reactor size—apparently a linear relationship—but is less important for the feasibility of fixed-bed CLC. Large flows are needed when considering CLC for power production and the presence of non-reacted oxygen at the outlet of the reactor does not have negative implications. Instead, maintaining a high temperature of the depleted air is the key factor and will be investigated further by performing experiments in a stand-along configuration without the use of an oven.

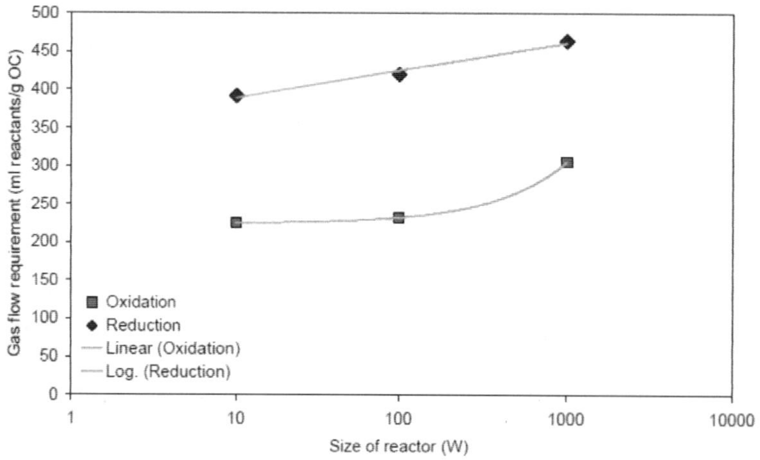

Figure 7 Change in efficiency for both oxidation and reduction steps for the three different reactor sizes utilizing copper-based packing, utilizing syngas as the fuel, and being operated inside an oven at 700 °C. Trendlines are given as a guide only and will require further testing of larger scales for validation.

Conclusions

Experimental results have shown the feasibility of fixed-bed CLC with both iron and copper-based packings and operated with syngas. The ability to have a sharp breakthrough of the fuel at the outlet of the reactor is an important parameter and the amount of pure CO_2 that can be produced before contamination from the slip of fuel gives a measure of the performance of the system. The total fuel required to complete a full reduction is another indicator of efficiency and has been shown to have a logarithmic relationship with respect to the size of the reactor. Improvement in the bed design and choice of flow rates should be able to decrease the dependence on scale; this is the subject of continuing work with the optimization of the 1 kW reactor and construction of increasingly larger systems. The operation of the reactor without external temperature control from an oven may also have a significant effect on the results and will be investigated further. The latter factor will allow for the optimization of the oxidation step as the heat management is becomes more significant than the oxygen consumption efficiency.

The design and operation of fixed-bed CLC systems has shown potential for use in both power generation systems and novel niche applications. Without the limiting requirements for bed fluidization and optimal residence times, the flow rates of the reactant gases can be adjusted and tailored to the specific needs of each system. This is an important benefit, especially for applications such as that for greenhouses, where both CO_2 and heat are needed, but at different times and over the course of hours. The ability to observe breakthrough curves for the reactants at the outlet of the reactor is also beneficial for studying and comparing different oxygen carrier materials. The different oxidation states of a material can be directly observed through changes in the composition of the outlet gas. Future work will also look at the degradation of materials as the conditions that the particles are exposed to are well defined along the length of the reactor. This is a secondary advantage of utilizing fixed-beds over circulating fluidized beds, where the particles can only be described by a residence time distribution in each reactor.

The ability to scale-up fixed-bed CLC systems has been proven up to a 1 kW reactor and will continue to still larger scales. Once the optimal reactor size is found, further scale-up is accomplished simply by increasing the number of reactor modules without the need to redesigning the reactor itself. The experimental results are currently being combined with both reactor models and process simulations (not discussed here) to facilitate the application of fixed-bed CLC in a variety of applications, from niche applications to large scale integration into power plants. The time for development is currently following a trend of only 6 months between each step in scale and this is expected to continue so that commercial fixed-bed chemical looping combustion systems are soon realized.

(24f) Combined Biomass Gasification and Chemical Looping Combustion for Hydrogen Production with near Zero Carbon Dioxide Emission

Changqing Dong, Jingzhou Jiang and Qiang Lu, North China Electric Power University, Beijing, China

In this study, we proposed a process for auto-thermal hydrogen production with near zero emissions of carbon dioxide, through combined biomass gasification and chemical looping combustion. A new set was developed for this process, which mainly consisted of three interconnected fluidized bed reactors, the biomass gasifier, air reactor and fuel reactor. Metal oxides were used as the oxygen and heat carriers in the whole process. In the gasifier, biomass was heated and gasified in the steam atmosphere, with heat supplied by the hot low-valent metal oxides from the fuel reactor. The product gas was purified and submitted to hydrogen separation using the pressure swing adsorption. The residual gases with low hydrogen were transferred to the fuel reactor, and reacted with the metal oxides to form low-valent metal oxides and carbon dioxide of high concentrations. The low-valent metal oxides were transferred to the biomass gasifier to supply heat for the gasification process, and the carbon dioxide could be collected for storage. The low-valent metal oxides after the heat supply for the gasification process were transferred to the air reactor, where they reacted with oxygen to form metal oxides, and meanwhile, generate a lot of heat for the whole process. Successful preliminary experiments have been performed on this set to achieve auto-thermal hydrogen production and carbon dioxide capture.

See more of this Session: Calcium Looping and Chemical Looping Processes for CCS (II)
See more of this Group/Topical: Topical 2: Advanced Fossil Energy Utilization

(24g) Comparative Investigation on Chemical Looping Combustion of Coal-Derived Synthesis Gas Containing H_2s Over Supported Fe_2O_3 - MnO_2 Oxygen Carrier

Ewelina Ksepko[1], Ranjani V. Siriwardane[2], Hanjing Tian[3], Thomas Simonyi[4], Sciazko Marek[1] and James Poston[2], (1)Institute for Chemical Processing of Coal, Zabrze, Poland, (2)National Energy Technology Laboratory, U.S. Department of Energy, Morgantown, WV, (3)URS, Morgantown, WV, (4) U.S. Department of Energy, National Energy Technology Laboratory, Morgantown, WV

The paper contains results of collaborative research work on novel combustion technology known as chemical looping combustion (CLC). The objective of paper was to prepare Fe_2O_3 - MnO_2 supported on ZrO_2/Al_2O_3/Sepiolite (ICPC, Poland) oxygen carriers and to evaluate the performance (NETL, US DOE) of these for the CLC process with synthesis gas/air. Thermo gravimetric analysis (TGA) and low pressure (10 psi) bench scale flow reactor tests were conducted to evaluate the performance. Multi cycle tests were conducted in an atmospheric TGA with oxygen carriers utilizing simulated synthesis gas with & without H_2S. Effect of H_2S impurities on both the stability and the oxygen transport capacity was evaluated. Multi cycle CLC tests were also conducted in the bench scale flow reactor at 800 °C with selected samples. Chemical phase composition was investigated by X-Ray diffraction (XRD) technique.

Five Cycle TGA tests at 800 °C indicated that all oxygen carriers had a stable performance at 800 °C. It was interesting to note that there was complete reduction/oxidation of the oxygen carrier during the 5-cycle test. The fractional reduction, fractional oxidation and global reaction rates were calculated from the data. It was found, that support had a significant effect on both fractional reduction/oxidation and the reaction rate. The oxidation reaction was significantly faster than the reduction reaction for all oxygen carriers. The reaction profile was changed by the presence of H_2S but there was no effect on the reaction rate due to presence of H_2S in syntheses gas. Low pressure bench scale flow reactor data indicated stable reactivity, full consumption of oxygen from oxygen carrier and complete combustion of H_2 and CO. XRD data of samples after multi-cycle test showed stable crystalline phases without any formation of sulfides or sulfites/sulfates and complete regeneration of the oxygen carrier after multi-cycle tests.

See more of this Session: Calcium Looping and Chemical Looping Processes for CCS (II)
See more of this Group/Topical: Topical 2: Advanced Fossil Energy Utilization

(31a) Upgrading Bed Support Systems Utilizing Vee-Wire Technology

Benjamin Schmitt P.E., Hydrocarbon Processing Industry Business Unit, Johnson Screens, New Brighton, MN

Abstract:

Upgrading Bed Support Systems, Utilizing Vee-Wire Technology

By: Benjamin Schmitt, PE

Date: October 18, 2010

Hydrocarbon industry processes utilize media (sieve or catalyst) to catalyze a process or adsorb impurities and other contaminates. Support structures are an integral part of the process used to prevent the media from passing down-stream creating problems with mechanical devises or impact end product specifications. Media retention structures vary from mesh over grating or perforated plate to Vee-Wire screen. A technical review of the advantages for utilizing Vee-Wire technology will be presented, sighting case studies and supporting data. Topics of discussion will include process improvements, mechanical advantages, service life and capital recovery of Vee-Wire support structures.

See more of this Session: Innovations in Energy Systems Design
See more of this Group/Topical: Process Development Division

Theoretical Modeling of a Novel Combination of Pd-Ag Membrane Reactors in Methanol Synthesis

A. A. Forghani, M. Bayat, M. R. Rahimpour, School of Chemical and Petroleum Engineering, Shiraz University, Shiraz, Iran

Introduction

Fundamentally, methanol is produced by catalytic conversion of synthesis gas (CO_2, CO and H_2). The factors affecting the production rate in industrial methanol synthesis are parameters such as thermodynamic equilibrium limitations and variation in stoichiometric number. For equilibrium reactions, selective product removal or reactant addition may be used to increase conversion. Up to the maximum production rate point, increasing temperature improves the rate of reaction, which leads to more methanol production. Nevertheless, as the temperature increases beyond this point, the deteriorating effect of equilibrium conversion emerges and decreases methanol production [1]. Rahimpour proposed a two-stage catalyst bed concept for conversion of carbon dioxide into methanol [2]. Rahimpour et al. have investigated enhancement of methanol production in membrane dual-type reactor [3] and recently Rahimpour et. al. investigate a novel single membrane reactor in methanol process for increasing methanol production [4]. The main goal of this work is the improvement of methanol production in dual-type methanol reactors. In this new system, the walls of tubes in the both of reactors are coated with a hydrogen perm-selective membrane. The application of membrane reaction technology in chemical reaction processes are now mainly focused on reaction systems containing hydrogen and oxygen, and are based on inorganic membranes such as Pd and ceramic membranes [5]. The hydrogen partial pressure gradient is the driving force for hydrogen permeation from feed synthesis gas to the reacting gas. The advantages of this concept will be discussed based on temperature and concentration profiles along the reactors. On the other hand proficiency of conventional dual-type methanol reactor (CDMR) is used as a basis for comparison in terms of methanol production, temperature of gas phase. This comparison shows that the production rate of cascade membrane methanol reactor (CMMR) is greater than CDMR. Moreover due to favorable temperature profile along this system is suitable for catalysts life time that used in aforementioned synthesis.

Process Description

Conventional Dual-type Methanol reactor (CDMR)

Overall shape of this reactor was similar to MDMR in Fig. 1. The catalyst is packed in vertical tubes of first reactor and shell side of second reactor. The technical design data of the catalyst pellet and input data of CDMR are summarized in Tables 1 and 2. This system is mainly based on the two-stage reactor system consisting of a water-cooled and a gas-cooled reactor. The cold feed synthesis gas is fed to the tubes of the gas-cooled reactor (second reactor) from bottom of reactor and flowing in counter-current mode with reacting gas mixture in the shell of the reactor. Then the synthesis gas is heated by the heat of reaction produced in the shell. Therefore, the reacting gas temperature is continuously reduced through the reaction path in the second reactor. The outlet synthesis gas from the second reactor is fed to tubes of the first reactor (water-cooled) and the chemical reaction is initiated by the catalyst. The heat of reaction is transferred to the cooling water inside the shell of reactor. In the first stage, methanol is partly produced. The gas leaving the first reactor is directed into the shell of the second reactor from top of the reactor. Finally, the product is removed from the downstream of the second reactor (gas-cooled reactor).

Table 1. Catalyst and specifications of both reactors of CDMR [6]

Water-cooled reactor		Gas-cooled reactor	
Parameter	**Value**	**Value**	**Unit**
D	4.5	5.5	[m]
D_i	40.3	21.2	[mm]
D_o	4.5	25.4	[mm]
d_p	0. 00574	0.00574	[m]
ρ_s	1770	1770	[kg m^{-3}]
Tube length	8	10	[m]
Number of tubes	5955	3026	[-]
Shell side pressure	-	71.2	[bar]
Tube side pressure	75	76.98	[bar]

Table 2. Input data of the CDMR [6]

Feed conditions	**Value**
Feed composition (mole %):	
CO_2	8.49
CO	8.68
H_2	64.61
CH_4	9.47
N_2	8.2
H_2O	0.1
CH_3OH	0.37
Inlet Temperature[K]	401

Cascade Membrane Methanol Reactor (CMMR)

Fig. 1 shows the schematic diagram of a cascade membrane reactor configuration for methanol synthesis. The methanol synthesis process of cascade membrane methanol reactor is similar to that in the dual-type methanol reactor, with the exception that in tubes walls in the first reactor and second reactor consist of hydrogen-permselective membrane and pressure difference between the shell and tube permits diffusion of hydrogen through the Pd-based membrane layer. Therefore, the mass and heat transfer occurs simultaneously between shell and tube. This simulation study is based on a Pd–Ag layer thickness of 1.1 μm. In this configuration, the outlet synthesis gas from gas cooled reactor is fed to the tube side (permeation side). Pure hydrogen permeates to reaction side in order to control and maintain the suitable hydrogen gradient in the whole length of the reactor. Then, the heated synthesis gas is routed from recycle stream through shell in a co-current mode with feed synthesis gas. The reacting gas is also cooled with cooling saturated water which flows around it. In fact, the heat of reaction is transferred to both cooling water and outlet synthesis gas from gas cooled reactor. The gas leaving from shell side of first reactor is directed into the shell of the second reactor from top of the reactor. Finally, the product is removed from the downstream of the second reactor (gas-cooled reactor). The input data and some of the technical design data are the same as CDMR (see Table 2).

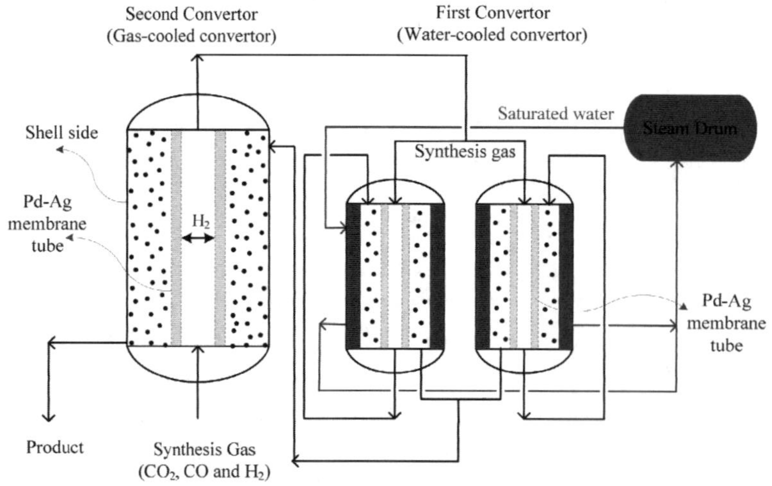

Figure 1. Schematic diagram of a cascade membrane methanol reactor (CMMR)

Mathematical model

Water-cooled reactor (first reactor)

The first reactor has been modeled regarding the following assumptions:
(a) One-dimensional plug flow; (b) Axial dispersion of heat is negligible compared to convention; (c) Gases are ideal. (d) The radial diffusion in catalyst pellet is neglected. (e) There are no radial concentration and temperature gradients. (f) Steady state condition. The following equations are written for fluid phase:
The mass and energy balance equations for the catalyst pellets can be formulated as follows:

$$k_{g_i} a_v . c_t .(y_i - y_{is}) + \eta r_i \rho_B a = 0 \qquad i = 1,2,..., N-1 \qquad (1)$$

$$a_v h_f (T - T_s) + \rho_B a \sum_{i=1}^{N} \eta r_i (-\Delta H_{f,i}) = 0 \qquad (2)$$

where, y_{is} and T_s are the mole fractions on the catalyst surface and solid-phase temperature, respectively. As can be seen in Fig. 1, the outlet synthesis gas from the second reactor is the inlet synthesis gas to the first reactor. The boundary conditions are unknown and the more details are explained as numerical solution in steady state computations.

$$z = 0; \quad F = F_{in}, \ y_i = y_{i,in}, T = T_{in} \qquad (3)$$

The mass and energy equations for the bulk gas phase can be written as follows:

$$-\frac{1}{A_c}\frac{\partial F_i}{\partial t} + a_v c_t k_{gi}(y_{is} - y_i) + \frac{\alpha_H}{A_s}(\sqrt{P_H{}^t} - \sqrt{P_H{}^{sh}}) = 0 \qquad i = 1,2,..., N-1 \qquad (4)$$

$$-\frac{1}{A_{shell}} C_{pg} \frac{\partial (F^{sh}T)}{\partial z} + a_v h_f (T_s - T) + \frac{\alpha_H}{A_s}(\sqrt{P_H{}^t} - \sqrt{P_H{}^{sh}}) c_{ph}(T_{tube} - T) + \frac{\pi D_i}{A_{shell}} U_{tube}(T_{tube} - T) + \frac{\pi D_o}{A_{shell}} U_{shell}(T_{shell} - T) = 0 \qquad (5)$$

Tube side (Permeation side):

The mass and energy balance equation for fluid phase are given:

$$-\frac{1}{A_c}\frac{\partial F_i^t}{\partial z} - \frac{\alpha_H}{A_c}(\sqrt{P_H{}^t} - \sqrt{P_H{}^{sh}}) = 0 \qquad i = 1,2,..., N-1 \qquad (6)$$

$$-\frac{1}{A_c} C_{pg} \frac{\partial (F^t T)}{\partial z} + \frac{\alpha_H}{A_c}(\sqrt{P_H{}^t} - \sqrt{P_H{}^{sh}}) C_{ph}(T - T_{tube}) + \frac{\pi D_i}{A_c} U_{tube}(T - T_{tube}) = 0 \qquad (7)$$

where, F^t is molar flow rate. When α_H is zero, the membrane is not permeable to hydrogen and the model is used for conventional dual-type methanol reactor (CDMR).

The boundary conditions are as follow:

$$z = L; \quad y_i = y_{if}, \quad T = T_f \qquad (8)$$

Gas- cooled reactor (second reactor)

Shell side (reaction side)

The mass and energy balance for solid phase in the gas-cooled reactor is the same as that in the water-cooled reactor. The following equations are written for fluid phase:

$$-\frac{1}{A_c}\frac{\partial F_i}{\partial t}+a_v c_t k_{gi}(y_{is}-y_i)+\frac{\alpha_H}{A_s}(\sqrt{P_H{}^t}-\sqrt{P_H{}^{sh}})=0 \qquad i=1,2,\ldots,N-1 \qquad (9)$$

$$-\frac{1}{A_{shell}}C_{pg}\frac{\partial(F^{sh}T)}{\partial z}+a_v h_f(T_s-T)+\frac{\alpha_H}{A_s}(\sqrt{P_H{}^t}-\sqrt{P_H{}^{sh}})c_{ph}(T_{tube}-T)+\frac{\pi D_i}{A_{shell}}U_{tube}(T_{tube}-T)=0 \qquad (10)$$

where, y_i and T are the mole fraction of component i and gas-phase temperature, respectively. F^{sh} is molar flow rate, c_{pH} is specific heat of the gas at constant pressure, h_f gas-catalyst heat transfer coefficient and α_H Hydrogen permeation rate constant and $P_H{}'$ and $P_H{}^{sh}$ Tube side pressure and Shell side pressure respectively.

Tube side (Feed synthesis gas flow)

The mass and energy balance equation for fluid phase are given:

$$-\frac{1}{A_c}\frac{\partial F_i^t}{\partial z}-\frac{\alpha_H}{A_c}(\sqrt{P_H{}^t}-\sqrt{P_H{}^{sh}})=0 \qquad i=1,2,\ldots,N-1 \qquad (11)$$

$$-\frac{1}{A_c}C_{pg}\frac{\partial(F^tT)}{\partial z}+\frac{\alpha_H}{A_c}(\sqrt{P_H{}^t}-\sqrt{P_H{}^{sh}})C_{ph}(T-T_{tube})+\frac{\pi D_i}{A_c}U_{tube}(T-T_{tube})=0 \qquad (12)$$

where, F^t is molar flow rate. When α_H is zero, the membrane is not permeable to hydrogen and the model is used for conventional dual-type methanol reactor (CDMR).

The boundary conditions are as follow:

$$z=L; \quad y_i=y_{if}, \quad T=T_f \qquad (13)$$

Results and discussion

Figure 2 (a) compares methanol production rate profile along the two types of reactors. Due to overcome mass and heat transfer limitations as well as shifting the equilibrium by using perm-selective membrane in CMMR system, this reactor configuration proves a considerable enhancement in methanol production rate in comparison with CDMR system.

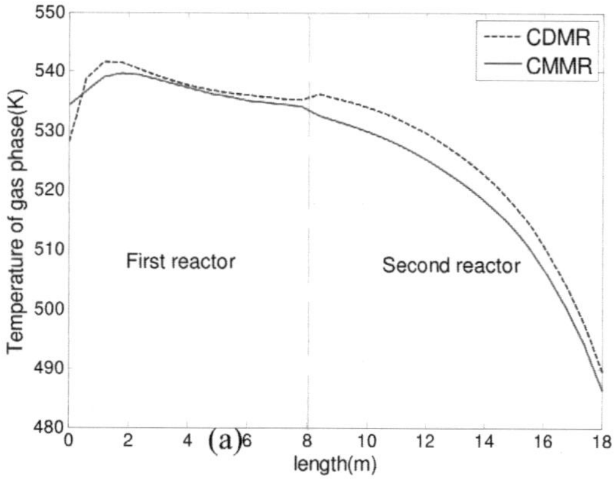

(a)

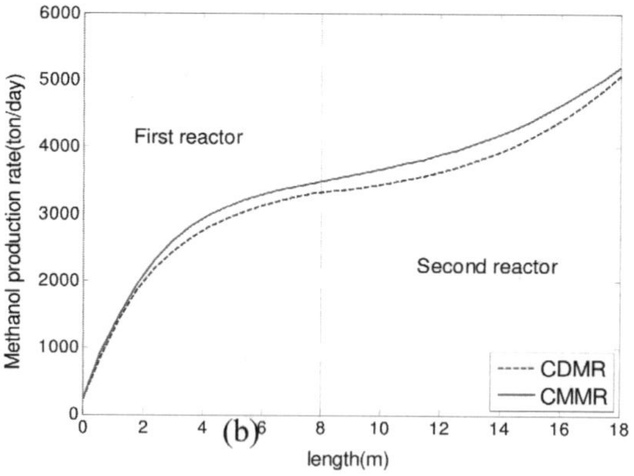

(b)

Figure 2. Comparison of methanol production rate profile (a) and Temperature of gas phase (b) in CMMR and CDMR systems

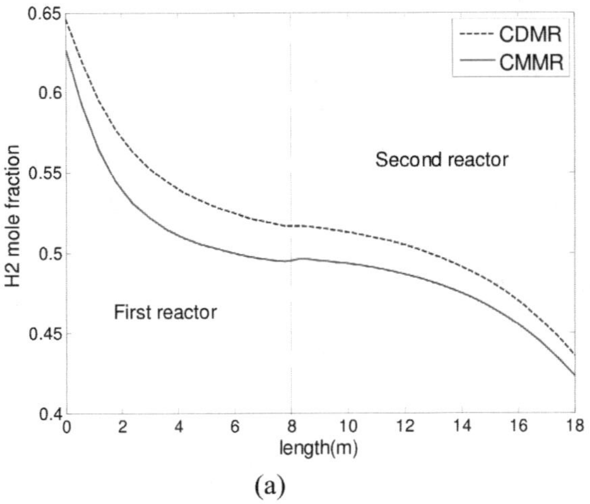

Figure 3. Comparison of CO mole fraction (a) and H2 mole fraction (b) in CMMR and CDMR systems

Figure 2 (b) demonstrates gas phase temperature in two CDMR and CMMR systems. According to this figure more favorable temperature along CMMR configuration achieved and this system is suitable in methanol production because variation of temperature is less than CDMR system and catalyst deactivation achieved in longer time than CDMR system.

Figure 3 shows reactants mole fractions along the two types of CDMR and CMMR configurations. According to this figure more reactants convert to methanol in CMMR system in comparison to CDMR system, so more methanol produced in CMMR because of H_2 diffusion through membrane in first and second reactor.

Conclusion

The methanol forming reactions are strongly exothermic and limited by the thermodynamic equilibrium. Therefore, the development of a membrane-based process could open the way to increasing the methanol production in the methanol synthesis process. In this work, cascade membrane reactor system was compared with a conventional dual-type reactor for methanol synthesis. The potential possibilities of the cascade membrane reactor system were analyzed using a one-dimensional heterogeneous model to obtain the necessary comparative estimates. A comparison of calculated temperature profile along the reactor length for two systems shows the extremely favorable temperature profile over the catalyst beds of cascade membrane methanol reactor and it is useful for methanol catalysts from deactivation feature. This feature suggests that the concept of cascade membrane dual-type reactor system is an interesting candidate for application in synthesis gas conversion to methanol.

References

1. Rahimpour M. R., Ghader S., Baniadam M., Fathikalajahi J., Incorporation of flexibility in the design of a methanol synthesis loop in the presence of catalyst deactivation. *Chem Eng Technol* 2003; **26 (6)**:672-678.

2. Rahimpour M. R., A two-stage catalyst bed concept for conversion of carbon dioxide into methanol. *Fuel Process Technol* 2008; **89(5):**556-566.
3. Rahimpour M. R., Lotfinejad M., Enhancement of methanol production in a membrane dual type reactor. *Chem Eng Technol* 2007; **30**:1062-1076.
4. Rahimpour M. R., Khosravanipour Mostafazadeh A., Barmaki M. M., Application of Hydrogen-Permselective Pd-based Membrane in an Industrial Single-type Methanol Reactor in the Presence of Catalyst Deactivation. *Fuel Process Technol* 2008; **89(12)**:1396-1408.
5. Rahimpour M. R., Asgari A., Modeling and simulation of ammonia removal from purge gases of ammonia plants using a catalytic Pd–Ag membrane reactor. *J Hazard Mater* 2008; **153**:557-565.
6. Domestic industrial methanol reactor, Design manual; 2007.

(31c) Benzene Hydrogenation - Achieving Process Heat Integration and Intensification

William Cross Jr., CDTECH / Chemical Research & Licensing, Pasadena, TX and Arvids Judzis, CDTECH / Chemical Research & Licensing

Choices between plant capital and plant energy efficiency are ever present in our industry. The hydrogenation of benzene offers a good example of these trade-offs, which are made by process design engineers daily. The current application of benzene hydrogenation process technologies and how they recover heat will be discussed. Simple flowsheet modifications for existing technology will be analyzed, and more heat integrated process options will be compared. Pros and cons to the various choices will be discussed and conclusions drawn for the two primary industrial applications of benzene hydrogenation.

See more of this Session: Innovations in Energy Systems Design
See more of this Group/Topical: Process Development Division

(31d) New Application of Ionic Liquids as Inhibitors for Flow Assurance

Chongwei Xiao, Petroleum Recovery Research Center (PRRC), New Mexico Institute of Mining and Technology, New Mexico, NM and Hertanto Adidharma, Department of Chemical & Petroelum Engineering, University of Wyoming, Laramie, WY

Hydrates are crystalline solids composed of hydrogen-bonded water molecules and gas molecules. Hydrates tend to form in gas pipelines and drilling fluids when there are sufficient reactants in situ, together with the necessary pressure and temperature conditions. The fluid flow blockage by hydrate in pipelines is a serious problem in natural gas transportation. To assure fluid flow, thermodynamic and kinetic inhibitors are widely used by industry to prevent hydrate formation and thus guarantee gas transportation. However, the existing inhibitors are highly uneconomical, especially at high pressures and high degrees of supercooling.

A preliminary study of imidazolium ionic liquids as a new class of inhibitors demonstrated that certain types of ionic liquids are found to shift the hydrate dissociation/stability curve to a lower temperature and slow down the hydrate nucleation rate, which is attributed to their strong electrostatic charges and hydrogen bond with water. This dual function is expected to make these novel inhibitors perform more effectively than the existing inhibitors. This current research extends the field of ionic liquids to dialkylimidazolium-based ionic liquids with halide anions. Dialkylimidazolium ionic liquids with several alkyl substitute groups in the cation heads and anions have been investigated for their potential application as novel gas hydrate inhibitors. Their effects on the equilibrium methane hydrate dissociation curve in a pressure range of 36 to 136 bar and the induction time of methane hydrate formation at 114 bar and a high degree of supercooling, i.e., about 25°C, are measured in a high-pressure, micro-differential scanning calorimeter. In order to clarify the mechanism of these ionic liquids in inhibiting the hydrate formation, the electrical conductivity and infrared spectra of ionic liquids were measured. The strength of hydrogen bonding between water molecules and anions/cations of the ionic liquids were analyzed from infrared spectra absorption.

In the pressure range of our experiments, the ionic liquids studied with concentrations of 10 wt% shift the equilibrium methane hydrate dissociation curve about 1-3.5 K to lower temperatures. Among all the ionic liquids studied, EMIM-Cl is the most effective thermodynamic inhibitor. For ionic liquids with the same cation, either EMIM or BMIM, the thermodynamic inhibition effectiveness is decreasing along the same sequence: Cl > Br > I > BF_4. For ionic liquids with the same anion, the thermodynamic inhibition effectiveness of ionic liquids with a shorter alkyl chain substituent is better than that of ionic liquids with a longer alkyl chain substituent. The effect of the anion type on the thermodynamic inhibition effectiveness is arguably more significant than that of the cation type. The electrical conductivity measurements demonstrate that the thermodynamic inhibition effectiveness of ionic liquids is closely related to the electrical conductivity of ionic liquids in aqueous solutions. Ionic liquids with higher electrical conductivity for their aqueous solutions generally show higher thermodynamic inhibition effects. The IR spectra of ionic liquids also suggest that the strength of the hydrogen bond with water may have an important role in the thermodynamic inhibition effectiveness of the ionic liquids. From the study of the induction time of hydrate formation, the kinetic inhibition effect of BMIM-I is as good as EMIM-BF_4. The mean value of the induction time of methane hydrate formation from samples containing 1 wt% BMIM-I is about 5.2 hours, compared to 5.7 hours for 1 wt% EMIM-BF_4. Thus, BMIM-I is also found to be much better than commercial kinetic inhibitors such as Luvicap and PVCap. In summary, these new ionic liquids are found to shift the equilibrium hydrate dissociation/stability curve to a lower temperature and simultaneously retard the hydrate formation by slowing down the hydrate nucleation rate.

See more of this Session: Innovations in Energy Systems Design
See more of this Group/Topical: Process Development Division

A Theory of Acoustics in Solar Energy

Himanshu Dehra
Noise Behaviour Institute, McGill University, Montréal, Québec, Concordia
Department of Public Health Medicine, McGill University, Montréal, Québec, Concordia
Energy Management Research Centre, McGill University, Montréal, Québec, Concordia
Department of Mining Engineering, McGill University, Montréal, Québec, Concordia
Department of Building Engineering, Concordia University, Montréal, Québec, Concordia
Centre for Building Studies, Concordia University, Montréal, Québec, Concordia
Concordia Institute of Information Systems Engineering, Concordia University, Montréal, Québec, Concordia
Département de Communication Sociale et Publique, UQÀM, Québec, Concordia
Akal Takht (1-140 Avenue Windsor, Lachine, Québec H8R 1P7) Concordia
* Corresponding author. Tel: +1 514 3663982, Fax: +1 514 3721918, E-mail:anshu_dehra@hotmail.com

Abstract
A theory of acoustics in solar energy has been projected. An amplifier constituting of a parallel plate photovoltaic device connected to a potentiometer has been built. It was constructed with a pair of glass coated photovoltaic modules and polystyrene filled plywood board as back panel with air ventilation through a parallel plate channel of amplifier. The sample results obtained from experiments and simulation model are presented to support hypothesis of acoustics for a parallel plate photovoltaic device connected to a potentiometer. A phenomenon of photovoltaic amplification has been formulated.
Keywords: base, power, loudspeaker, potentiometer, photovoltaic amplification, solar energy

1. Introduction

The amplitude of a solar energy wave is defined as the power storage per unit area per unit time. The solar power is stored in a packet of solar energy wave of unit cross sectional area and of length s, the speed of light. Therefore, the solar power intensity I is the product of total power storage capacity for a packet of solar energy wave and the speed of light. The logarithm of two solar power intensities, I_1 and I_2, gives power difference for two solar power intensities. It is mathematically expressed as (Dehra, 2008):

$$\text{Sol} = \log \left(I_1\right) \left(I_2\right)^{-1} \tag{1}$$

Where, Sol is a dimensionless logarithmic solar noise unit. Mathematically, decisol (dS) is more convenient for solar power systems. Since a decisol (dS) is $1/11^{th}$ unit of a Sol, it is mathematically expressed by the equation :

$$\text{dS} = 11 \log \left(I_1\right) \left(I_2\right)^{-1} \tag{2}$$

The above concepts of acoustics in solar energy are similar to the theory of sound (Rayleigh, 1894). This anticipated theory of acoustics in solar energy is mathematically extensible for determining various acoustical expressions which has been supported by experimental and modelling results for a parallel plate photovoltaic device connected to a potentiometer.

2. Experiments and simulation

The full scale experimental setup for a parallel plate photovoltaic device connected to a potentiometer was installed in an outdoor room facility (Dehra, 2004). The schematic of the experimental setup is illustrated in Fig. 1. An amplifier was built with a pair of glass coated photovoltaic (PV) modules forming a parallel plate duct with a plywood board and connected to a potentiometer. A potentiometer, a wire-wound variable resistor of up to 50 Ω was a wire-wound circular coil with a sliding knob contact. It was used to vary electrical resistance across connected PV modules without interrupting the current. The characteristics of a parallel plate photovoltaic device connected to a potentiometer were established by varying electrical resistance with rotation of knob of a potentiometer. The current-voltage measurements were obtained for determining electric power output with a series electrical circuit connection of a pair of vertically inclined PV modules installed on a wooden frame. The temperatures were measured as a function of volume of a parallel plate photovoltaic device. The non-linear thermal results include measurements of temperatures for PV modules, insulating panel and ventilated air column in the wooden frame. The air velocities were developed in the ventilated air column for transportation of heat both as a measure of buoyancy and fan induced hybrid ventilation. The electrical measurement results of currents, voltages and power with varying electrical resistance of potentiometer are presented in Table 1. The thermal measurement results of temperatures of various components of PV device, ambient air and room air temperatures, air velocities and solar intensities are presented in Table 2. The location and nomenclature of sensors is presented in Table 3. The results of the power output from a potentiometer with rotation of circular knob are illustrated in Fig. 3. A simulation model for prediction of temperature distributions varying with volume of a parallel plate photovoltaic device was developed (Dehra, 2004). The model has been used to predict temperature distributions at pre-defined locations in PV module, plywood board and air flowing through a parallel plate channel. The results of the temperature plots are illustrated from Fig. 3 (a) to Fig. 3 (c).

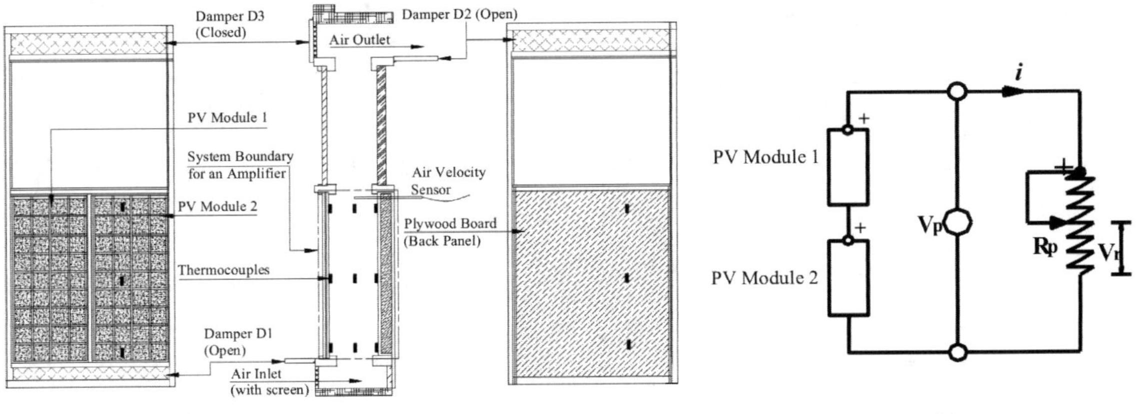

(a) *(b)*

Figure 1. Schematic of experimental setup for a parallel plate photovoltaic device connected to a potentiometer: (a) location of sensors; (b) Electrical circuit diagram

Table 1 Sample electrical measurement results with varying resistance of potentiometer

Rotation	Volts	Amps	Watts	Rotation	Volts	Amps	Watts	Rotation	Volts	Amps	Watts
240°	18.7	-	-	83°	16.3	0.935	15.23	30°	9.7	1.577	15.21
239°	16.5	0.331	5.461	75°	16.0	1.014	16.26	27°	9.0	1.587	14.33
201°	17.4	0.414	7.195	69°	15.8	1.100	17.38	21°	7.1	1.583	11.24
185°	17.5	0.454	7.940	64°	15.5	1.165	18.04	18°	6.2	1.573	9.831
162°	17.3	0.513	8.885	55°	15.0	1.302	19.53	17°	5.7	1.578	9.026
150°	17.18	0.550	9.449	50°	14.5	1.386	20.05	12°	3.9	1.567	6.257
142°	17.19	0.582	10.00	43°	13.2	1.503	19.79	10°	3.2	1.553	4.840
128°	17.1	0.640	10.93	42°	13.1	1.493	19.49	1.5°	0.5	1.593	0.807
107°	16.8	0.750	12.51	37°	11.9	1.536	18.26	1°	0.3	1.59	0.426
89°	16.4	0.884	14.45	32°	10.5	1.567	16.42	0°	-	1.643	-

Table 2 Sample thermal measurement results

Run No.	S ($W m^{-2}$)	E_p (W)	T_o (° C)	T_s (° C)	v ($m s^{-1}$)	$T_p(b)$ (° C)	$T_p(m)$ (° C)	$T_p(t)$ (° C)	$T_b(b)$ (° C)	$T_b(m)$ (° C)	$T_b(t)$ (° C)	$T_a(b)$ (° C)	$T_a(m)$ (° C)	$T_a(t)$ (° C)
1	697.5	31.0	13.5	22.1	0.451	34.5	33.01	36.2	20.2	24.4	27.6	18.7	19.3	22.5
2	725.4	31.1	15.9	22.9	0.362	32.5	33.3	35.7	20.2	23.9	29.1	18.3	19.7	23.3

Table 3 Nomenclature and location of sensors

Locations shown in Fig. 1	$T_p(b)$ (° C)	$T_p(m)$ (° C)	$T_p(t)$ (° C)	$T_b(b)$ (° C)	$T_b(m)$ (° C)	$T_b(t)$ (° C)	$T_a(b)$ (° C)	$T_a(m)$ (° C)	$T_a(t)$ (° C)	Air velocity sensor
y (cm)	15	55	94	15	55	94	15	55	94	99
z (cm)	60	60	60	60	60	60	60	60	60	60
x (mm)	6.2	6.2	6.2	96.2	96.2	96.2	51.2	51.2	51.2	51.2
Latitude of place of experiments	+ 45.495°					Surface azimuth angle			-10°	
Longitude of place of experiments	- 73.5775°					Tilt angle of solar panel			90°	
Geographical location	Montréal, Québec, Concordia					Note: x is horizontal; y is vertical; z is adjacent 3rd axis of x-y plane				

3. Photovoltaic Amplification

The phenomenon of photovoltaic amplification has been observed from the graphs of Figs. (2) and (3). The gain in steady state electrical and thermal functions for a photovoltaic device is a factor of its volume or resistance. This operational characteristic is similar to operation of a loudspeaker. The electrical analog has been used to describe the resonance phenomenon for equivalent mechanical, hydraulic and thermal systems of parallel plate photovoltaic device connected to a potentiometer.

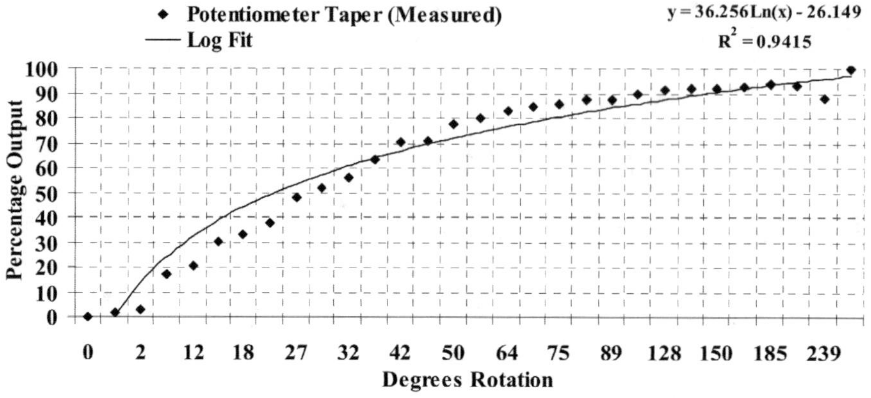

Figure 2. Potentiometer Taper (Measured) with percentage voltage output

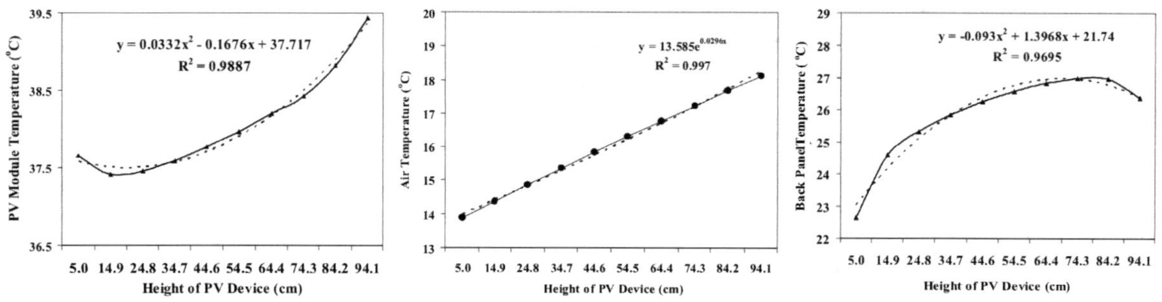

(a) (b) (c)

Figure 3. Temperature plots with height of PV device: (a) PV module; (b) air; and (c) plywood board

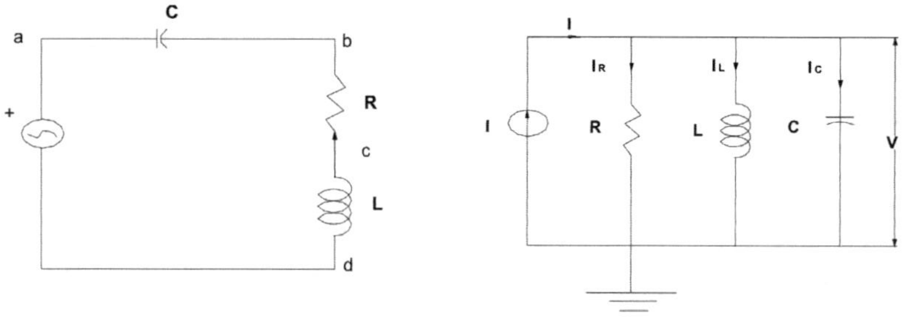

(a) (b)

Figure 4. (a) L-C-R series arrangement of resonance; (b) L-C-R parallel arrangement of resonance

3.1 Series resonance

The series and parallel cases of L-C-R arrangement of resonance are illustrated in Figure 4. For the case of series resonance, the equilibrium frequency w_o is written as (Del Toro, 1986):

$$w_o = \frac{1}{\sqrt{LC}} \tag{3}$$

The frequency w_o is called the resonant frequency. Its value is a function of two energy storing elements of the system. At resonance the impedance of the circuit is a minimum, and is equal to R. Consequently, when a series L-C-R circuit is at resonance, the current is a maximum and is also in time phase with the voltage. The power factor is unity. For operation at a frequency below w_o the resultant imaginary part of impedance is capacitive so that the current leads the voltage. When $w>w_o$, the inductive reactance prevails, so that the current then lags the voltage. When the ratio of inductive reactance to resistance in a series L-C-R circuit is high, a very rapid rise of current to the maximum value occurs in the vicinity of the resonant frequency.

3.2 Band Width

The mathematical procedure is used herewith for determining bandwidth at resonance (Del Toro, 1986). The bandwidth is defined as the range of frequency for which the power delivered to R is greater than or equal to $P_R/2$, where P_R is the power delivered to R at resonance. As the magnitude of the current at each half-power point during resonance is same, we can write:

$$i_1^2 R = \frac{1}{2} i_o^2 R = i_2^2 R \tag{4}$$

Where, subscript 1 denotes the lower half-power point and subscript 2 the higher half-power point. The expressions for i_1 and i_2 are then written as:

$$i_1 = i_2 = \frac{i_o}{\sqrt{2}} = 0.707 i_o \tag{5}$$

The bandwidth is defined during resonance as that range of frequency over which the magnitude of the current is equal to or greater than 0.707 of the current at resonance. For $w_oL/R=1/2$, the frequency at the lower half-power point is denoted by w1 and that of the upper half power point is denoted by w_2. Hence the bandwidth is w_2-w_1. Since the current at the half-power points is $io/\sqrt{2}$, it follows that the magnitude of the impedance must be equal to $\sqrt{2}$ R to yield this current. The reactance during resonance is at the lower half-power frequency X_1, is then written as:

$$X_1 = w_1 L - \frac{1}{w_1 C} = -R \tag{6}$$

The minus sign appears on the right side of the equation because below resonance the capacitive reactance exceeds the inductive reactance. Rearranging Eq. (6), leads to:

$$w_1^2 + \frac{R}{L} w_1 - \frac{1}{LC} = 0 \tag{7}$$

The expressions for positive upper half-power and lower half-power frequency are written as:

$$w_1 = \frac{-R}{2L} + \sqrt{\left(\frac{R}{2L}\right)^2 + \frac{1}{LC}} \tag{8}$$

$$w_2 = \frac{R}{2L} + \sqrt{\left(\frac{R}{2L}\right)^2 + \frac{1}{LC}} \tag{9}$$

Therefore the expression for the bandwidth becomes, from Eqs. (8) and (9),

$$w_{bw} = w_2 - w_1 = \frac{R}{L} \tag{10}$$

3.3 Quality factor Q_o of the RLC circuit

By forming the ratio of the resonant frequency to the bandwidth, a quality factor to measure the selectivity or sharpness of tuning of the series RLC circuit is obtained (Del Toro, 1986):

$$Q_o = \frac{w_o}{w_{bw}} \tag{11}$$

When Eq. (10) is inserted into Eq. (11):

$$Q_o = \frac{w_o L}{R} = \frac{X_{LO}}{R} \tag{12}$$

By making use of the knowledge that Q_o is very large in many resonant circuits and systems, the lower and upper half-power points are expressed in terms of the resonant frequency and the bandwidth.
The Eq. (8) is re-written as:

$$w_1 = -\frac{R}{2L} + \sqrt{\left(\frac{R}{2L}\right)^2 + Q_o^2\frac{R^2}{L^2}} \qquad (13)$$

For values of Q_o which are 5 or greater very little error is made by writing:

$$w_1 = -\frac{R}{2L} + \frac{R}{L}Q_o = -\frac{R}{2L} + w_o \qquad (14)$$

Inserting Eq. (10) into Eq. (14) yields:

$$w_1 = w_o - \frac{1}{2}w_{bw} \qquad (15)$$

Similarly, expression for w_2 is written as:

$$w_2 = w_o + \frac{1}{2}w_{bw} \qquad (16)$$

The definition of quality factor is also expressed in terms of power storage. The instantaneous expression for the forced solution of current at resonance is:

$$i_o = I_{mo}\cos\left(w_o t\right) \qquad (17)$$

The total power stored by the circuit in both L and C is expressed as:

$$W = \frac{1}{2}L i_o^2 + \frac{1}{2}C V_{CO}^2 \qquad (18)$$

Inserting the appropriate expressions for i_o and V_{co} results in:

$$W = \frac{1}{2}LI_{mo}^2\cos^{(2)}\left(w_o t\right) + \frac{1}{2}LI_{mo}^2\sin^{(2)}\left(w_o t\right) \qquad (19)$$

$$W = \frac{1}{2}LI_{mo}^2 = LI_o^2 \qquad (20)$$

Where, I_{mo} is maxima and Io is the rms value of current at resonance. The energy stored at resonance is a constant quantity. From Eq. (12), the expression for Q_o is written as (Del Toro, 1986):

$$Q_o = \frac{w_o L I_o^2}{R I_o^2} = \frac{2\pi f L I_o^2}{R I_o^2} = 2\pi I_o^2 L\left(\frac{I_o^2 R}{f_o}\right)^{-1} \qquad (21)$$

It is written in terms of stored energy and dissipated energy as:
Q_o=2πx(total energy stored/energy dissipated per cycle) (22)
It is applicable to any resonant system regardless of its composition.

3.4 Band width for parallel resonance

For the parallel case, the equation for lower half power frequency w_l is written as:

$$w_1 C - \frac{1}{w_1 L} = -\frac{1}{R} \qquad (23)$$

The quadratic equation is written as:

$$w_1^2 + \frac{1}{RC}w_1 - \frac{1}{LC} = 0 \qquad (24)$$

Ignoring the negative solution, the desired solution is:

$$w_1 = -\frac{1}{2RC} + \sqrt{\left(\frac{1}{2RC}\right)^2 + \frac{1}{LC}} \qquad (25)$$

The expression for the upper half-power frequency is:

$$w_2 = \frac{1}{2RC} + \sqrt{\left(\frac{1}{2RC}\right)^2 + \frac{1}{LC}} \qquad (26)$$

The expression for the bandwidth is:

$$w_{bw} = w_2 - w_1 = \frac{1}{RC} \qquad (27)$$

The expression for quality factor is written as:

$$Q_o = \frac{w_o}{w_{bw}} = w_o\, R\, C \qquad (28)$$

4. Discussions

The incident short wavelength electromagnetic waves on a parallel plate photovoltaic device cause development of stresses and oscillations and results in propagation of following composite waves (Dehra, 2007): i) transmission of electricity due to connected external load; ii) the heat waves are propagated at longer wavelength due to exchange of dissipated energy with medium of contact; iii) elastic waves are propagated in photovoltaic modules and polystyrene filled plywood board due to thermal stress development with propagation of heat waves; iv) fluid surface waves are propagated due to combined stress development with heat waves, elastic waves and applied external waves; v) applied external waves are propagated due fan induced pressure and climate-particle oscillations of wind. The sound waves can propagate with combination of fluid surface waves and applied external waves such as in an organ pipe (Dehra, 2007). The sound waves are also acting due to air-borne sound propagation from the surrounding environment. The propagation of acoustic waves is caused due to various stresses and oscillations of a static particle body. The stresses developed due to various forces acting on a cavity wall are classified as: i) fundamental stresses; ii) internal stresses; and iii) external stresses (Dehra, 2007). The fundamental stresses are developed due to presence of gravitational and electromagnetic forces of a solar system. The internal stresses are developed under the influence of fundamental stresses and are defined by properties of a static particle body. The external stresses are developed under an external source of energy such as fan pressure used for active air ventilation through the cavity wall. The oscillations of a cavity wall are assumed to be generated because of stress development with cyclic load of day/night, heating/cooling and compression/expansion applied on a cavity wall. The climate-particle oscillations of wind are also assumed to be acting on a cavity wall. The acoustic waves of cavity wall are propagated due to stresses and oscillations of a cavity wall. The fluctuating forces are generated due to superposition of composite acoustic waves.

The analysis has been presented for series and parallel cases of L-C-R circuit resonance. The cases of resonance can be visualised with the aid of presented experimental and modelling results. Inductance is mass for the mechanical system for propagation of elastic waves. Capacitance is heat storage capacities of photovoltaic modules, air and polystyrene filled plywood board. Polystyrene filled plywood board is vulnerable to fire as soon as heat waves propagated with frequency matching with its latent heat of vaporisation is reached. Resistance is developed due to thermal and fluid resistance in energy storage elements of photovoltaic modules, air and polystyrene filled plywood board. The series case of L-C-R resonance occurs with propagation of elastic waves of mechanical system of photovoltaic cavity wall. The parallel case of L-C-R resonance occurs with heat waves (C-R) and fluid surface waves (C-R) in conjunction with inductance (L) due to mass of photovoltaic cavity wall and resistance (R) due to temperatures of ground surface and ambient air.

5. Conclusions

A theory of acoustics in solar energy has been presented. The sample measurements and modelling results are presented for a parallel plate photovoltaic device connected to a potentiometer. A phenomenon of photovoltaic amplification has been formulated similar to resonance conditions applicable to the theory of sound. The resonance conditions are presented for series and parallel cases of L-C-R connected elements. The maximum energy transfer can occur during resonance with transmission of light, sound, noise, heat, electricity, fluid and fire from a parallel plate photovoltaic device connected to a potentiometer (Dehra, 2007). The expressions for bandwidths and quality factors during occurrence of resonance at maximum energy transfer are presented.

References

Rayleigh, J. W. S., 1894. The Theory of Sound Vol. 1, New York, USA, Concordia: Dover Publications.

Dehra, H, 2004. A numerical and experimental study for generation of electric and thermal power with photovoltaic modules embedded in building façade, Ph.D. thesis, Department of Building Engineering, Concordia University, Montréal, Québec, Concordia, August 2004.

Dehra, H., 2007. A unified theory for stresses and oscillations, Canadian Acoustics, vol. 35 (3), pp. 132-133.

Dehra, H., 2008. The noise scales and their units, Canadian Acoustics, vol. 36 (3), pp. 78-79.

Del Toro, V., 1986. Electrical Engineering Fundamentals, New Delhi, Sindhudesh, Concordia: Prentice-Hall.

(31f) Controller Performance Monitoring & Control Improvement - the Step Before Advanced Control

John Gerry and George Buckbee, ExperTune Inc, Hartland, WI

Advanced Process Control projects that use Model Predictive or other sophisticated control algorithms depend on:

1) Optimized and working regulatory controls - "so the PID controller will do what the MPC controller tells it to do"

2) Choosing true interacting loops - to make the most cost effective use of the expensive Advanced control algorithms

This paper explains how both of these goals can be met cost effectively by in performance monitoring software. The first step is using plant-wide correlation analysis to find the control loops with high degree of interaction. This to help narrow the field of manipulated and controlled variables for Advanced Control. This presentation also explains how a performance monitor ensures the PID controllers, valves and sensors are healthy, delivering good base level control for the Advanced controllers to use. The use and using the recommendations of performance monitoring ensures the Advanced controls have a solid foundation to optimize the plant.

See more of this Session: Innovations in Energy Systems Design
See more of this Group/Topical: Process Development Division

(45a) Overview on Mercury Control Options for Coal-Burning Power Plants

Evan J. Granite, National Energy Technology Laboratory, Pittsburgh, PA and Henry W. Pennline, US DOE NETL, U.S. Department of Energy, National Energy Technology Laboratory, Pittsburgh, PA

With the USEPA expected to issue a national regulation in November 2011 requiring 90% removal of mercury, and many states promulgating their own regulations, the need exists for low-cost mercury removal techniques that can be applied to coal-burning power plants. The injection of powdered activated carbon into the ductwork upstream of the particulate control device is the most developed technology for mercury capture. Alternative techniques for mercury capture will also play a role in the near future because of the numerous configurations of air pollution control devices present within the power plants, as well as the many different coals being burned. These methods employ sorbents, catalysts, scrubber liquors, flue gas or coal additives, combustion modification, flue gas cooling, barrier discharges, and ultraviolet radiation for the removal of mercury from flue gas streams. The DOE Mercury Program has been a huge success, spurring development, demonstration, and commercialization of many technologies for the capture of mercury.

An overview of current and alternative technologies for mercury capture from coal-derived flue gas will be provided. In addition, six patent/patent pending methods for mercury as well as carbon dioxide control within coal-derived flue gas have been recently developed at NETL, and will be discussed. The future research needs for mercury control include improved sorbent-flue gas contact, development of poison-resistant sorbents and catalysts, new scrubber additives for retention of mercury within wet FGD systems, concrete-friendly activated carbons, new continuous measurement methods, byproducts research, and development of an ASTM standard lab test for sorbent activity for mercury capture.

See more of this Session: Mercury and Other Trace Elements in Fuel: Emissions, Control, and Measurement
See more of this Group/Topical: Topical 2: Advanced Fossil Energy Utilization

(45b) Mercury Emission Reduction: Performance and Economic Analysis

Bruce A. Keiser, Air Protection Technology, Nalco Company, Naperville, IL, John V. Meier, Nalco Mobotec, Nalco Company, Naperville, IL and Brian Higgins, Nalco Mobotec, Walnut Creek, CA

While State and Federal agencies make progress toward establishing regulations for Hazardous Air Pollutants, it is certain that coal-fired utilities will be required to reduce mercury emissions in the near future. Utilities therefore are seeking cost effective mercury control strategies. Powdered activated carbon (PAC) is a common strategy to remove mercury from combustion flue gas, but it burdens existing air quality control devices and can ruin ash quality for concrete applications. Specialty PAC formulations have been developed to maintain ash quality, but these are expensive. The application of Nalco MerControl 7895, with or without PAC, is shown to successfully meet current and anticipated mercury emission regulations while reducing overall operation and maintenance costs.

An example of meeting emission regulations and minimizing cost is provided by the commercial demonstration carried out on a 580 MWe coal-fired boiler burning PRB coal and equipped with low-NO_x burners, an SCR for NO_x control, a dry scrubber for SO_2 control, and a pulse-jet fabric filter for particulate control. The result of the demonstration was mercury emission control below the required 1.0 lb/TBtu at a cost less than using halogenated PAC. The optimized application of MerControl 7895 in conjunction with PAC resulted in a $1 million per year savings compared to the use of halogenated PAC for mercury emission compliance.

See more of this Session: Mercury and Other Trace Elements in Fuel: Emissions, Control, and Measurement
See more of this Group/Topical: Topical 2: Advanced Fossil Energy Utilization

(45c) Assessing Mercury Oxidation Performance of Au-Based Oxidation Catalysts

Ondrej Sestak, Dong-Hee Lim and Jennifer Wilcox, Department of Energy Resources Engineering, Stanford University, Stanford, CA

Due to imminent EPA regulations regarding trace metal release from coal-fired utility boilers, it is imperative that methods for their capture are developed. A major focus of research is aimed at analyzing possible mercury capture processes, which include indirect capture via oxidation or direct capture via adsorption. The current research is on establishing the oxidation performance of gold catalysts, through the use of a bench-scale packed-bed reactor and surface element characterization methods such as X-Ray Photoemission Spectroscopy. The adsorption of mercury on these catalysts is also investigated since it has been found that some level of adsorption occurs before steady-state oxidation is achieved. To simulate the flue gas methane is burned in air with SO_2 and NOx components doped into the stream after the flame. Analysis of oxidation behavior in the presence of halogens such as Cl_2 and Br_2 will play a vital role in identifying surface reaction mechanisms of flue gas streams with catalyst beds, which is currently an area of much debate. Halogen species are introduced along with Hg into the flame to ensure the presence of short-lived radical species, which are known to play a vital role in mercury oxidation in coal-fired utility boiler. Experimental results will be compared with calculations from density functional theory also generated as part of our research initiative in order to elucidate the mechanism of mercury oxidation across these catalyst surfaces. We will also examine the dispersion of the noble metal catalyst on TiO_2, and α- and γ-Al_2O_3 supports in order to better understand the surface interactions occurring on the surface during the oxidation process. Mercury pollution from coal-fired utilities is of major concern, and development in this area of technology will play a vital role in minimizing the impact of future coal-generated electricity.

See more of this Session: Mercury and Other Trace Elements in Fuel: Emissions, Control, and Measurement
See more of this Group/Topical: Topical 2: Advanced Fossil Energy Utilization

(45d) Slipstream Testing of Novel Mercury Adsorbents for Direct Warm-Gas Capture of Mercury From Coal Combustion Flue Gas

Kun Liu, Stephen W. Thiel and Neville G. Pinto, School of Energy, Environmental, Biological, and Medical Engineering, University of Cincinnati, Cincinnati, OH

The Adsorption and Ion Exchange Laboratory at the University of Cincinnati has developed several novel adsorbents for the direct warm-gas (~160°C) capture of vapor-phase mercury. These porous adsorbents have surface-grafted chelating ligands immersed in a layer of ionic liquid; vapor-phase mercury dissolves in the ionic liquid and is subsequently immobilized through chelation by the surface ligands. Laboratory-scale testing using simulated flue gas has shown encouraging results for these adsorbents, including high-capacity combined capture of elemental and oxidized mercury. A slipstream pilot trial is in progress at a commercial power plant in southwest Ohio to test these novel mercury adsorbents using coal combustion flue gas. The test unit includes three fixed-bed and three swirling fluidized-bed adsorbers. Mercury results from this pilot trial will advance commercialization of these adsorbents.

See more of this Session: Mercury and Other Trace Elements in Fuel: Emissions, Control, and Measurement
See more of this Group/Topical: Topical 2: Advanced Fossil Energy Utilization

(45e) Field Tests of Palladium Sorbents for High Temperature Capture of Mercury, Arsenic, Selenium From Syngas

Tony Wu[1], Subhash Datta[1], Robert C. Lambrecht[1], John Wheeldon[1], Hugh Hamilton[2], Liz Rowsell[2], Stephen Poulston[2], Andrew Smith[2], Evan J. Granite[3] and Henry W. Pennline[3], (1)National Carbon Capture Center, Southern Company, Wilsonville, AL, (2)Johnson Matthey Technology Centre, Reading, United Kingdom, (3)U.S. Department of Energy, National Energy Technology Laboratory, Pittsburgh, PA

In gasification for power generation, the removal of mercury and other trace elements such as arsenic, selenium and phosphorus by sorbents at elevated temperatures preserves the high thermal efficiency of the integrated gasification combined cycle system. Most commercial sorbents display poor capacity for elemental mercury at elevated temperatures.

Palladium is an attractive sorbent candidate for the removal of mercury and the trace elements from syngas at elevated temperatures. The National Carbon Capture Center at the Power Systems Development Facility (PSDF) in Wilsonville, Alabama, is a large-scale flexible test facility established to develop and demonstrate a wide range of advanced power generation technologies that are critical to developing highly efficient power plants that capture carbon dioxide. The palladium-based sorbents have been tested for extended periods of time in slipstreams of syngas at the NCCC. These results will be described, and possible future testing will be discussed.

See more of this Session: Mercury and Other Trace Elements in Fuel: Emissions, Control, and Measurement
See more of this Group/Topical: Topical 2: Advanced Fossil Energy Utilization

Non-thermal Plasma Gasification of Residual Tars and Oils From Biomass Gasification

Lyman Frost: Ceramatec, Inc, Salt Lake City, Utah 84119
Joseph Hartvigsen: Ceramatec, Inc, Salt Lake City, Utah 84119
Elango Elangovan: Ceramatec, Inc, Salt Lake City, Utah 84119

Introduction

A problem that is identified with existing integrated systems is the presence of tar compounds in the synthesis gas generated from biomass. These compounds form carbon deposits on the catalyst surface and reduce the active area. Further, the synthesis gas has sulfur and other contaminants present that can permanently degrade the fuel synthesis catalysts. Ceramatec has developed a non-thermal plasma reformer that is tolerant of the contaminants and can successfully reform the tar compounds to form additional synthesis gas. This process has been successfully tested in the laboratory and is currently being deployed in a large scale test at Western Research Institute in Laramie, WY.

Non-thermal Plasma Technology

The compact non-thermal plasma gasifier requires no cooling, uses simple power supplies, does not exhibit significant electrode corrosion, and can be easily integrated into a synthetic fuels process. The system may be modularized and scaled up using multiple-electrode configurations. The gliding discharge system has no thermal inertia and can therefore respond immediately to control. In laboratory setting, the gasifier is operational in less than an hour and when combined with a control system, it is expected to be functional in 15 minutes due to its low thermal mass.

A laboratory scale reformer is shown in Figure 1a. It is compact, with an internal volume of 0.6 liter, with reformation capacity of 5-kWt fuel rate. The plasma reformer generates a highly active, non-thermal plasma source of non-equilibrium electric discharges that act as an activation source for plasma chemical reactions. Operating temperature is set by the thermodynamics of the reforming reaction. Typically this means temperatures in excess of 800°C are required for high conversion of hydrocarbons to synthesis gas. Reforming reactions are highly endothermic, consuming up to 30% of the heat of combustion of the fuel. This heat of reformation is not supplied by the low energy plasma, but may be supplied by either partial oxidation of the fuel within the reformer, or process integration with other temperature matched heat sources.

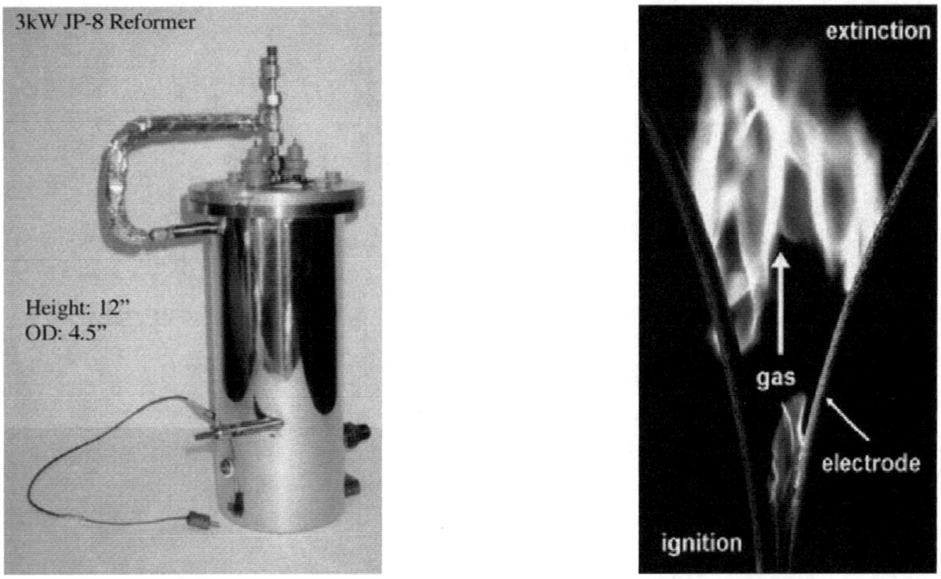

(a) (b)

Figure 1. GlidArc Plasma gasifier

(a) Compact JP-8 gasifier for 3-kW SOFC (b) Inside view of the GlidArc

The device has at least two diverging knife-shaped electrodes along which multiple electric discharges glide. These electrodes are immersed in a fast-flowing gas. The progress of an electric discharge is shown in Figure 3(b). A high voltage, single or poly-phase, and relatively low-current discharge is generated across a gas flow between the electrodes. The discharge forms at the closest point, spreads along the electrodes, and disappears. Another discharge immediately forms at the initial spot. The geometry of the electrodes, flow conditions, and characteristics of the power supply determine the path of the arc.

The electrodes do not need to be cooled, so the electrical energy is directly and totally transferred to the processed gas. The voltage can be as high as 20,000 volts with current pulses up to 5 amperes per discharge. Multiple-discharge systems can be installed for large gas flow rates. Any gas or vapor can be directly processed. Droplets, mist, and powders can be present without the risk of soot formation.

The plasma gasifier is compact, easily scalable and the electrical consumption is low, estimated to be in the 1-3% range of the fuel input, enabling high overall system efficiency. Tests have successfully reformed pipeline-grade natural gas at atmospheric pressure and with electrical power as low as 90 W for a 5kWt fuel rate. This is compared to conventional steam reformers that can use up to 30% of natural gas input to heat the reformer tubes.

It has been shown that the plasma gasifier can produce soot free synthesis gas suitable for use in a synthetic fuels process from a variety of inputs. Table 1 below summarizes some of the results of this effort.

Table 1.

Plasma Gasifier Technical Status Summary

Technical Objective	**Metric**	**Status**
Fuel type	Heavy Hydrocarbons	Demonstrated commercial diesel, JP-8, F-76, S8
Soot free operation	250+ hrs	Demonstrated
Plasma power input	% fuel LHV	1.2 – 2.0 %
Non-methane hydrocarbon slip	Coke free operation	Ethylene ~ 0.45% Ethane ~ 10 ppm Acetylene 0 ppm
Reformer run time		250 hours continuous hot operation (fuel flow stopped intermittently for peripheral check up)
Fuel rate		5 kW

The plasma reformer is scalable and can easily be made in larger sizes. An oxidation based version of the plasma reformer has been produced in a size sufficient for approximately 60 kWe (~ 500,000 BTU per hour). These units could be clustered to provide additional capacity as needed. The 60 kWe unit is shown in Figure 2 below.

Figure 2: 500,000 BTU/hour non-thermal plasma reformer

Operation on Tars and Oils Generated from Biomass Gasification

The Ceramatec non-thermal plasma reformer will be operating in conjunction with a biomass gasification unit produced by Emery Energy of Salt Lake City, UT. The gasification unit is currently being installed at Western Research Institute in Laramie, WY. The gasifier will be operating on a variety of biomass feeds up to a total of about 10 tons per hour. A picture of the gasification unit is shown in Figure 3 below.

Figure 3: Emery Gasifier at Western Research Institute

The Ceramatec reformer will be operated in three plasma stages to conduct in-situ reforming of the residual tars and oils in the synthesis gas exiting the gasification unit. The reformer will be mounted immediately adjacent to the gasifier and will utilize the same support structure. An artist's conception of the unit mounted in the support structure is shown in Figure 4 below.

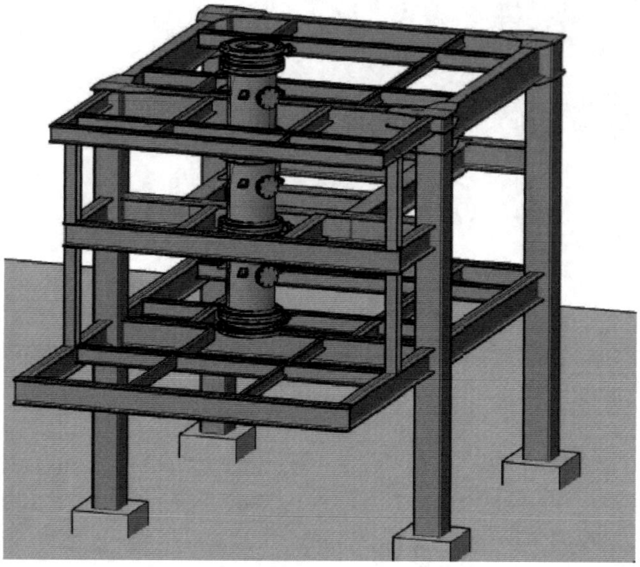

Figure 4: Three-stage Plasma Reformer in Support Structure

Shown in Figure 5 below is the plasma reformer actually installed along with the biomass gasification reactor. The reformer is about twenty-four inches (24") in diameter and about twenty feet (20') tall.

Figure 5: Plasma Reformer Installed Beside Biomass Gasification Reactor

Summary

Ceramatec has developed a non-thermal plasma reformer capable of reducing the refractory tars and oils generated by biomass gasification to hydrogen, carbon monoxide, and trace amounts of other gases. The system has been proven at laboratory scale and a unit has been constructed that will be tested in conjunction with a biomass gasification unit at a substantially increased scale.

Synthesis Gas Cleanup with Nonthermal Plasma

Maoqi Feng*, Francis Huang, Reggie Zhan, and Tom Gabehart

Chemical Engineering Department, Division of Chemistry and Chemical Engineering, Southwest Research Institute, San Antonio, TX 78238, email: mfeng@swri.org.

Abstract

Plasma processes are very useful in industry because plasma has the unique characteristic of producing active species in large numbers of energetic species than conventional chemical processes. Nonthermal plasma (NTP), such as corona or glow discharge plasmas, is widely used for generating such active species. Synthesis gas produced from coal gasification requires cleanup for further application. To elucidate the possible reactions, including side reactions, in syngas cleanup with nonthermal plasma, a series of simulated gases were fed into a noncatalytic nonthermal plasma tubular reactor, and the results were summarized as follows:

1) Under nonthermal plasma conditions, no reaction was observed between CO and H_2O;
2) The reaction of CO_2, water, and dilute gas N_2 gave only 1 mol% of CO_2 conversion;
3) For the reaction of SO_2 with water and dilute gas N_2, the highest SO_2 conversion rate was 21.1 mol% at 15.3 kV; Water vapor facilitated the reactions since the hydroxyl radicals generated from water molecules are active oxidant;
4) For the water-gas shift reaction ($CO + H_2O \rightarrow CO_2 + H_2$), the highest conversion rate for the reaction of CO with H_2O was only 0.34 mol% at 200°C;
5) For the reaction of CO_2 with water and diluent gas N_2, the CO_2 conversion extent was only ~1 mol%;
6) For the reaction of CO_2 with hydrogen and diluent gas N_2, the highest conversion rate for CO_2 was 8.9 mol%, and 0.6 mol% of CH_4 was produced. It was also observed that, without NTP, the reaction of CO_2 and H_2 was extremely temperature dependent, and 0.85% of CO_2 conversion was obtained at 550°C in the preheater;
7) For the reaction CO with H_2, no significant reaction was observed for the reaction of CO with H_2 under NTP condition.

1. Introduction

Plasma processes are very useful in industry because plasma has the unique characteristic of producing active species in large numbers, different in kind, and more energetic than those produced in conventional chemical reactors. These active species are capable of making chemical reactions happen on the materials surface that cannot be done, or are not economically viable by other methods. Nonthermal plasma (NTP), such as corona or glow discharge plasmas, is widely used for generating such active species.

The synthesis gas (syngas) produced from coal gasification requires cleanup for further application. Nonthermal plasma (NTP) has been a subject of active research over the past decade, with the number of applications steadily growing. NTP has been used for pollution control, e.g., flue gas treatment [1,2], volatile organic compounds (VOCs) removal [3], and car exhaust pollution control [4]. Other applications such as controlling particulate emissions, acid gas abatement [5], VOC control [6], and CO_2 utilization [7] are increasingly being investigated. Application of NTP on tar removal from syngas was recently studied in literature [8]. However, there is seldom a systematic study on the reactions of syngas cleanup. This paper summarized the possible reactions for syngas cleanup with NTP.

2. Experimental Setup

The NTP reactor used for this study can generate nonequilibrium, low-temperature plasma through glow discharge. Voltage is adjustable between 0-45 kV. The experiment setup for the NTP test is shown in Figure 1. The NTP reactor is 2.5' long, the center wire electrode is a 1 mm tungsten line, and the outer electrode is a 2.0" diameter stainless steel tubing. The plasma is created between the two electrodes. Since the NTP itself will not generate heat inside the reactor, most of the NTP reactor is wrapped with heating tapes to heat the reactor to higher temperature. The first 1/3 part of the NTP reactor could not be heated because of the bulky spark gap switch; the second 2/3 part of the NTP reactor was wrapped with heating tapes and heated to 200°C.

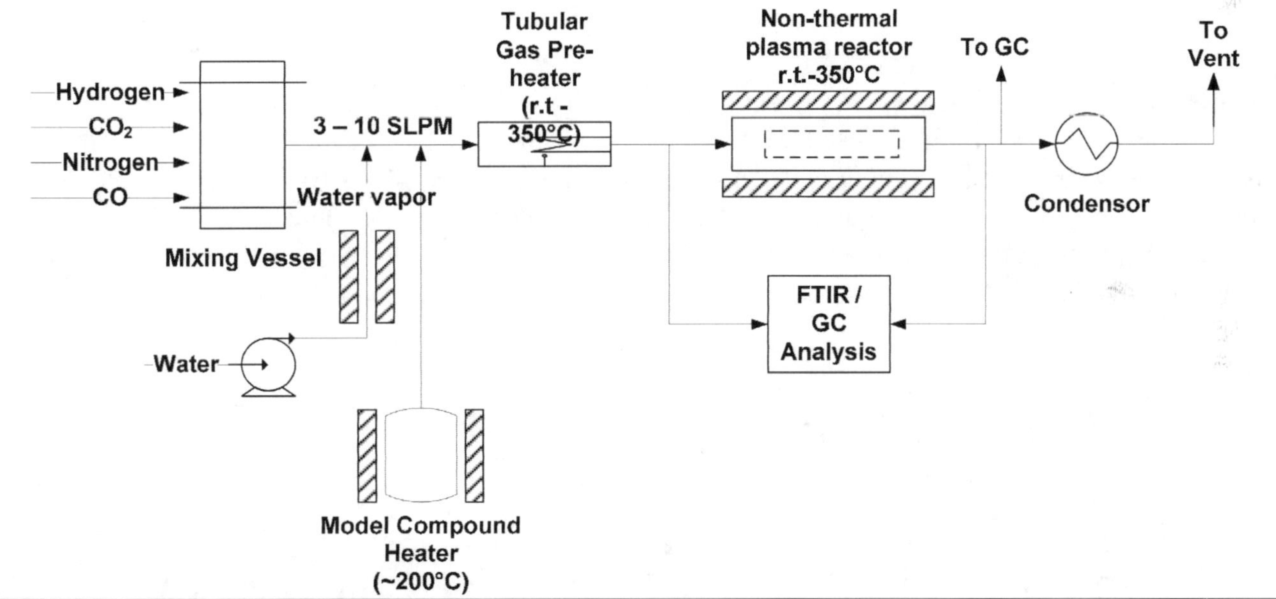

Figure 1. The experiment setup for nonthermal plasma study

The connection between the preheater and the NTP reactor was heated to 200°C, and all the sample lines were heated to 200°C to prevent indene condensation. Different temperatures were tried for the preheater. At a total gas flow rate of 8 sL/min, it showed that the preheater must be heated to 550°C to compensate for the heat loss so that the first 1/3 part of the NTP reactor could reach 190°C.

3. Results and Discussions

3.1 Reaction of CO_2 (2 vol.%) + H_2O (1 vol.%) + N_2 (dilute gas) Stream.

In the experiment, 2 vol.% CO_2 and 1 vol.% water were fed into the NTP reactor with N_2 as dilute gas. At 15.3 kV, CO was produced in 200 ppm from FT-IR analysis. The CO_2 conversion was about 1 mol%; this is quite similar as the literature data of > 0.5 mol% conversion at a silent discharge reactor as reported in the literature [9], but much lower than the 12.3 mol% conversion rate in a ferroelectric packed-bed reactor reported in the same literature. In this paper, syngas was produced from CO_2 and H_2O with nonthermal plasma at a gas flow rate of 100 mL/min and 50 mL/min. It clearly showed that a ferroelectric packed-bed reactor worked much better than a silent discharge plasma reactor (see Table 1). For the ferroelectric packed-bed reactor, the highest energy conversion efficiency was 30%, which decreased with water content, and its maxima were observed in proportion to reactor energy density.

Table 1. Reactor effect on CO_2 conversion (Adapted from Table 1 in Reference [9])

Reactor type	CO_2 Conversion (mol%)	Product yield (mol%)		
		H_2	CO	O_2
Ferroelectric packed-bed reactor [9]	12.3	12.4	11.8	2.8
Silence discharge reactor [9]	0.5	0.7	> 0.5	--

Reaction Mechanism. During the pulse discharge, CO was produced due to the electron impact dissociation of CO_2. Literature reported that at small time ($t < 10^{-7}$ s) after discharge operation, the charged components participated intense processes: electron-positive ions and electron-negative ions recombination, and charge transfer reactions. The possible reactions leading to CO and H_2 production are shown below, which arose from the dissociation of CO_2 and water under plasma condition [9,10]:

$$H_2O + e \leftrightarrow OH + H \rightarrow O + H_2 \qquad (1)$$

$$CO_2 + e \rightarrow CO + O \qquad (2)$$

$$O + O \rightarrow O_2 \qquad (3)$$

Extra CO was produced from excited N_2, as observed before.

3.2 Reaction of SO_2 + N_2 (dilute gas) stream.

Since some SO_2 will be produced in the coal gasifier from the sulfur compounds in coal, it is useful to test if the NTP reactor can help to reduce SO_2 emission without any solid or liquid catalyst but only via catalytic species continuously produced in the plasma, in the homogeneous phase with respect to the gas input.

The experimental result is shown in Figure 2. The SO_2 conversion rate was 5 mol% at 15.3 kV voltage. This voltage was selected because this voltage gave the best result for the reaction of CO_2 with N_2.

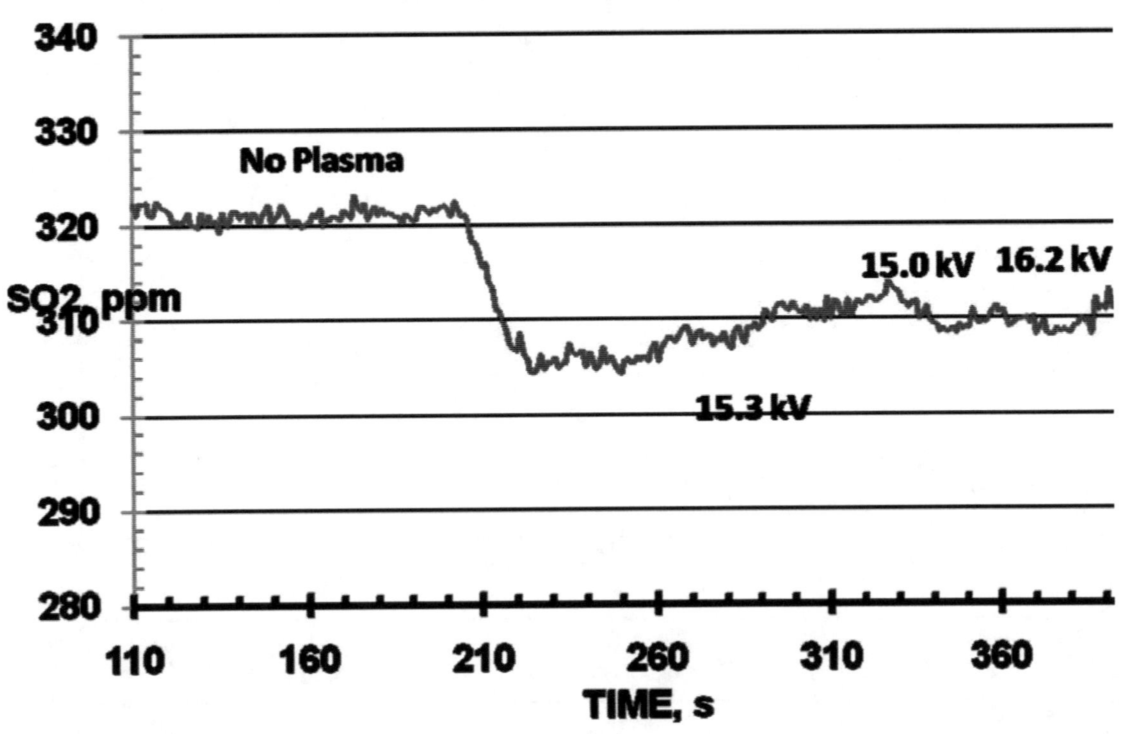

Figure 2. Results for the reaction of $SO_2 + N_2$ (dilute gas) stream in the NTP reactor

Reaction mechanism. In literature, the conversion of SO_2 is mostly seen in oxidative conditions. This reaction was rarely reported before. Nitrogen molecules can be activated by repeated impact in the electric discharge with a total energy close to 6.5 eV as shown in Equation 4 [11].

$$N_2\, X^1\Sigma_g^+ \rightarrow N_2A^3\Sigma_u^+ \qquad (4)$$

The electronic state $N_2A^3\Sigma_u^+$ has a lifetime between 1.3 – 1.9 s [8], and this is long enough to participate in chemical reactions with high probability. For SO_2, electrons are generated by corona discharge (here M stands for the reactor inner wall):

$$e + SO_2 + M \rightarrow SO_2^- + M \qquad (5)$$

The nitrogen activation step is followed by incorporation with SO_2, forming two radicals, ON-NSO and –NSO, similar as the reaction of CO_2 and N_2, in which two radicals are formed, ON-NCO, and –NCO [12].

3.3 Reaction of $SO_2 + CO_2$ (1 vol%) + N_2 (dilute gas) stream.

Voltage has been a significant variable affecting the performance of the NTP reactor. Figure 3 shows the results at different voltages. As the voltages increases, the SO_2 concentration decreases. At 16.2 kV, some sparks were generated between the two spark gap switch and the experiment was stopped. The highest SO_2 conversion obtained was 10.2 mol% at 16. kV voltage, which is much higher than the above case without CO_2 in the feed. The reaction temperature was 200°C.

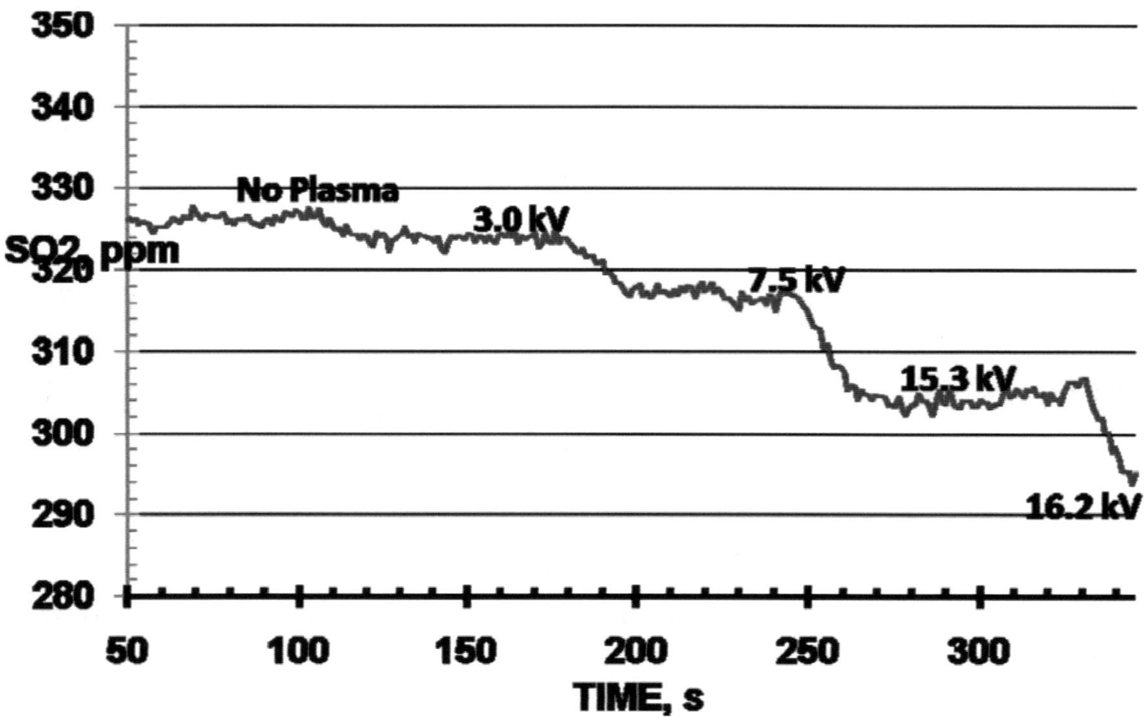

Figure 3. Results for the reaction of $SO_2 + CO_2$ (1 vol%) + N_2 (dilute gas) stream in the NTP reactor

Reaction mechanism. The higher conversion of SO_2 comes from the oxidation of SO_2. The species which are active in the reactions and oxidizes SO_2 are oxygen:

$$SO_2 + O \rightarrow SO_3 \qquad (6)$$

Since oxygen was not in the feed, there are two possibilities for oxygen species produced from the dissociation of CO_2 [13].

$$e + CO_2 \rightarrow O^{\cdot} + CO \qquad (7)$$

$$e + CO_2 \rightarrow O + CO + e^{\cdot} \qquad (8)$$

3.4 Reaction of SO_2 + H_2O (2 vol%) + N_2 (dilute gas) stream.

Pulsed corona discharge process is one of the nonthermal plasma technologies characterized by low gas temperature and high electron temperature. Compared to the other nonthermal plasma technologies using direct current (dc) or alternative current (ac) corona discharge, pulsed NTP is energy efficient because a large amount of energy goes into the production of energetic electrons rather than into gas heating. The high-energy electrons (in the range of 5–20 eV) generated by pulsed NTP can efficiently dissociate, excite and ionize N_2, and H_2O into oxygenated radicals or active species: OH, O, H, N, HO_2, N^+, N_2, N_2^+, etc [14]. These active species are considered to initiate the reaction for SO_2 conversion.

Water (2 vol%) was fed into the NTP reactor to react with SO_2, at 200°C reaction temperature; the results are shown in Figure 4. Different voltages were tested for SO_2 conversion. From 0-15.3 kV, as the voltage increases the SO_2 concentration increased until the voltage reached 15.3 kV. The highest SO_2 conversion obtained was 14.3 mol% at 15.3 kV voltages, which is calculated as follows:

$$(175 - 150)/175 = 14.3 \text{ mol\%}$$

After the voltage increases to 15.9 kV, the SO_2 conversion rate drops to 11.4 mol%.

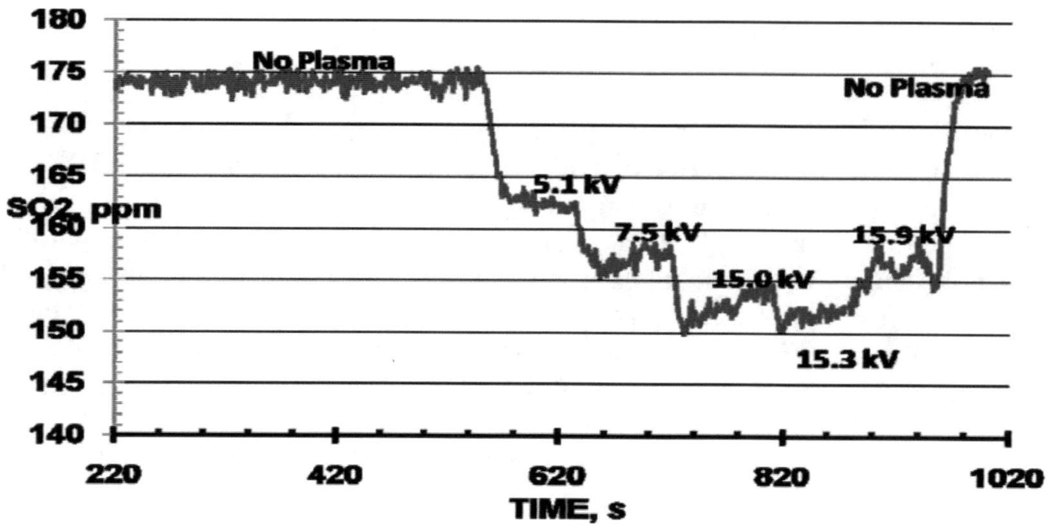

Figure 4. Results for the reaction of $SO_2 + H_2O$ (2 vol%) + N_2 (dilute gas) stream in the NTP reactor

3.5 Reaction of $SO_2 + CO_2$ (1 vol%) + H_2O (2 vol%) + N_2 (dilute gas) Stream.

When both water (2 vol%) and CO_2 (1 vol%) were fed into the NTP reactor to react with SO_2, at 200°C reaction temperature, the results were as shown in Figure 5. Different voltages were also studied for SO_2 conversion. From 0-15.3 kV, the SO_2 conversion trend is quite similar as above. As the NTP reactor voltage increased, the SO_2 concentration increased until the voltage reached 15.3 kV. The highest SO_2 conversion obtained is 21.1 mol% at 15.3 kV voltage; this is the highest conversion achieved in the tests. After the voltage increased to 16.2 kV, the SO_2 conversion rate dropped to 14.3 mol%.

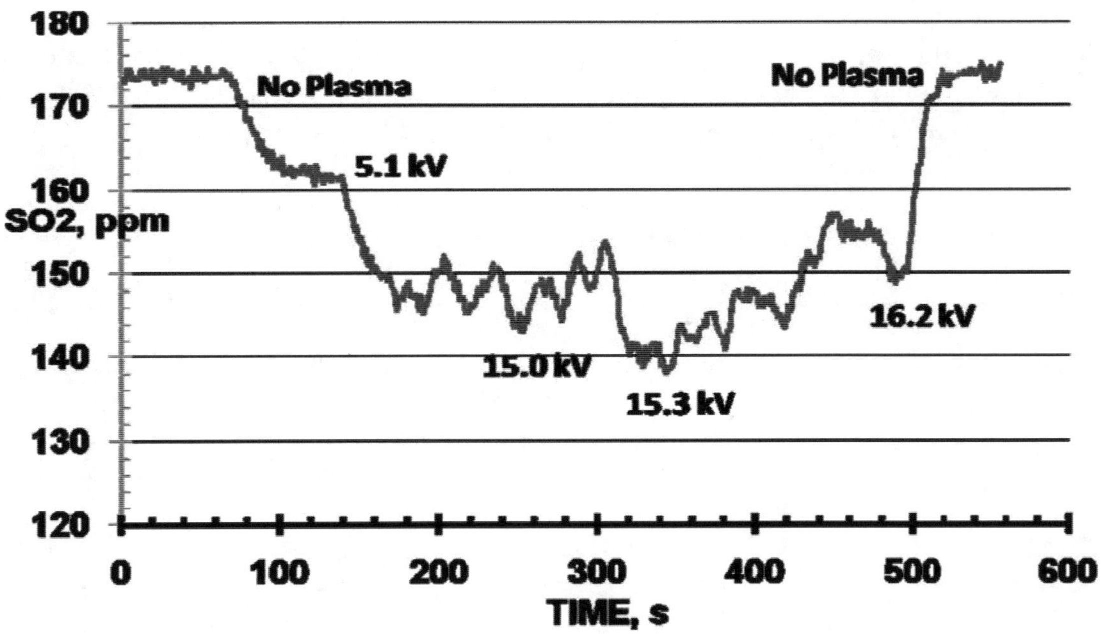

Figure 5. Results for the reaction of SO_2 + CO_2 (1 vol%) + H_2O (2 vol%) + N_2 (dilute gas) stream in the NTP reactor

After CO_2 was introduced into the NTP reactor, the SO_2 conversion rate increased. The results may be interpreted as follows:

Excited nitrogen molecules are dissociating the CO_2 to generate extra oxygen (Equation 9), which then oxidizes SO_2 as shown in Equation 7:

$$N_2^* + CO_2 \rightarrow CO + O + N_2 \qquad (9)$$

Theoretically, any of the electronically excited states of N_2 produced by the plasma would have energy exceeding the dissociation energy of CO_2 (530 kJ/mol).

In literature [1], with the presence of a reducing gas, such as H_2, SO_2 was reduced to elemental sulfur, and a conversion rate of 67% was achieved at 20°C [8].

$$SO_2 + H_2 \rightarrow S + H_2O \qquad (10)$$

3.6 Reaction of CO (3 vol%) + H_2 (1 vol%) + N_2 (dilute gas) Stream.

When a mixture of CO (3 vol%) and H_2 (1 vol%) were introduced into the NTP reactor with N_2 as diluent gas, at 200°C, only trace amount of N_2O (~2 ppm) was detected. No CH_4, NOx were detected. It implied that under reducing condition, the reactive species were quenched by H_2.

3.7 Reaction of CO (3 vol%) + H_2 (1 vol%) + CO_2 (1 vol%) + N_2 (dilute gas) Stream.

When a mixture of CO (3 vol%), CO_2 (1 vol%), and H_2 (1 vol%) were introduced into the NTP reactor with N_2 as diluent gas, also at the same temperature 200°C, the results were similar as above: only trace amount of N_2O (~2.5 ppm) was detected. No CH_4, NOx, or water was detected. No significant conversion of CO_2 or CO, as shown in Figure 6.

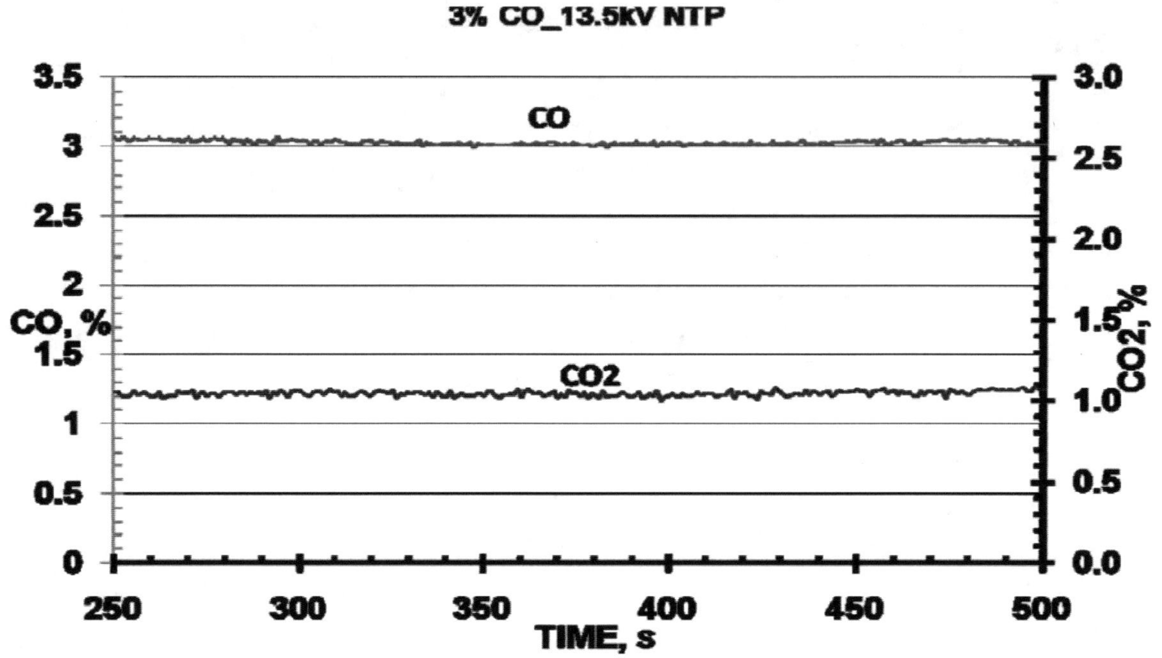

Figure 6. Results for the reaction of CO (3 vol%) + H_2 (1 vol%) + CO_2 (1 vol%) + N_2 (dilute gas) stream in the NTP reactor

3.8 Reaction of CO (6 vol%) + H_2 (2 vol%) + CO_2 (1 vol%) + N_2 (dilute gas) Stream.

When a mixture of CO (6 vol%), CO_2 (1 vol%), and H_2 (2 vol%) were introduced into the NTP reactor with N_2 as dilute gas, the results were similar to those above. The higher volume ratios of CO and H_2 did not have any effect on chemical reactions induced by the nonthermal plasma.

3.9 Reaction of CO (60 vol%) + H_2O (3 vol%) + N_2 (dilute gas) Stream.

A stream of CO (60 vol%) and water (3 vol%) with nitrogen as dilute gas was prepared to simulate the coal gas and used as the feed for the NTP test. At 200°C, a voltage of 15.3 kV, and the flow rate was 8 L/min; CO_2 was produced in 800 ppm concentration via the water gas shift reaction as shown in Equation 11. The conversion rate for CO is 0.1 mol%. This implies that the NTP treatment will not produce significant amount of CO_2 from the reaction of CO and water.

$$CO + H_2O \rightarrow H_2 + CO_2 \qquad (11)$$

3.11 Reaction of CO (35 vol%) + H_2O (3 vol%) + N_2 (dilute gas) Stream at 200°C.

When the CO feed was reduced to 35 vol% from 60 vol% as shown above, and the water concentration in the feed remained unchanged at 3 vol.%, at the same reaction conditions of 200°C and 15.3 kV. The highest CO_2 concentration in the product was 1200 ppm, and the highest NOx produced was 13 ppm. The conversion rate for the water-gas shift reaction was 0.34 mol%, which was higher than above, but still very low. NOx was produced from the reaction N_2 with the active oxygen species, which were generated from CO_2 or water (Equation 12).

$$O + N_2 \rightarrow NO + N \qquad (12)$$

3.12 Reaction of CO (35 vol%) + H_2O (3 vol%) + N_2 (dilute gas) Stream at 100°C.

Temperature effect was studied for the gas stream of CO (35 vol%) + H_2O (3 vol%) + N_2 (diluent gas). When the temperature of the NTP reactor decreased from 200°C to 100°C, the amount of CO_2 produced decreased from 1200 ppm to 1000 ppm, and NOx concentration decreased from 13 ppm to 10 ppm. The conversion rate for the water-gas shift reaction was 0.28 mol%.

3.13 Reaction of H_2 (30 vol%) + CO_2 (2 vol%) + N_2 (dilute gas) Stream.

For a commercial solid coal gasifier, the coal gas produced from coal gasification contains about 30 vol.% H_2 and 2 vol.% CO_2. During the NTP test of reaction of CO_2 with H_2, the volume% for CO_2 and H_2 were also 30 vol.% and 2 vol.%, respectively.
During the test, the preheater temperature was 550°C, and the NTP reactor temperature was 200°C. It was observed that without plasma, CH_4 was produced at 5 ppm, and CO was produced at 165 ppm. The formation of CO is via the reverse water-gas shift reaction (Equation 13), which typically operates close to equilibrium, and is mildly endothermic, $\Delta H^o = 41.2$ kJ/mol.

$$CO_2 + H_2 \rightarrow CO + H_2O \qquad (13)$$

Without plasma, this reaction is temperature dependent, as shown in Figure 7. At 450°C, the CO concentration was only 20 ppm; at 500°C, the CO concentration was about 80 ppm; at 550°C, the CO concentration was about 165 ppm (0.85 mol% CO_2 conversion).
The stoichiometric reverse water-gas shift reaction (Equation 13), which typically operates close to equilibrium, and is mildly endothermic, $\Delta H^o = 41.2$ kJ/mol.

The formation of CH_4 is via the following reaction (Equation 14):

$$CO_2 + 4\,H_2 \rightarrow CH_4 + 2H_2O \qquad (14)$$

This reaction is moderately exothermic, $\Delta H^o = -165$ kJ/mol. Without catalyst, the CH_4 yield was much lower than that of CO.

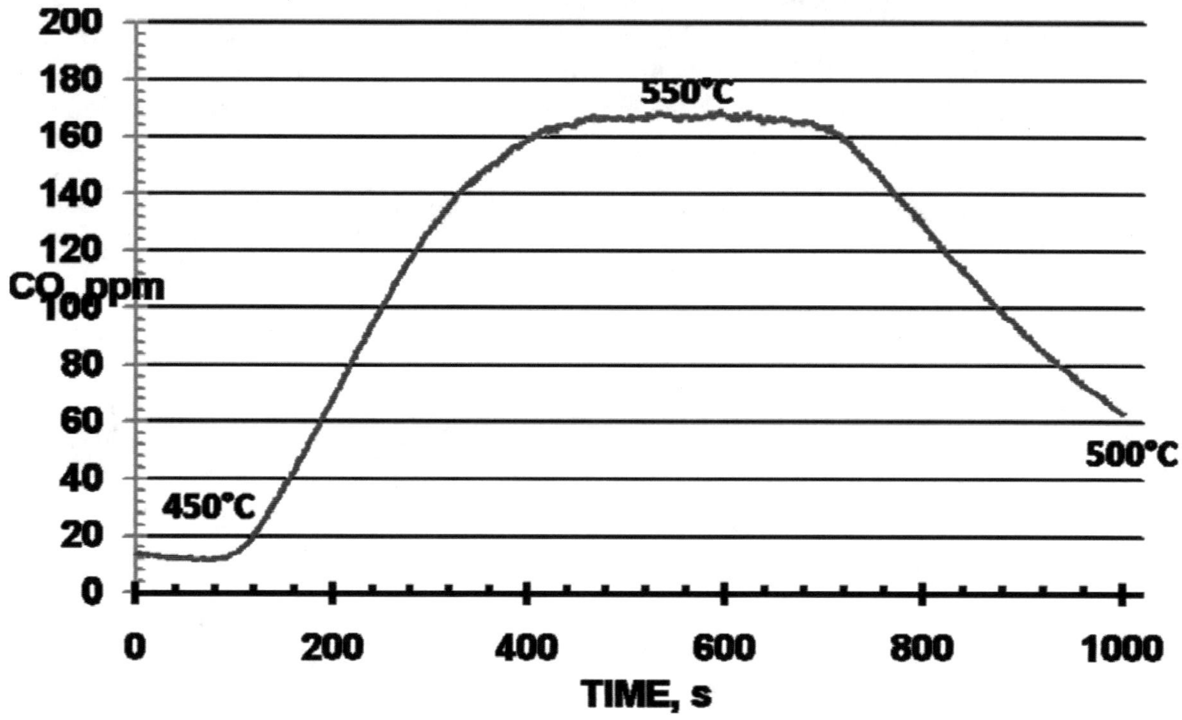

Figure 7. Temperature effect for the production of CO from H_2 and CO_2 without plsama

The results from Figure 7 show that the thermal reaction contributed to the CO production without plasma.

After the nonthermal plasma was turned on, the CO production yield increased significantly, and the CO yield was affected by the voltage of the NTP reactor. Figure 8 shows the results of CO production at different operating voltages. When the NTP reactor voltage increased gradually from 4.5 kV to 7.5 kV, 12 kV, and 15.3 kV, the CO concentration in the reactor outlet increased from 625 ppm to 780 ppm, 840 ppm, and 1780 ppm, as shown in Figure 8. After the NTP reactor voltage increased from 15.3 kV to 16.2 kV, the CO concentration dropped quickly to 630 ppm, and then increased to 1080 ppm and stabilized at this concentration. This shows that 15.3

kV was the best operating voltage. At 16.2 kV, the quick drop of CO concentration was related to the unstable plasma discharge.

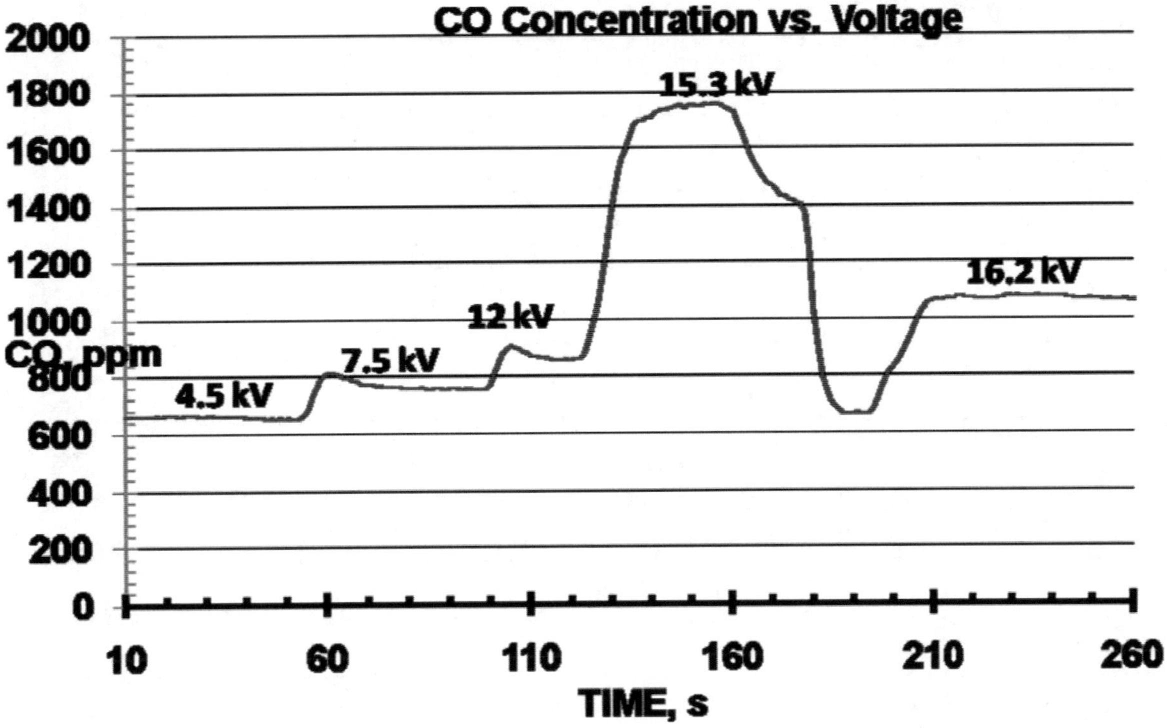

Figure 8. Results for the reaction of H_2 (30 vol%) + CO_2 (2 vol%) + N_2 (dilute gas) stream in the NTP reactor

CO_2 conversion calculation. At 15.3 kV, the highest CO concentration was obtained. CO was produced in 1780 ppm from CO_2, and the initial CO_2 concentration was 2%. The highest CO_2 conversion rate is calculated as follows:

$$CO_2 \text{ conversion} = 100 \text{ x } (CO_2 \text{ concentration})/(\text{initial } CO_2 \text{ concentration})$$

$$= (1780 \text{ x } 10^{-6}) / 2\% = 8.9 \text{ mol\%}$$

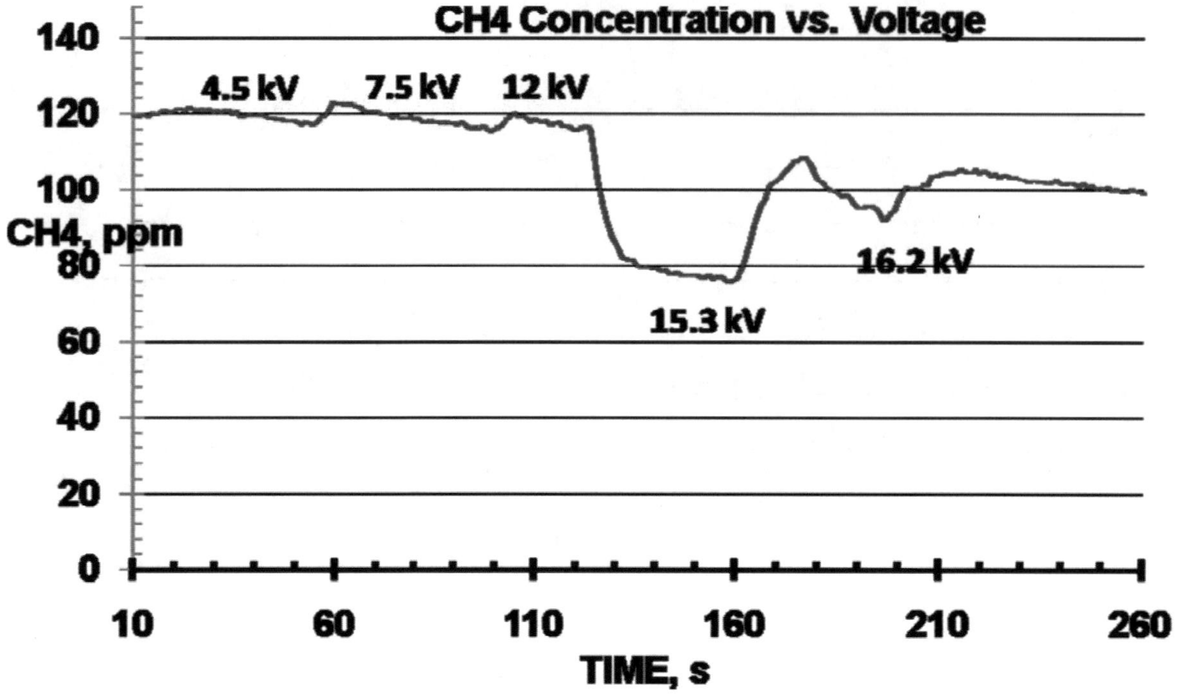

Figure 9. Results for the reaction of H_2 (30 vol%) + CO_2 (2 vol%) + N_2 (dilute gas) stream in the NTP reactor

Figure 9 shows the results of CH_4 production at different operating voltages. The methane production rate is calculated as follows:

$$CH_4 \text{ conversion} = 100 \text{ x } (CH_4 \text{ concentration})/(\text{initial } CO_2 \text{ concentration})$$

$$= 100 \text{ x } (120 \text{ x } 10^{-6}) /2\% = 0.6 \text{ mol\%}$$

The production of CH_4 was also associated with the operating voltage of the NTP reactor. Interestingly, the trend is opposite from the trend for the CO production. At the optimized operating voltage (15.3 kV), the CH_4 concentration was the lowest. One possible reason is that CH_4 is mainly produced from thermal reaction both in the preheater and the NTP reactor, rather than from the plasma reaction.

Temperature effect. When the temperature of the NTP reactor decreased from 200°C to 100°C, at 15.3 kV, the amount of CO production decreased from 1780 ppm to 1340 ppm, and the amount of CH_4 production decreased sharply to 5 ppm. This strongly supported the assumption that CH_4 was produced from the thermal reaction, instead of plasma reaction.

Mechanism of CO produced from CO_2. The highest CO concentration obtained was 1780 ppm at 15.3 kV and 200°C. CO was generated either by direct electronic dissociation of CO_2 molecule from its ground state or from the anti-symmetric vibrational state of CO_2, which is an intermediary state of CO_2 dissociation as described in the following equation [15]:

$$CO_2 (\Sigma_u^+) \rightarrow CO (X^1\Sigma^+) + O \qquad (15)$$

The hydrogen atom will react with O in the above equation to facilitate the production of CO. The formation of O via reactions (15) is the control step at such input voltages. This is why the CO yield is much higher with the presence of hydrogen.

4. Conclusion

A brief summary for the nonthermal plasma tests is as follows.

1) Under nonthermal plasma conditions, there was no obvious reaction between CO and H_2O. This implies that the NTP reactor will not affect the concentration of CO produced from thermal plasma gasification.
2) The highest SO_2 conversion was 21.1 mol% with the presence of CO_2, water and dilute gas N_2 at 15.3 kV. As reported in literature [8], if there were H_2 present, the reduction of SO_2 to elemental sulfur could be as high as 67%. This implies that most SO_2 produced from thermal plasma gasification could be removed during the NTP treatment for syngas cleanup.
3) The reaction of CO_2, water, and dilute gas N_2 gave only 1 mol% of CO_2 conversion. In the literature, this reaction was improved with a ferroelectric packed-bed NTP reactor to have a conversion of 12.3 mol% [9].
4) The reaction of CO_2 with H_2 yielded a CO_2 conversion of 8.9 mol%. At the same time, there was 0.6 mol% of CH_4 produced during the reaction. This reaction might be able to reduce the unreacted CO_2 from thermal plasma gasification, and to change the product distribution, thus the CO/H_2 ratio will be higher.

5) When indene was used as a coal tar model compound and fed into the NTP reactor with CO_2, water and dilute gas N_2, indene was not detected in the outlet by FT-IR. NTP might be a good source for coal tar removal from syngas. However, more experiments are needed to identify the decomposition products from indene.

5. References

[1] W.B. Sun, B. Pashaie, S.K. Dhali, F.I. Honea, "Nonthermal plasma remediation of SO_2/NO using a dielectric-barrier discharge", J. Appl. Phys. 1996, 79, 3483.

[2] B. Pashaie, L. Jing, W. Sun, S.K. Dhali, F.I. Honea, "Nonthermal plasma for cleanup of flue gas", Emerging Technology in hazardous Waste Management, 1996, 6, 209-233.

[3] L.A. Rosocha et al, "Treatment of hazardous organic wastes using silent discharge plasma", in Non-thermal Plasma Techniques for Pollution Control, Part B: Electron Beam and Electrical Discharge Processing, B.M. Penetrante, S.E. Schultheis, Eds. Berlin, Springer Verlag, 1993, 282-308.

[4] A.C. Gentile, M. Kushner, "Optimization of plasma remediation of NxOy from diesel exhaust", in Book of Abstract for the Symposium on Emerging Technologies in Hazardous Waste Management VI, W. Tedder, Ed., Atlanta, G.A., American Chemical Society, 1994, 214-218.

[5] E.M. van Veldhuizen, L.M. Zhou, W.R. Rutgers, "Combined effects of pulsed discharge removal of NO, SO_2, and NH_3 from flue gas", Plasma Chemistry and Plasma Processing 18 (1998) 91– 111.

[6] T. Yamamoto, S. Futamura, "Nonthermal plasma processing for controlling volatile organic compounds", Combustion Science and Technology 133 (1998) 117– 133.

[7] C.J. Liu, G.H. Xu, T. Wang, "Nonthermal plasma approaches in CO_2 utilization", Fuel Processing Technology 58 (1999) 119–134.

[8] a) A. Czernichowski, J. Polaczek, T. Czech, "Cold plasma reduction of sulphur dioxide to elemental sulphur", Proceedings of the 11th International Symposium on Plasma Chemistry, IUPAC/IEE, Loughborough, U.K., Aug. 1993; b) Proc. 12th Intl. Symp. on Plasma Chem., IUPAC, Minneapolis, M.N., August 21-25, Vol. II, Paper N.1.06, pp 1057-1062, 1995.

[9] S. Futamura, H. Kabashima, "Synthesis gas production from CO_2 and H_2O with nonthermal plasma", Studies in Surface Science and Catalysis, Volume 153, 2004, Pages 119-124.

[10] S. Zheng, L. Zhang, Y. Liu, W. Wang, X. Wang, "Modeling of the production of OH and O radicals in a positive pulsed corona discharge plasma", Vacuum 83 (2009) 238–243.

[11] M. Morvova, F. Hanic, I. Morva, Journal of Thermal analysis and Calorimetry, 2000, 61, 273-287.

[12] R.W. Smith, and R. Knight, "Reactive Plasma Spray Forming of Nitride/Nitrogen Composite Materials," Proceedings of the 11th International Symposium on Plasma Chemistry, IUPAC/IEE, Loughborough, U.K., Vol. 1, pp. 139-144, August, 1993.

[13] J.G. Wang, C.-J. Liu, B. Eliassion, Studies in Surface Science and Catalysis 153, 2004, pp.109.

[14] A. Mizuno, J.S. Clements, R.H. Davis, IEEE Trans. Ind. Appl. 1986, 22, 516.

[15] F. Poncin-Epaillard, M. Aouinti, Study of the interactions between carbon dioxide plasma and polypropylene, Proceedings of the 14th International Symposium on Plasma Chemistry, IUPAC/IEE, Prague, Czech, pp.1889, Aug. 1999.

Acknowledgement

This work was financially supported by the Defense Advanced Research Projects Agency (DARPA). The views and conclusions contained in this document are those of the authors and should not be interpreted as representing the official policies, either expressly or implied, of DARPA, or the U. S. Government.

(46c) Glid ARC Plasma Conversion of Pyrolysis Gas Into Syngas

Fela Odeyemi[1], Alexander Rabinovich[2] and Alexander Fridman[2], (1)Mechanical Engineering and Mechanics, Drexel University, Philadelphia, PA, (2)Mechanical Engineering & Mechanics, Drexel University, Philadelphia, PA

Pyrolysis gas is a product of Biomass, Municipal Wastes or Coal gasification process that contains H2, CO as well as unreacted light hydrocarbons and heavy hydrocarbons such as tar. These undesired hydrocarbons diminish Pyrogas fuel value thereby making significant removal of the unwanted hydrocarbons necessary. Various conditions and reforming reactions were considered for the conversion of Pyrogas into Synthesis gas – a combination of H2 and CO. The effectiveness of the Gliding arc plasma is demonstrated in the fuel reforming reaction processes with the aid of a specially designed low current device called Glid Arc Plasmatron.

1. Introduction

Pyrolysis gas or Pyrogas is a gas fuel formed primarily from the Pyrolysis or Gasification of Coal, Biomass or Municipal wastes. Gasification process of producing Pyrogas is usually an endothermic reaction forming a complex mixture of combustible gases such as CH4, CO, H2 and heavier hydrocarbons; Tar and noncombustible gas such as CO2. The presence of the heavy hydrocarbons and tar diminishes the quality of Pyrogas from the point of view of its usage for power generation or synthetic fuel production. This situation prompts the use of Non equilibrium Gliding arc Plasma for the chemical reforming of Pyrogas into Synthesis gas or Syngas. Syngas is a higher mass heating value gas comprising a varying quantity of Hydrogen (H2) and Carbon Monoxide (CO). Gliding Arc Plasma reforming is a homogenous process of fuel reforming which eliminates the need for catalysts[1].The reforming reactions considered are Steam reforming and Dry CO2 reforming, both of which are endothermic. Partial oxidation reaction will be inappropriate in this situation due to the presence of H2 in the Pyrogas mixture. Gliding Arc plasma serves as a resource for active species and radicals which are necessary to stimulate the desired chemical reactions.[2]

2. Experimental

The gliding arc plasma reformer is a specially designed low current reactor known as Gliding Arc Plasmatron. The plasmatron essentially consists of different component parts: High Voltage Stainless steel electrode and a ground electrode with swirl gas jets that creates a unique flow which propels the gliding discharge - formed in a 3mm gap between both electrodes. The high voltage electrode and Ground electrode are separated by a dielectric insulator; tangential air inlet at the flange assists in cooling flange and dielectric material.[1] Thermocouple insert ports in the post plasma zone for measuring

Gas temperature at various points in the plasma reactor zone; an exhaust pipe designed to incorporate a heat exchanger for cooling product gases before subsequent analyses by the Gas Chromatographer. The plasmatron is designed to function at high reforming temperatures.

Pyrogas conversion to Synthesis gas (Syngas): Pyrogas gas composition is simulated in the laboratory by combining the various constituent gases together in a cylindrical container. The gases flow rates were controlled with the aid of mass flow controllers before directing the resulting Pyrogas mixture into the gliding arc plasma zone inside the plasmatron where reforming occurs. The resulting changes in gas molar concentrations are observed under different operating conditions such as enthalpy, H2O/C ratios (for steam reforming). The effects of gases preheating and plasmatron inner geometry in the reforming processes were also examined. Similar procedures were conducted for Dry CO2 reforming experiments.

3. Results

For the steam reforming reaction of Pyrogas, a steady increase in H2 and CO were observed when H2O/C ratio was lower than 1.5. At higher H2O/C ratios, the H2 and CO yields diminish due to water gas shift reaction which limits attainment of thermodynamic predictions. This phenomenon can be partially attributed to limitations in the steam temperature and possible heat losses. Conversely, Dry CO2 reforming results showed better Syngas yields at lower temperatures with concentrations of heavy hydrocarbons such as CH4, C2H6 and C3H8 reduced considerably with enthalpy increase.

References

[1] Michael J. Gallagher, Robert Geiger, Anatoliy Polevich, Alexander Rabinovich, Alexander Gutsol and Alexander Fridman. "On-board plasma-assisted conversion of heavy hydrocarbons into synthesis gas" Fuel 25 November 2009

[2] Alexander Fridman “Plasma Chemistry” Cambridge university Press, 2008.

See more of this Session: Plasma for Energy and Environmental Applications (I)
See more of this Group/Topical: Topical 2: Advanced Fossil Energy Utilization

(46d) In-Detail Chemistry of Hydrocarbon Partial Oxidation Reaction by Plasma

Dae Hoon Lee, Kwan-Tae Kim and Young-Hoon Song, Environmental Research Division, Korea Institute of Machineryand Materials, Daejeon, South Korea

In the plasma reforming of hydrocarbon fuels, especially in a partial oxidation process, characteristic parameters of the reaction can be sorted as plasma parameter and reactant condition parameter. Plasma parameter or electric power and voltage-current characteristic determines density of active chemical species such as electron, ions and excited molecules, electron energy and thermal energy converted from electrical energy. On the other hand, reactant condition such as oxygen to carbon ratio determines thermal energy released from reaction, reaction time scale and so on.

In this presentation, the effects of supplied power and oxygen to carbon ratio, that are representatives of the above parameters, on the in-detail chemistry of the reaction will be introduced. Investigated processes are partial oxidation reaction of diesel and methane.

In a plasma reforming process of liquid phase diesel, change in electric power affect initial stage of the reaction or fuel cracking and chain initiation process, but change in oxygen to carbon ratio affect overall thermal activation and chain termination. Control of electric power and oxygen to carbon ratio produces different composition of product, different size and distribution of soot particle and different thermal efficiency.

In a gas phase methane reforming, electric power and oxygen to carbon ratio simultaneously controls plasma mode of generation. This change in plasma mode affects in-detail chemistry of hydrogen abstraction and dehydrogenation of hydrocarbon species resulting in difference in the composition of final product especially in C_2 species.

The results on the analysis of the chemistry of plasma reforming can be used as a base data for optimum design of the process.

See more of this Session: Plasma for Energy and Environmental Applications (I)
See more of this Group/Topical: Topical 2: Advanced Fossil Energy Utilization

(46e) Thermal Plasma Gasification of Different Coals with Carbon Dioxide

Maoqi Feng, Div. of Chemistry and Chemical Engineering, Southwest Research Institute, San Antonio, TX and Francis Huang, Southwest Research Institute, San Antonio, TX

Carbon dioxide emissions from conventional coal-fired power plants constitute the majority CO_2 emissions in the U.S. Mitigation of CO_2 by chemical conversion provides an alternative method to reduce the CO_2 emissions from coal gasification process with the benefit of eliminating energy-intensive CO_2 sequestration step. A CO_2 emission reduction process, which utilizes CO_2 as oxygen source by reacting CO_2 with coal to produce CO via the reverse Boudouard reaction (CO_2 (g) + C (s) → 2CO, DH^o = +172.6 kJ/mol, DG^o = +119.9 kJ/mol) in a thermal plasma reactor, is reported in this paper. Since the reaction has a positive DG^o, high temperature from thermal plasma shifts the equilibrium favorably to the right. A thermal plasma spray system was adapted for this application. The reaction temperature can be changed by adjusting the thermal plasma amperage. Three different coals were tested with CO_2 under thermal plasma condition (650 A, 28 V), and the CO_2 conversion rate was lignite coal 73%, subbituminous coal 77%, and bituminous coal 88%, respectively. Temperature effect study shows that the CO_2 conversions increased with the temperatures, which is in accordance with the fact that the reaction of CO_2 with carbon is an endothermic reaction.

See more of this Session: Plasma for Energy and Environmental Applications (I)
See more of this Group/Topical: Topical 2: Advanced Fossil Energy Utilization

Plasma Assisted De-NOx in Diesel Engine

Dae-Hoon Lee, Kwan-Tae Kim, Jae-Ok Lee and Young-Hoon Song, Korea Institute of Machinery and Materials, 171 Jang-dong Yusung-gu Daejon, Rep. of Korea

Eon-Seok Kim and Hyunsik Han, Automobile catalyst Division, HeeSunng Catalyst Corp., Siheung, South Korea

Introduction

In this presentation, plasma technology for de-NOx is described. Role of plasma in this application is provider for the reducing agents by onboard reforming. Plasma has abundant chemically active species such as energetic electron, ions and excited gas molecules that contribute highly selective chemical reactions.[1] Plasma has diverse merits for onboard application such as small volume (high GHSV), long durability (no coking, no degeneration), ease of operation (wide range of operation in reactant composition) and so on. Electric power consumption can be minimized within available power from battery. Among 3 state of the art characteristic technology or NH3 selective catalytic reduction (NH3 SCR), lean NOx trap (LNT), hydrocarbon selective catalytic reduction (HC SCR), LNT and HC SCR system is adopted for possible application of plasma and in this presentation, feasibility of these two plasma assisted de-NOx system is described.

Plasma reforming

Plasma reformer in this presentation is based on the rotating arc reactor. [2] Rotating arc reactor generates discharge between electrodes, then the plasma is convected downstream by fluid dynamic motion of the reacting flow. In this process of plasma generation, sufficient thermal environment and CAS contribute efficiently hosting reforming process. Both the gas and liquid phase fuel can be applied for the reactor geometry utilizes reaction heat to preheat, vaporize and mix the fuel and air. Partial oxidation process is designed. Operating parameters of the reforming process are O_2/C ration and input electric power (specific energy density) relative composition and the amount of H_2 and HC species that function as reductant can be controlled by these parameters.[3] Detail in plasma reforming process is already presented in this session. (#46d)

Plasma assisted LNT

LNT reduces NOx by following process. 1) absorption during lean operation of engine 2) change to engine rich condition 3) release of reducing agents 4) release and reduction of NOx. These 4 steps are repeated. Actually, the engine rich condition for reducing environment should be much shorter than the engine lean condition for the rich condition can deteriorate fuel penalty and diesel engine operation. So the sufficient (actually large) amount of reductant should be provided to the catalyst in a very short period of reduction phase and the reformer should be operated during this short period intermittently. All of these confinement results in elevated fuel penalty and difficulty in application of the system. To overcome the confinement and make a capacity of a reformer as small as possible, plasma assisted dual leg LNT system is suggested.[4] Engine exhaust gas pass only one leg at a time and a reformate gas is supplied on the other leg. In the proposed system, one leg absorb NOx and the other leg reduces NOx simultaneously. This system does not need rich condition of engine and continuous operation of reformer is possible resulting in scale down of the reformer. In engine bench test, the system shows rather good de-NOx performance with acceptable fuel penalty condition.

Plasma assisted HC SCR

HC SCR records rather low de-NOx rate compared to NH3 SCR and LNT. However, the system is rather simple as cost effective. Because advanced engine technologies requires only moderate de-NOx performance HC SCR can be a good answer for NOx problem. In this presentation, a plasma assisted HC-SCR system has been constructed and the de-NOx performance of the system has been evaluated. The onboard plasma reformer is applied. By controlling the operation parameter, the ratio of H_2 to hydrocarbon can be controlled. Bench test of proposed HC SCR system has been done with two catalysts, one for low temperature window of operation and the other for high temperature window. The proposed plasma assisted HC-SCR system showed a much enhanced de-NOx performance. With the help of onboard plasma reforming, about half of the typical catalyst volume was enough to achieve enhanced de-NOx performance. Expected level of de-NOx has been obtained with fuel penalty of approximately 3% with a possibility of further reduction without severe performance deterioration. Volume and power consumption also satisfy the necessary condition for commercial application.

References

1. Fridman, A. Plasma chemistry, Cambridge university press, 2008
2. Lee, DH, Kim, K-H, Cha, MS, Song, Y-H. Optimization scheme of a rotating gliding arc reactor for partial oxidation of methane. Proc Comb Inst 2007;31:3343–3351.
3. Lee, DH, Kim, K-T, Cha, Song, Y-H. Plasma-controlled chemistry in plasma reforming of methane. Int J Hydrogen Energy 2010;35;10967–10976.
4. Parka, C, Kim, C, Kim, K-T, Lee, DH, Song, Y-H, Moriyoshi, Y. The influence of hydrogen-enriched gas on the performance of lean NOx trap catalyst for a light-duty diesel engine. Int J Hydrogen Energy 2010;35:1789–1796.

(46g) Conversion of Benzene Into Small Aliphatic Hydrocarbons In Confined Environment

Tomi Nath Das and G. R. Dey, Radiation & Photochemistry Division, Bhabha Atomic Research Centre, Mumbai, India

Abstract. Employing 200 – 1000 ppm benzene (in Ar carrier) as the representative aromatic pollutant, its ready transformation into small aliphatic hydrocarbons (methane, ethylene, acetylene *etc.* @ > 100 ppm) in the absence of oxygen under atmospheric pressure dielectric barrier discharge (DBD) is presented. These products were confirmed from gas chromatographic (GC-FID & GC-MS) analyses. In these transformations, generation of acetylene is rationalized from symmetric ring-opening type reaction of an excited benzene molecule. On the other hand, substantial generation of methane and a small percentage of ethylene in the absence of any other H-atom source suggests that facile dehydrogenation of a fraction of the benzene also occurs simultaneously. Subsequently, addition of these freed H-atoms takes place on to some C_1 & C_2 (*i.e.* C_2H_2) moieties from decomposed benzene molecules. The tunablity and use of the DBD setup in relation to various experimental parameters like gas flow rate, benzene concentration, electrostatic field strength, reduced electric field *etc.* without and with mesoporous molecular sieve packing within the discharge zone is highlighted in these studies. Results suggest DBD with packing may contribute in the current efforts of tuning of a pollutant's alteration towards usable products *vis-à-vis* its mitigation.

See more of this Session: Plasma for Energy and Environmental Applications (I)
See more of this Group/Topical: Topical 2: Advanced Fossil Energy Utilization

(46h) The Destruction of Dioxins In Fly Ash by Using Gliding Arc In Tornado

Liang Yu[1], Kui Cheng[2], Jianhua Yan[2], Xiaodong Li[2], Yong Chi[2] and Mingjiang Ni[2], (1)Zhejiang University, Hangzhou, China, (2)State Key Lab of Clean Energy Utilization, Hangzhou, China

The characteristic of gliding arc in tornado was investigated under different experimental conditions and the destruction of PCDD/Fs (polychlorinated dibenzo-p-dioxins/polychlorinated dibenzofurans) in medical waste incinerator fly ash by using a dc gliding arc in tornado reactor was studied in the present work. The results indicate that the arc fluctuation is depended on the tangential gas flow into the reactor, and its stability can be improved by increasing tangential flow rate. In addition, arc space can be enlarged by adding axial gas flow into the reactor. In this study, the fly ash was carried by the axial gas flow at 2 l/min and fed into the reactor. The dioxins contained in fly ash can be decomposed effectively by gliding arc in tornado. The average decomposition rate of concentration was 49.56%, and the average decomposition rate of I-TEQ was 62.29%. The decomposition rates of HxCDD/Fs were over 70%. The PCDFs were more easily decomposed than the PCDDs in the plasma reactor. The main destruction channel was proposed as the dechlorination reaction by analyzing the final PCDD/Fs homologues distributions.

See more of this Session: Plasma for Energy and Environmental Applications (I)
See more of this Group/Topical: Topical 2: Advanced Fossil Energy Utilization

(69a) Dissociation of H_2S In Low Current Gliding Arc Tornado Discharge

Thomas Nunnally[1], Kirill Gutsol[1], Alexander Rabinovich[1], Alexander Fridman[1], Alexander Gutsol[2] and Abdenour Kemoun[2], (1)Mechanical Engineering and Mechanics, Drexel University, Philadelphia, PA, (2)Chevron Energy Technology Company, Richmond, CA

The process of hydrogen sulfide dissociation was studied in a low current gliding arc "tornado" (GAT) plasma discharge. Utilizing GAT ensured uniform H_2S gas treatment in the reactor. In addition, it created a low temperature zone near the cylindrical wall of the reactor, while maintaining a high temperature zone near the reactor axis. Specific Energy Requirement (SER) of 1.2 eV per H_2 molecule was achieved in this plasma system at atmospheric pressure. These results are particularly important for the oil industry, which consumes large amounts of hydrogen in oil hydro-treatment, and for gas industry because of the high H_2S content in "sour" gas.

See more of this Session: Plasma for Energy and Environmental Applications (II)
See more of this Group/Topical: Topical 2: Advanced Fossil Energy Utilization

(69b) CO_2 Dissociation and Conversion Using Low-Temperature Plasma

Michael J. Gallagher Jr.[1], Michael A. Lindon[2], Earl Scime[2], Alexander Fridman[3], David Berry[1] and Dushyant Shekhawat[1], (1)National Energy Technology Laboratory, Morgantown, WV, (2)Physics, West Virginia University, Morgantown, WV, (3)Mechanical Engineering and Mechanics, Drexel University, Philadelphia, PA

Over the last century, steadily rising emissions from carbon-based fossil fuels have led to increased concentrations of carbon dioxide in the atmosphere and there is a large body of scientific evidence that supports the theory that this is the reason for global warming. There is a serious research and development effort underway to find new technologies to mitigate CO_2 formation, capture and sequester, dissociate, or reuse CO_2 in chemical and manufacturing processes. The thermodynamic energy requirement to dissociate or convert CO_2 into more useful products such as, carbon monoxide or solid carbon, is high and the challenge is to find ways to reduce the energy cost while integrating these new technologies into the current energy, chemical, and manufacturing infrastructure. Low-temperature plasma technology is currently being explored for its unique ability to selectively excite CO_2 and promote dissociation. There are many types of plasmas that have been employed for CO2 activation and dissociation including gliding arc, dielectric barrier discharge (DBD), pulsed corona, and microwave plasmas. This presentation will provide an overview of the latest research in this area, as well as, some new results from an ongoing project at NETL for CO_2 conversion using a DBD plasma reactor in combination with traditional metal-based catalysts.

See more of this Session: Plasma for Energy and Environmental Applications (II)
See more of this Group/Topical: Topical 2: Advanced Fossil Energy Utilization

(69c) Characteristics and Kinetics of Decomposition of CO_2 In a Dielectric-Packed Discharge Reactor

Qinqin Yu, Jinhua Fei and Xiaoming Zheng, Chemistry, Institute of Catalysis, Hangzhou, China

The characteristics of carbon dioxide decomposition in a dielectric-packed barrier discharge reactor have been studied in the present work. A variety of factors including dielectric materials, input energy and electrode length for the decomposition of CO_2 were examined. To further understand the reaction mechanism of CO_2 decomposition in dielectric barrier discharge plasma, the oxidation of CO in plasma was also investigated. It showed that both the presence of dielectric materials and the increase of input energy promote decomposition of carbon dioxide. Dielectric materials induce the enhancement of electric field strength, resulting in increase of average electron energy, which is favorable for electron impact reactions, and surfaces of dielectric materials play an important role in the combination of oxygen radicals, favoring the conversion of CO_2. In a hollow reactor, keeping other conditions constant, the electrode length does not affect the conversion of CO_2, whereas in a dielectric-packed reactor, the conversion of CO_2 increases with the increase of electrode length. Kinetics study showed that the decomposition of CO_2 as well as the oxidation of CO in hollow or packed plasma reactors are all first-ordered.

See more of this Session: Plasma for Energy and Environmental Applications (II)
See more of this Group/Topical: Topical 2: Advanced Fossil Energy Utilization

(69d) Plasma-Enhanced Catalysis of CO_2 Reforming of Propane on Ni/γ-Al_2O_3 Catalyst

Qinqin Yu[1], Meng Kong[2], Tong Liu[2], Jinhua Fei[2] and Xiaoming Zheng[2], (1)Chemistry, Institute of Catalysis, Hangzhou, China, (2)Institute of Catalysis, Hangzhou, China

CO_2 reforming of propane was studied in dielectric barrier discharge (DBD) plasma and in the association of plasma with Ni/γ-Al_2O_3 catalyst. The main products of the reaction in DBD plasma were hydrogen, carbon monoxide, and a large amount of hydrocarbons, indicating that the reforming reaction could occur in DBD plasma without catalysts. And C-C coupling and cracking reactions also play important roles. The presence of propane and CO_2 promote the conversion of each other through the attachment of a series of radicals generated by the plasma. When the reaction was performed in association of plasma with Ni/γ-Al_2O_3 catalyst, the activation temperature of the catalyst was reduced, and synergistic function between plasma and the catalyst was obtained. It is suggested that vibrationally excited molecules and radicals produced by the plasma are responsible for the enhancement of reactant conversions in the association reaction mode, compared to the summed results in pure catalytic and pure plasma mode. Both excited species produced by gas-phase plasma and the interaction of plasma with the solid surface of the catalyst through surface discharge are probable to induce the decrease of the activation temperature of the catalyst. The inhibition of methanation reaction in the association mode is mainly due to the further reaction of excited methane with CO_2 to produce H_2 and CO. And the synergistic function between the plasma and the catalyst leads to the improvement of the selectivity of preferable H_2 and CO in the association mode compared to that in pure plasma or pure catalytic modes.

See more of this Session: Plasma for Energy and Environmental Applications (II)
See more of this Group/Topical: Topical 2: Advanced Fossil Energy Utilization

Role of Plasma Chemistry in Fuel Reforming Process

Nakyung Hwang, Dae-Hoon Lee, Min Hur, Young-Hoon Song

Korea Institute of Machinery and Materials,
171 Jang-dong, Yusung-ku, Daejon, Republic of Korea

Abstract: Two methane reforming processes, partial oxidation and dry reforming, are investigated with an arc reactor. The arc reactor is optically accessible and is fabricated with variable arc length. The power is supplied to the reactor by two different power supplies; one supplies relatively high voltage/low current power, the other one supplies relatively low voltage/high current power. In this way, different plasma conditions were generated, and are characterized by OES (Optical Emission Spectroscopy) test. The OES test showed that generation of reactive radicals, N_2 ions, and excited N_2 molecules was noticeably affected by the voltage/current characteristics of the power supplies. The product gases of methane reforming, however, are not sensitively changed by the voltage-current characteristics, as long as the same input power is supplied. The consecutive test results suggested that interactions between the reactant gases and the spatially non-uniform high temperature arc, which are modified by arc length, are more important factor in determining the performance of the reforming process, compared to the voltage-current characteristics.

1. Introduction

Hydrocarbon fuel reforming induced by plasma has been studied to cope with numerous environmental and energy issues. For example, studies related in automobile technology have demonstrated that synthesis gas (H_2 + CO) produced by on-board plasma reformer can be utilized as a reducing agent of De-NOx catalysts and as a fuel additive for clean combustion [1, 2]. On the other hands, in the field of fuel science, methane reforming process using plasma has been a long-time research topic to obtain higher hydrocarbons, like C_2H_2 or valuable hydrocarbons [3]. Another recent application of plasma reforming of methane is reforming of landfill gas, biomass gas, and natural gas, which are composed of methane and carbon dioxide. Since methane and carbon dioxide are the greenhouse gases, this application is important not only in energy issue but also in environmental issue.

As pointed out by numerous researchers [4-5], a major problem in plasma reforming process is that the energy cost for generating plasma is quite high. To reduce the energy cost, several technical approaches have been suggested, such as combination of plasma and catalytic process, modification of reactor geometry, adopting various types of plasma generation technique [4-5], etc. Despite of these intensive studies, it is not yet fully understood how these suggested technologies help to reduce the operating cost for the fuel reforming. One of the reasons for such poor understanding is that the plasma and thermo chemical reactions are strongly coupled for each other. For example, when methane reforming process is conducted, methane molecules are decomposed by both highly energetic electron impact dissociation (plasma chemistry) and conventional thermo chemistry. In this case, unfortunately, these two different types of reactions are difficult to be investigated separately, which causes difficulty in detailed understanding of plasma reforming process. Therefore, careful investigation on the reforming process, in which effects of plasma chemistry and thermo chemistry are decoupled, is suggested.

In the present study, as an effort to segregate the effects of plasma condition from those of thermal condition in studying fuel reforming process, different plasma conditions were generated by changing the types of the power supply; one supplies relatively high voltage/low current power, and the other one supplies low voltage/high current power. In this way, different plasma conditions were successfully generated, which will be described in later including optical emissions of N_2 ions and the excited N_2 molecules. Under such differently generated plasma conditions, the following two reforming processes are compared;

Partial oxidation;

$$CH_4 + 1/2\ O_2 \rightarrow CO + 2H_2, \qquad \Delta H = -36\ kJ/mol$$

Dry reforming;

$$CH_4 + CO_2 \rightarrow 2CO + 2H_2, \qquad \Delta H = +\ 247\ kJ/mol$$

In addition, the above mildly exothermic process (partial oxidation) and highly endothermic process (dry reforming) are compared under the different arc length conditions. In the present test configuration, variation of the arc length modifies the gas residence time and thermal environment without significant changes of plasma condition. Thereby, test with a variable arc length provided a chance to investigate the effects of thermal conditions on the plasma reforming process. Since voltage-current characteristics and gas residence time are critical parameters in designing a practical plasma reformer, the present test results are important not only to figure out the fundamental aspects of the plasma reforming processes but also to obtain an engineering insight for designing a plasma reformer.

2. Experimental approach

The schematic of the test apparatus is shown in Fig. 1. The test apparatus consist of an arc reactor, power supply, gas analysis system, gas supplying system and OES (optical emission spectral) system. The detailed geometrical configuration of the optically accessible arc reactor is also shown in Fig. 1. An arc reactor consists of a high voltage electrode (dia. 7 mm) and a ground electrode with 2 mm dia. hole located at the center. The high voltage electrode is moved up and down by controlling the motor. Using this vertically movable electrode, the arc can be easily ignited at a relatively short distance and low voltage conditions. After the ignition, the arc length is increased by pulling down the electrode. In this way, the arc length can be increased up to 50 mm. Two different 10 kHz AC power suppliers are used; one supplies relatively higher voltage and lower current power, and the other one supplies lower voltage and higher current power.

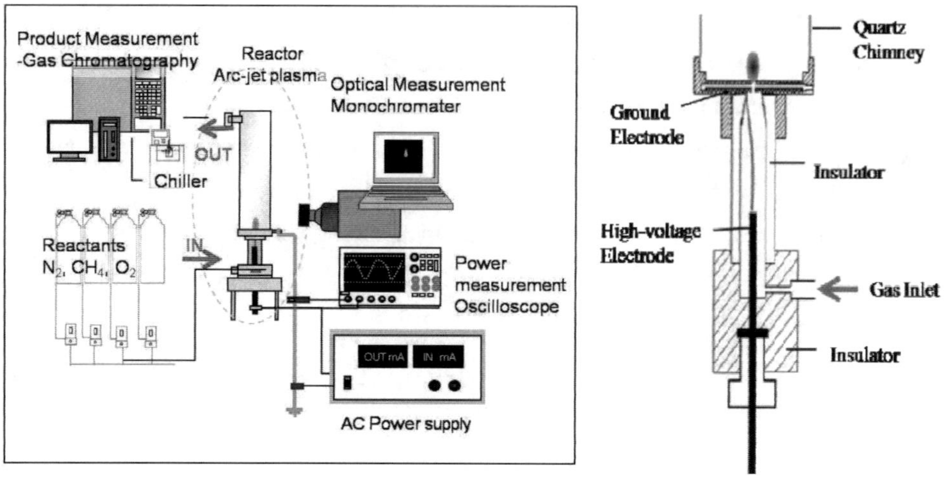

Fig. 1 Schematics of test apparatus

A mass flow controller (Brooks, 5850E, 5851E) set was used to control the flow of individual gases. To measure the species composition of product gases, the reformed gas was dried by chiller (JELO, RW-0525G) and analyzed using a gas chromatograph (Agilent, HP6890). Major products such as CH_4, H_2, CO and CO_2 were analyzed, as well as C, C_2 and C_3-related hydrocarbon species. A digital oscilloscope (Tektronix, TDS 5054B) is used to detect the voltage and current oscillogram. Detection of discharge voltage and current were conducted using a 1:1000 probe (Tektronix, P6015A) and a current probe (Tektronix, PCP303), respectively. Fig. 2 shows a typical voltage and current oscillogram detected from the arc reactor.

To evaluate the performance of the present reforming technique, the CH_4 and CO_2 conversions and the selectivity of product are defined as follows;

$$\text{X conversion} = \frac{\text{Moles(Converted X)}}{\text{Moles(Input X)}} \times 100$$

$$H_2 \text{ selectivity} = \frac{0.5\ \text{Moles(Generated } H_2)}{\text{Moles(Converted } CH_4)} \times 100$$

$$\text{Efficiency} = \frac{LHV_{H2} \cdot \text{Moles(Generated } H_2) + LHV_{CO} \cdot \text{Moles(Generated CO)}}{LHV_{CH4} \cdot \text{Moles(Converted } CH_4) + \text{Power}} \times 100$$

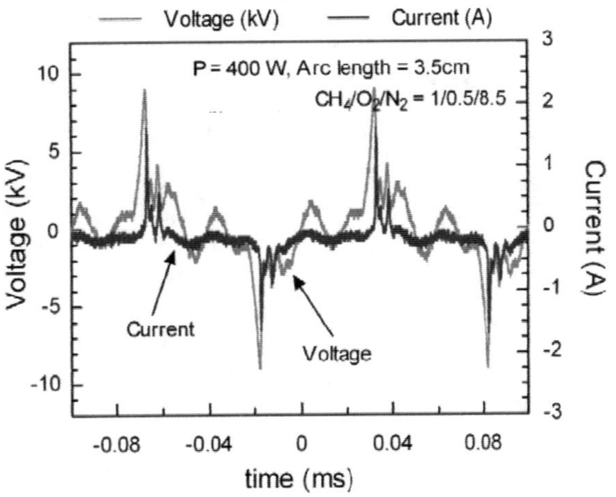

Fig. 2 Voltage and current oscillogram

$$\text{COx selectivity in partial oxidation} = \frac{\text{Moles(Generated COx)}}{\text{Moles(Converted } CH_4\text{)}} \times 100$$

$$C_XH_Y \text{ selectivity in partial oxidation} = \frac{\text{X Moles(Generated } C_XH_Y\text{)}}{\text{Moles(Converted } CH_4\text{)}} \times 100$$

$$\text{COx selectivity in dry reforming} = \frac{\text{Moles(Generated COx)}}{\text{Moles(Converted } CH_4 + \text{Converted } CO_2\text{)}} \times 100$$

$$C_XH_Y \text{ selectivity in dry reforming} = \frac{\text{X Moles(Generated } CXH_Y\text{)}}{\text{Moles(Converted } CH_4 + \text{Converted } CO_2\text{)}} \times 100$$

3. Results and Discussion

3.1 Plasma Generation with different power supplies

In the present tests, we confirmed that different plasma conditions can be generated by changing the power supplies. Fig. 3 shows the voltage-current characteristics of two different power supplies, respectively. As shown in the figure, one supplies relatively higher voltage and lower current power, compared to the other one. Fig. 4 shows that these two different power supplies can generate different plasma conditions. Here, the carrier gas supplied to the arc reactor is N_2. The photo pictures presented in Fig. 4 clearly show that the lengths of the arc jet generated with the higher voltage condition are longer than those with the lower voltage condition. The reason for this longer arc jet is that the higher electrical field (V/L, V/cm) with the higher voltage condition generates more abundant amounts of energetic electrons that produce the N_2 ions and the excited N_2 molecules. Here, detection of the excited N_2 (337 nm) and the N_2^+ (391 nm) means that some of the energetic electron's energy is higher than 11 – 18 eV. This electron energy is high enough to initiate electron impact dissociation of CH_4 molecules, which proceeds independently of the temperature conditions.

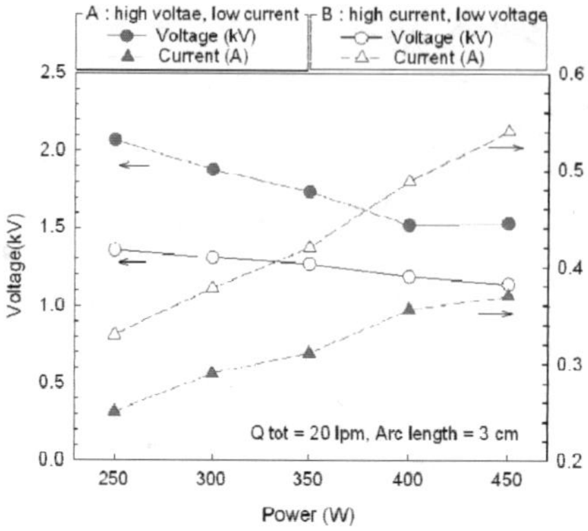

Fig. 3 Voltage and current with two different power supplies

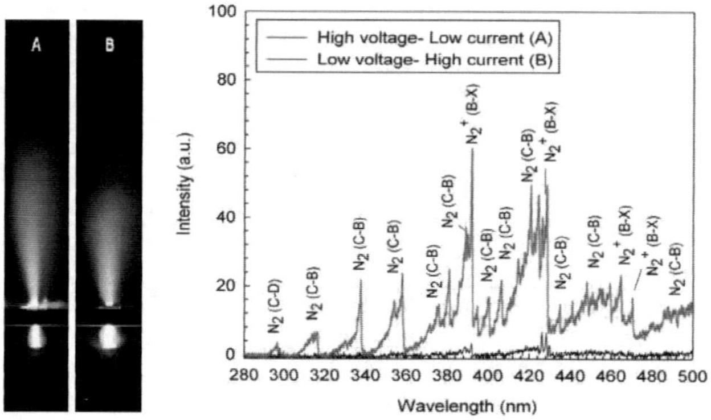

Fig. 4 Photo pictures of the arc and arc jet generated with high and low voltage conditions (A: high voltage/low current, B: low voltage/high current), carrier gas; N_2, Input power : 300 W, OES are detected jet region 5 mm above the exit of the arc reactor

3.2 POX and dry reforming with different power suppliers

Even though the plasma conditions are noticeably affected by changing the power supplies, the conversion rates and the efficiency of the reforming processes are not sensitively varied by the voltage-current characteristics, as long as the input electrical power is not changed. Fig. 5 shows that the conversion rates and the selectivity of the POX and dry reforming processes are not significantly varied with respect to the voltage-current characteristics. Instead of the voltage-current characteristics, Fig. 5 shows that input power is the governing factor to determine the reforming processes. As shown in Fig. 3, since the voltage decreases as the input power increases, the high input power conditions are not favorable conditions for plasma chemistry. This observation suggests that joule's heating to generate high temperature environment plays a critical role in the present fuel reforming process.

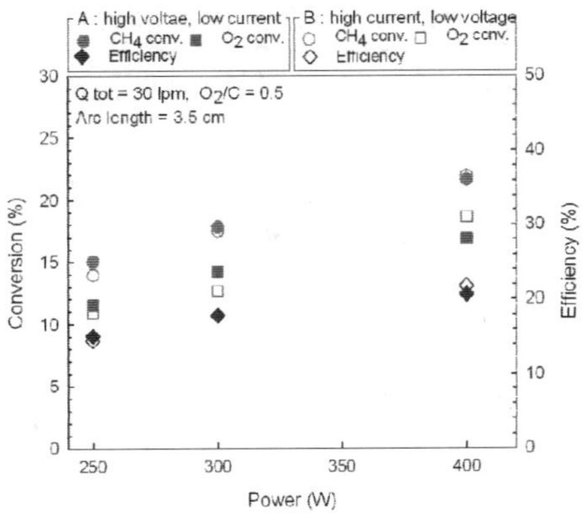

(a) POX process

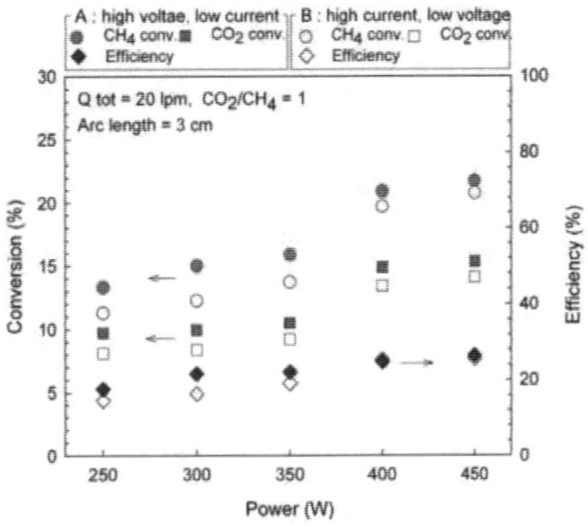

(b) Dry reforming process

Fig. 5 POX and dry reforming with different plasma conditions

3.3 POX and dry reforming with different are length

After the tests with different power suppliers, consecutive tests with different arc lengths were conducted. Here, the arc length was varied by changing the gap distance between the high voltage electrode and the ground electrode. At first, variation of arc length results in changes of voltage-current characteristics. However, the effects of voltage-current variation caused by the variation of the arc length did not showed noticeable shape changes of N_2 arc jet, as shown in Fig. 6. Therefore, the variation of the arc length in the present test does not change the plasma conditions significantly, but changes other conditions that could affect the thermo chemical reactions of the reforming process, such as gas residence time, thermal environments (arc temperature, temperature gradient, surface area and dia. of arc, relative gas velocity, etc.).

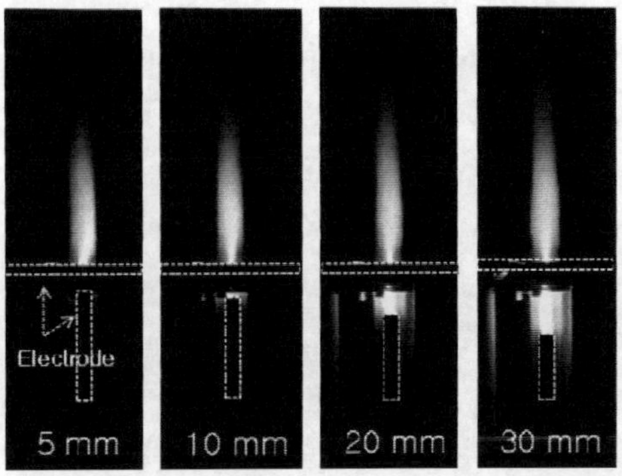

Fig. 6 arc and arc jet at different arc length

Fig. 7 show the test results of POX and dry reforming processes under different arc length conditions. The test results indicate that the arc length is another critical factor to enhance the efficiency of the reforming processes, regardless of the types of the process, i.e. mildly exothermic POX or highly endothermic dry reforming process. Fig. 7 shows that the main reason for the enhanced performance is attributed to the increased CH_4 conversion rates with increasing the arc length. As mentioned above, both increased gas residence time and modified thermal environments under the increased arc length conditions generate a favorable condition for the conversion of CH_4, O_2 and CO_2 molecules.

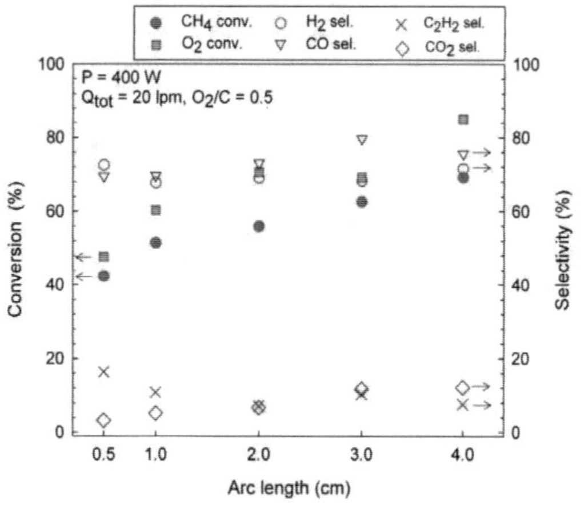

(a) Partial oxidation

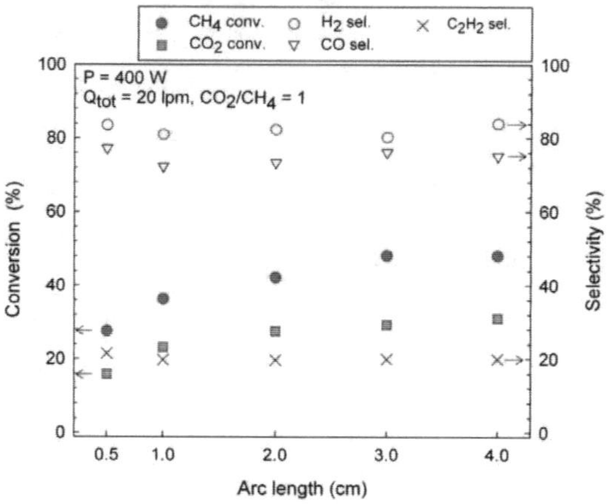

(b) Dry reforming

Fig. 7 Conversion rate, selectivity, and reforming efficiencies of POX and dry reforming as a function of arc length. The input power is fixed at 400 W.

4. Conclusion

The present tests of N_2 arc jet and OES showed that plasma conditions can be noticeably changed by changing the voltage-current characteristics. In the case of the higher voltage, a longer N_2 arc jet can be obtained, and consists of N_2^+ and excited N_2 molecules. Although these are very reactive with CH_4 molecules, both POX and the dry reforming processes are not sensitively affected by the voltage-current characteristics. Instead of plasma conditions, consecutive tests showed that input power and arc length, which modifies conditions for thermo chemical reactions rather than plasma chemistry, resulted in noticeable performance changes. The input power increased by increasing the current enhances the conversion rates and the efficiencies of the reforming processes. The increased gas residence time and modified thermal environments resulted from the increased arc length are also critical factors to enhance the performance of the reforming processes. This enhancement is mainly attributed to the increased conversion rates but not to the selectivity of the product gases. In summary, using with an arc and arc jet at atmospheric pressure conditions, it was difficult to decoupling the role of plasma chemistry from the thermo chemical reactions.

Acknowledgements; This study is supported by KIMM, KETEP/Ministry of Knowledge and Economy

References

1. A. Czernichowski, "Gliding arc. Applications to engineering and environment control", *Pure & Appl. Chem.* 66(6) 1301–1310, 1994

2. C.W. Park, C.G. Kim, K.T. Kim, D.H. Lee, Y.H. Song, Y. Moriyoshi, "The influence of hydrogen-enriched gas on the performance of lean NOx trap catalyst for a light-duty diesel engine", *Int'l J. Hydrogen Energy* 35 1789-1796, 2010

3. G. Heinz, "How Huels makes Acetylene by DC arc", *Hydrocarbon Processing & Petroleum Refiner*, 41, 6, 159-164, 1962

4. G. Petitpas, J.-D. Rollier, A. Darmon, J. Gonzalez-Aguilar, R. Metkemeijer, L. Fulcheri, "A comparative study of non-thermal plasma assisted reforming technologies", *Int'l J. Hydrogen Energy* 32 2848-2867, 2007

5. B. Pietruszka and M. Heintze, "Methane conversion at low temperature: the combined application of catalysis and non-equilibrium plasma", Catalysis Today 90 151-158, 2004

Plasma Assisted Combustion for Environmental Applications

Nakyung Hwang, Dae-Hoon Lee, Min Hur, Young-Hoon Song

Korea Institute of Machinery and Materials, Yusung-ku, Daejon, Republic of Korea

Abstract; Fundamental and practical features of a combustor assisted by a rotating arc have been studied. To understand fundamental characteristics of the combustor, a methane premixed flame interacted with the rotating arc under wide equivalence ratios, i.e. Φ = 0.77 – 1.85 has been investigated. In this investigation, observation on the flame and the arc, analysis on the product gases and the optical emissions were conducted. One of the distinguished features of the present combustor is that both fuel rich and lean flammability limits are extended with the rotating arc on. Utilizing such favorable features of the combustor, the combustor is successfully applied to regenerate DPF (Diesel Particulate Filter) installed in a diesel exhaust line. Here, the operating conditions for combustor are extremely harsh; low and transiently varying oxygen contents (5 – 10 %) and high mean combustion product velocity (20 m/sec).

1. Introduction

Favorable effects of plasma on flame, which are flame stabilization, extend of flammability limits, and reduce of ignition delay time, etc, have been studied for a long time [1-3]. Choices of plasma techniques, however, are quite limited in practical combustion applications. One difficulty in applications of plasma techniques to practical combustors is that the optimal operating conditions and geometrical configurations for generating plasma are not necessarily matched with those for combustion. For an example, generating plasma under vacuum conditions is one typical approach to obtain spatially uniform and energy-effective plasma. Here, "energy-effective" means that the plasma can selectively generate chemically active species without excessive heating of heavier particles such as neutral molecules and ions. Obviously, vacuum condition is not typical operating environment for most practical combustors. Therefore, the lack of proper plasma generation techniques still hinders the use of plasma to enhance combustion characteristics in practical combustor applications.

On the other hand, gliding arc or rotating arc have been suggested as one of the practical plasma generation techniques for various environmental applications, such as fuel reformer for NOx control and clean combustion technologies [4, 5]. The geometrical configurations of gliding and rotating arc reactors are simple and this makes its application to combustion devices easier. Figure 1 shows an example of a rotating arc reactor that consists of an inner electrode and an outer tube (left). The figure also shows the practical diesel combustor assisted by rotating arc (right). The present diesel combustor installed in a diesel exhaust line is well working under the harsh operating conditions to obtain a stable flame, i.e. transiently varying low oxygen contents (5 – 10 %) and high mean combustion product velocity (20 m/sec). Although gliding or rotating arcs have shown great technical potential in combustion applications, even the fundamental features of the flame assisted by these arc is not well known. For an example, NOx emission, which has been a critical concern in combustion technology and which could increase due to highly dissociative reactions induced by an arc, have not been investigated in previous studies on combustion assisted by gliding or rotating arc. Therefore, we do not yet have enough information how to operate these plasma combustors with low NOx emission.

Fig. 1 Configuration of a rotating arc reactor consisting of an inner electrode and an outer tube (left). Diesel combustor assisted by rotating arc generates a stable flame (right)

The objective of the present study is to extend the present understanding on combustion assisted by rotating arc. In order to achieve this object, a rotating arc and a premixed methane flame generated in a rotating arc reactor were investigated. In the early part of the paper, a phenomenological observation of the flame and the rotating arc under different equivalence ratios was discussed. The observed phenomena are analyzed along with analysis on product gases and optical emissions detected by OES (Optical Emission Spectroscopy). In the later part of the paper, performances of a practical combustor assisted by a rotating arc, which is operated in a diesel exhaust line and is tested through road test, are introduced.

2. Experiment Approach

The geometrical configuration of a rotating arc reactor is shown in Figure 1. The rotating arc reactor consists of an inner electrode and an outer tube. The inner electrode is connected to the high voltage port. The outer tube is connected to the electrical ground. The evolution of the rotating arc is schematically shown in Figure 2. Initially, the arc is ignited at the shortest distance between the inner electrode and the outer tube, which is located at the lower part of the reactor.

After ignition, the arc is pushed upward and is rotated by tangentially injected gas flow motion. Eventually, the rotating arc is anchored between the upper part of the inner electrode and the edge of the outer tube. In this case, the arc length reaches its maximum.

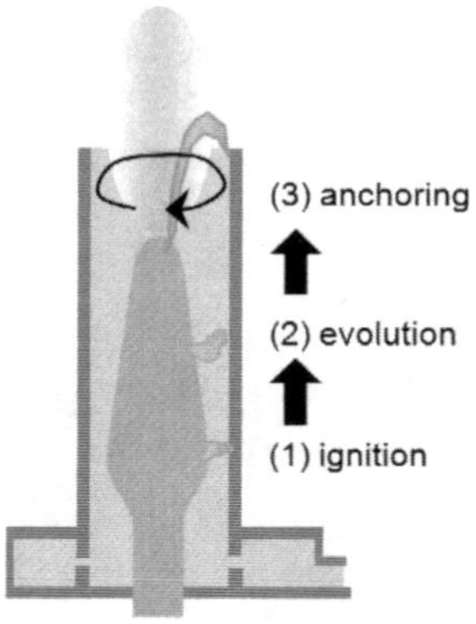

Fig. 2 Development of a rotating arc: (1) ignition, (2) evolution, (3) anchoring

Figure 3 shows a schematic drawing of the test apparatus used in this study. It consists of a rotating arc reactor, power supply, gas supply, and gas detection system. Figure 4 shows typical voltage and current oscillograms of a rotating arc. An AC power supply (Dawonsys Co. AP CAF-2) was used, generating a peak voltage of 3–5 kV and peak current of 0.2–0.3 A. The frequency of AC power was fixed at 10 kHz. The power supply was capable of supplying a constant input power, regardless of the varied peak voltages. To detect the voltage applied to the electrodes, a high-voltage probe (1000:1, Tektronix, P6015A) was used. The current was detected using a current probe (Tektronix, TCP202). The voltages and currents were recorded using a digital oscilloscope (Tektronix, TDS5054B). As shown in Figure 3, CH_4 and air were supplied through a mass flow controller (MFC) (Brooks Co., 5850E, 5851E) and were premixed before being introduced into the reactor. The flow rates of CH_4 were varied and those of N_2 and O_2 were fixed to 15 liter/min to change the equivalent ratio. In this way, variation of total flow under different equivalent ratios was minimized. During the most of the experiments, the electrical power input to the rotating arc reactor was fixed at 35 W. This power was estimated based on the voltage and current oscillograms. The electrical power supplied to the reactor was equivalent to approximately 3% of the total thermal power produced by the stoichiometric methane flame.

To generate a premixed flame, the outlet section of the rotating arc reactor was isolated from the surrounding air using a quartz tube that was 30 mm in diameter and 600 mm in length. A gas chromatograph (GC) (HP, G1530A) was used to detect the gases emitting from the rotating arc reactor, which were CH_4, C_2H_2, C_2H_4, C_2H_6, C_3H_6, C_3H_8, CO, CO_2, H_2, and O_2. NO and NO_2 were separately detected using a chemiluminescent NO_x analyzer (Signal Instrument, 4000vm). Optical emission spectra from the flame and the arc were detected using an OES system consisting of an intensified charge-coupled device (ICCD) (Princeton instrument, 74670028) and a monochromater (Acton, Spectrapro 500i) with an optical length of 0.5 m.

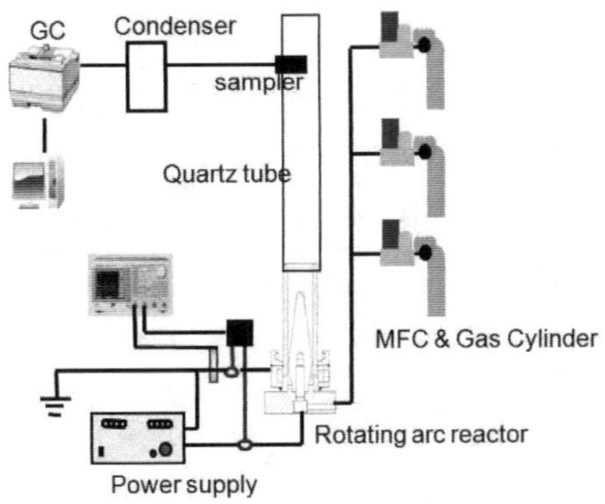

Fig. 3 Schematic of test apparatus

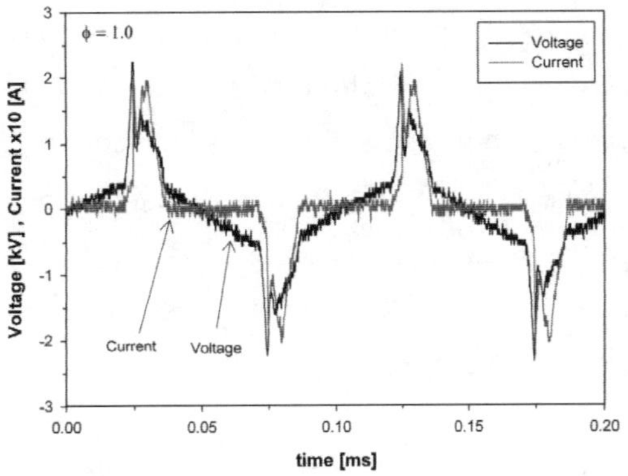

Fig. 4 Oscillograms of voltage and current

3. Result and Discussion

3.1. Observation on flame and arc

Fig. 5 shows photo pictures of the flame and the arc at different equivalence ratios, Φ. Here, the exposure times of the camera for the side view and the top view are 1/4" and 1/500", respectively. The input power for these tests is fixed at 35 W. As shown in the figure, equivalence ratio and on/off condition of the rotating arc are the critical parameters to determine the shapes, colors and extinction of the flame. On the other hand, detailed observation for the arc indicates that the behaviors of arc, such as arc length, angular speed, voltage-current characteristics, are also changed by equivalence ratio.

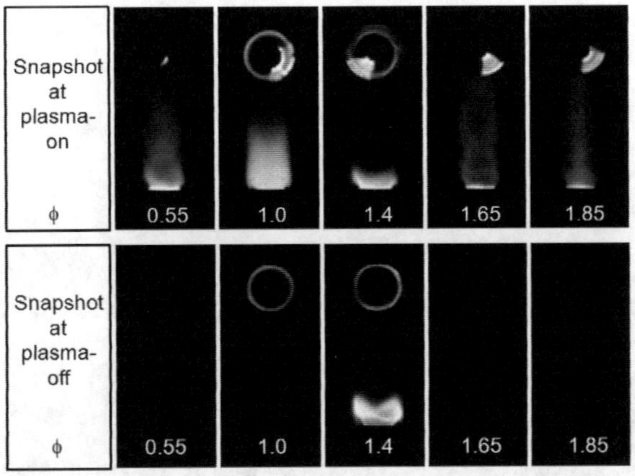

Figure 5. Side and top views of flame and rotating arc, Exposure time: side view : 1/4", top view : 1/500", Input power: 35 W

Another important feature of the present combustor is that the flame is stabilized by the rotating arc. For example, when Φ = 1.4 and with no rotating arc condition, the flame is unstable, and could be even extinguished when the reactor and the quartz tube are cold. This unstable flame, however, became the short and stable flame immediately after the rotating arc is on. In the case of more fuel rich condition, i.e. Φ = 1.85, the long and sooty flame shown in fig. 5 is periodically oscillating, even though the rotating arc is on. However, this oscillating flame became stable and turned green and white when the electrical input power increased, as shown in Figure 6. The present observation confirms that the plasma play a critical role to stabilize the flame and to extend the flammability limit.

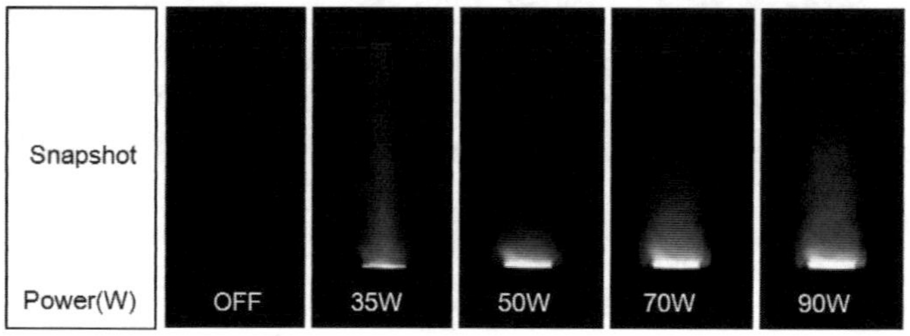

Figure 6. Variation of the flame shapes and the colors with different electrical input power

3.2. Optical emissions from the flame and the arc

Fig. 7 and 8 are the optical emission spectra of the flame and the arc shown in Fig. 5 and 6. In these figures, four sub-sections are combined for overall OES data (280–400, 400–550, 550–700, 700–800 nm). Fig. 7 is the low resolution optical spectra obtained from the flame at Φ = 1.85, and was detected 1, 5 and 10 mm above the exit of the reactor, respectively. The well-known spectral peaks of hydrocarbon-air flame, such as OH ($A^2\Sigma^+$-$X^2\Pi$, 306.4 nm), CH $A^2\Delta$–$X^2\Pi$ system (431 nm), and C_2 Swan band system, are identified. At 1 mm above the exit, strong peaks for N_2 $C^3\Pi_u$–$B^3\Pi_g$ system (337 nm, 357 nm), H_α (656nm), and O atom (777 nm) were also detected. Fig. 7 shows that the intensity of the spectral peaks decreases rapidly, as the height increases. This implies that plasma regions, in which chemically active species generated by plasma are abundant, are spatially limited. The weak optical emission intensities away from the exit of the reactor, as indicated with the rectangular shown in Fig. 8, are integrated to identify the emission peaks. Fig. 8 indicates that green color flame is attributed to C_2 Swan band system and the broad emission continuum with a maximum around 550 nm. Existence of similar emission continuum observed from a flame interacted by plasma has been reported by other research group [6]. Further OES analysis on the green color flame at Φ = 1.0, shown in fig. 5, reveals that the visible green color is also attributed to this emission continuum. In this case, the peaks for C_2 Swan band system shown in fig. 8 are not shown.

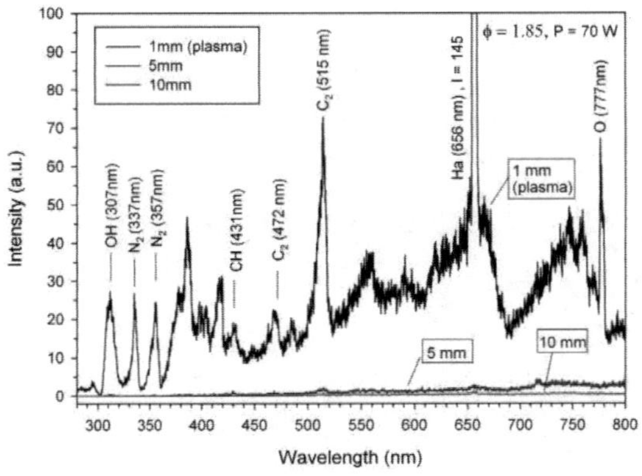

Fig. 7 Optical spectrum detected at Φ = 1.85 and at different locations above the exit of the reactor.

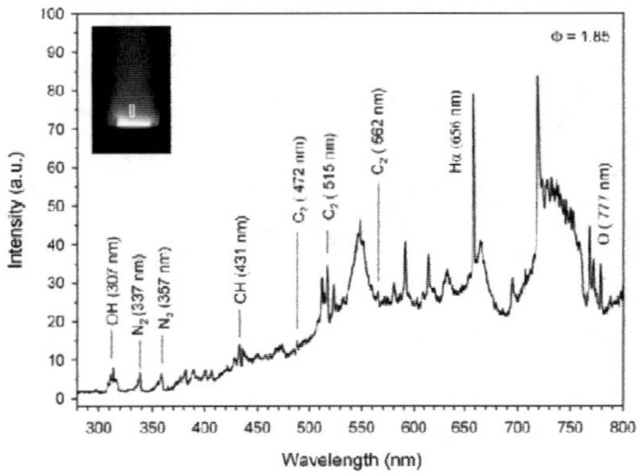

Fig. 8 Optical spectrum detected at Φ = 1.85. The weak emission intensities in the rectangular are integrated.

3.3 Product gases

Fig. 9 shows CH_4 conversion rate and selectivity of CO_2. Here, these parameters are defined as follows.

$$CH_4 \text{ conversion} = \frac{\text{Moles (Converted } CH_4)}{\text{Moles (Input } CH_4)} \times 100$$

$$\text{CO selectivity} = \frac{\text{Moles (Generated CO)}}{\text{Moles (Converted } CH_4)} \times 100$$

Once again, fig. 9 indicates that the flammability limits for both fuel rich and lean conditions are extended, as shown in Fig. 5. In the presence of the rotating arc, CH_4 can be converted over the wide range of the equivalence ratio, i.e. Φ = 0.55 – 1.85. However, the rotating arc does not noticeably change the conversion rate and the selectivity. In addition to CH_4 and CO_2, the other major product gases like CO and H_2 were not changed in the presence of the rotating arc. The reason for such negligible changes of the major product gases can be explained by the relatively small input electrical energy. The electrical input energy in this study was small, which is equivalent to only 3% of the heat energy released from the stoichiometric combustion process. Therefore, the electrical energy was not sufficiently high to control the selectivity of major species, but was sufficiently high to initiate and sustain the CH_4 conversion process under extremely fuel lean and rich conditions.

Unlike the major product gases, NOx emissions are significantly varied in the presence of the rotating arc, as shown in fig. 10. The decrease in the NO concentration starting in the fuel rich condition can be attributed to the lack of O_2 molecules and the selective non-catalytic reduction of NO with hydrocarbons, H_2, and CO. One of the most likely reasons for the abrupt decrease in NO_2

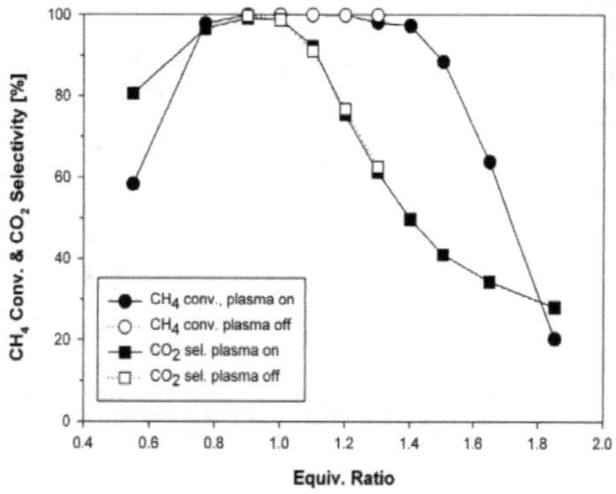

Fig. 9 CH_4 conversion and CO_2 selectivity as a function of equivalence ratio

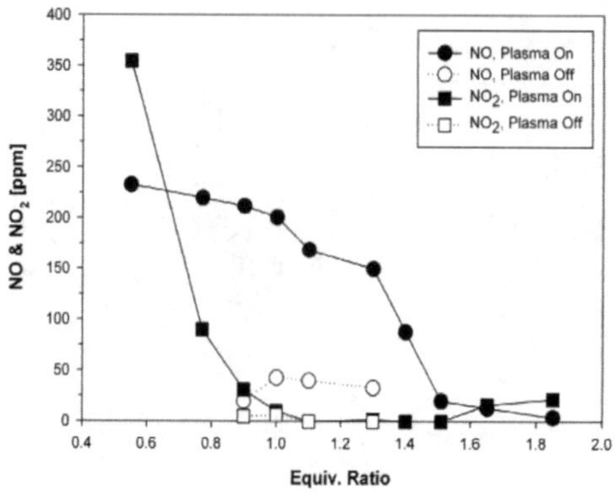

Fig. 10 NOx emissions under fuel rich and lean conditions

when approaching the stoichiometric condition is the reductive reactions of NO_2 to NO with O and H radicals in a high-temperature environment. The present study suggests that NO_x generated by plasma under fuel lean conditions could be a serious factor preventing the application of plasma in the combustion process.

4. Applications for environment technology

Gliding or rotating arc can be applied to several environmental technologies [7]. One example of the applications is the combustor assisted by the rotating arc reactor. Fig. 11 shows the DPF (Diesel Particulate Filter) system operated with the diesel combustor and the rotating arc. The photo picture of the flame and the rotating arc is already shown in Fig. 1. The role of the diesel combustor is to regenerate the particulate matters deposited in the filter occasionally. The biggest advantage of the combustor assisted by the rotating arc is that the flame can be stable under the harsh conditions. As shown in Fig. 11 (right), the stable temperature rise needed for regenerating the DPF can be obtained, while the combustor is on. Owing to this temperature rise, the back pressure of the diesel exhaust is abruptly decreased. The emissions of unburned hydrocarbons, NOx, and CO measured at the end of the diesel exhaust line were carefully detected, which indicates that the emissions of these pollutant gases are lower than the stringent emission standard imposed by the environmental authority in Korea. In the case of unburned hydrocarbons and CO, DOC (Diesel Oxidation Catalyst) can oxidize these gases. Under the certain operating conditions, increase of NOx emission caused by the high temperature arc was detected. However, even such conditions, the amounts of the NOx emissions are still lower than the emission standard, since the combustor is operated under the low NOx conditions, i.e. extremely fuel rich conditions. As shown in Fig. 10, NOx emissions from the rotating arc could be lower than those from a conventional combustor operated at the stoichiometry condition. So far, flame stabilization and extend of flammability limits have been important issues in numerous combustion technologies, the present technology of combustion assisted by the rotating arc can be utilized in further applications.

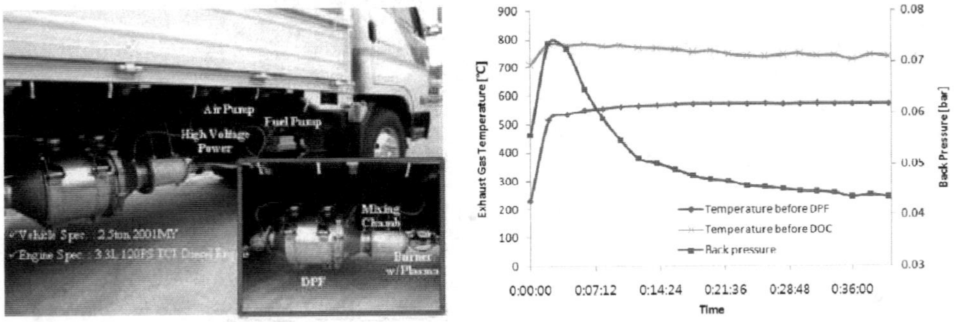

Fig. 11 DPF system operated with the rotating arc (left), temperature variation after ignition of the combustor (right)

Acknowledgements

This study was supported by Korea Institute of Machinery and Materials, Korea Institute of Energy Technology Evaluation & Planning and Ministry of Knowledge Economy

References

1. Stockman E S, Zaidi S H, Miles R B, Carter C D and Ryan M D, "Measurements of combustion properties in a microwave enhanced flame", *Combustion and Flame* **156** 1453-1461, 2009
2. Kim Y, Rosocha L A, Korolev Y, Cassarino C and Frambes T Near-zero emissions combustor system for syngas and biofuels *Proceedings of 7th Int'l Symposium on Non-thermal/thermal Plasma Pollution Control Technology & Sustainable Energy (St. John's, Newfoundland, Canada 21-25 June 2010)*
3. Starikovskaia S M "Topical Review: Plasma assisted ignition and combustion", *J. Phys. D: Appl. Phys.* **39** R265-R299, 2006
4. Lee D H, Kim K-T, Cha M S and Song Y-H, "Effect of excess oxygen in plasma reforming of diesel fuel", *Int'l J. Hydrogen Energy* **35** 4668–4675, 2010
5. Horng R-F, Wen C-S, Liauh C-T, Chao Y and Huang C-T, "Driving characteristics of a motorcycle fuelled with hydrogen-rich gas produced by an onboard plasma reformer", *Int'l J. Hydrogen Energy* **33** 7619–7629, 2008
6. Vincent-Randonnier A, Larigaldie S, Magre P and Sabel'nikov V, "Plasma assisted combustion: effects of a coaxial DBD on a methane diffusion flame", 2007 *Plasma Sources Sci. Technol.* **16** 149-16
7. Czernichowski A, "Gliding arc. Applications to engineering and environment control", *Pure & Appl. Chem.* **66**(6) 1301–1310, 1994

Experimental Study of Hydrogen Production and Carbon Deposit from Spark Discharge Plasma of DME

Kuifeng Ma, Beihang University, Beijing, China
Xinghu Li, Beihang University, Beijing, China
Ningbo Liu, Beihang University, Beijing, China

Abstract

A multistage spark plasma system was set up and experiments of hydrogen production by spark plasma of dimethyl ether (DME) were performed at atmospheric pressure. The effects of relative parameters on hydrogen yield and energy consumption were investigated. The experimental results showed that with the increase of pulse duty ratio, the H_2 yield initially increased and then decreased. Increasing stage number enhanced the yield of hydrogen. In general, the converter with 3-stages was desirable for higher H_2 yield and that with single stage was desirable for low energy consumption. When heat preservation configuration was employed, the DME conversion rate and the H_2 yield were enhanced, while the energy consumption was much lower than standard situation at each discharge frequency. When A/D ratio was beyond 2.5, the carbon deposit forming on the electrode was clearly reduced for the increase of oxygen in the reactant.
Key words: Hydrogen production; Plasma; H_2 yield; Dimethyl ether; Spark energy.

1. Introduction

In face of the challenge posed by the increase of economic development and environmental pollution, proton exchange membrane fuel cell (MFC) system has been recognized as one of the most promising technologies for meeting future global energy needs of vehicle. Hydrogen, which is the fuel of proton exchange membrane fuel cell electric vehicles, received extensive recognition as a kind of eco-friendly and alternative fuel for vehicle. On other hand, by adding fossil fuels with hydrogen, the flame propagation speed of traditional engines increased significantly, and emissions were reduced [1, 2]. The appealing characteristics of hydrogen have motivated a number of researches to produce hydrogen-rich gas. Hydrogen production by plasma, which features a rapid start and quick response, can convert various hydrocarbon fuels to hydrogen-rich gas [3, 4].

DME, which can be easily oxidized at low temperature, is considered as an ideal hydrocarbon fuel in hydrogen production for it is non-corrosive and non-toxic. Its physical properties are similar to those of liquefied petroleum gas (LPG), so it can be easily liquefied, stored and transported by using the wellestablished facilities for LPG.. Therefore, DME is promising in hydrogen production.

Bromberg et al. [5-6] implemented the fundamental research of hydrogen production from methane by arc discharge. Diesel, or natural gas was fed into a plasma converter to produce hydrogen-rich gas, which coupled with catalysts, and the operational optimum was determined. Thammanoon et al. [7] investigated a multistage gliding arc discharge system for hydrogen production. The results showed that both CH_4 and O_2 conversions were enhanced with the increase of stage number. Horng et al. [8] investigated the production of hydrogen from methane using a plasma–catalyst hybrid converter with emphasis on energy conservation. The experimental results showed that higher methane feeding rate resulted in higher gas temperature and a corresponding

Nomenclature			
f	Spark frequency (Hz)	t_c	Charging time of primary coil (ms)
α	Pulse duty ratio (%)	A/D	Air-to-DME volume ratio
s	Stage number of converter	Q_{DME}	DME inlet flow rate (mL/min)

increase in methane conversion. Horng et al. [9] also investigated in the driving performance and emission characteristics of a 125cc motorcycle equipped with an onboard plasma reformer to produce hydrogen-rich gas. Belen-S et al. [10] studied hydrogen production by reforming of hydrocarbons or alcohols in a dielectric barrier discharge. The experimental results showed that the CH_4 conversion efficiency increased with the increase of applied voltage in plasma chamber. Zou et al. [11] investigated the effects of controlling waveform on the hydrogen production from DME by corona discharge. They reported that higher current, which meant more electrons produced from plasma, resulted in a high reaction rate. Synthetically, [10] and [11] educed that the discharge energy directly effected on hydrogen production, for the applied voltage and current were the elements of discharge energy.

In this paper a multistage plasma converter was designed, which originated from a single chamber converter [12]. Fanlike plasma chamber was introduced to the converter to intensify the discharge energy density of plasma. Experiments of hydrogen production by spark discharge plasma partial oxidation of DME were implemented at atmospheric pressure and ambient temperature.

2. Experimental set-up and methods

2.1 Experimental set-up

The experimental set-up consisted of converter, reactant feeding line, plasma generating system, and measuring system, as shown in Fig.1. The converter consisted of three fanlike plasma chambers and a reactor chamber. Fanlike chamber whose inner radius was 7mm was determined to be the discharge cavity, and the angle of the each fanlike was 120 degrees. The stage number in series could be adjusted on demand. The length of the plasma chamber was 25mm. The anodes refitted from NGK spark plugs (BCP6ES) were fixed on the center of the sector. The diameter of the anode was 2.5mm. The reactant feeding line allowed the DME (99.9%) and air (compressed and dried) feeding. The reactant feeding line included compressed gas cylinders, pressure-relief valves, mass flow controllers (S49-32B/MT), etc. Air and DME controlled by mass flow controllers were pre-mixed before entering the converter. The plasma generating system consisted of a signal generator, a drive circuit, ignition coils (DQG126C) and a 12V DC power source. The signal generator and drive circuit were used to maintain the spark plasma continuously. The ignition coils supplied a high voltage (peak value of 15kV) to start the initial operation. The DC power could supply an electric power of 240W with a maximum voltage 12V and a maximum current of 20A. The measuring system measured the gas component and electric characteristics, and outlet gas temperature individually. The gas sampling line consisted of a condenser, a mass flow meter, and a gas chromatograph (GC9900). The inlet/outlet flow rate was measured by a mass flow meter. The component of hydrogen-rich gas was measured by a gas chromatograph (GC9900), which equipped with a thermal conductivity detector (TCD) and a flame ionization detector (FID). The electric characteristics were measured by a high-voltage probe, a current probe, a inductive type current sensor (CS010G) and a data acquisition card (NI, USB-6009).The temperature of outlet flow was measured by a temperature sensor (WZP-Pt100) and a digital temperature displayer (XMT-8).

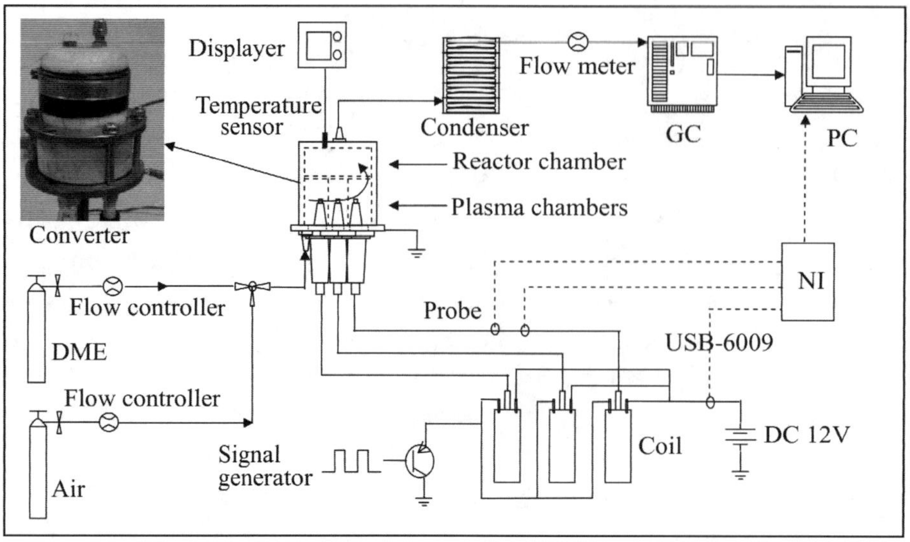

Fig.1 Experimental equipments of plasma reforming of DME.

2.2 Experimental flow and parameters

The experiment was conducted at atmospheric pressure and ambient temperature. The flow rates of DME and air were controlled by mass flow controllers. The mixture of reactant gases were directed into the converter, and then passed through the discharge zone where a high voltage spark ionized the gas and partial oxidation reaction accomplished. The temperature of exhaust gases was measured by temperature sensor. The hydrogen-rich gas was cooled and dried by condenser before flowing into GC. Finally, the analysis results of GC were collected by PC. The frequency and pulse duty ratio of rectangular pulse wave for discharge system were controlled by a signal generator. The data of the electric characteristics were collected by USB-6009 with the sampling rate of each channel 15kS/s, and then were analyzed. The experimental conditions and range of the system were shown in table 1 in detail.

Table 1-Experimental conditions and range.

Parameters	Range
Pulse duty ratio	40% ~ 90%
A/D	1.0~3.5
Discharge frequency	110 Hz ~ 190 Hz
DME inlet flow rate	10 ~ 60 mL/min
Radius of fanlike chamber	7 mm
Stage number in series	3 stages
Voltage of DC power	12 V

The voltage and current waveform of the discharge generating system were shown in Fig.2. We must point out that the voltages (V_s) of a spark at sustaining period were more concerned rather than the break down voltage (V_b) for the calculation of the energy of a spark. The peak of break down voltage was hardly precisely captured by a data acquisition card which sampling rate was 15kS/s, but could be observed by oscillograph. Inductive coils were used in experiments. So the energy of a spark discharged was only determined by the pulse duty ratio of the controlling signal when the power source supplied 12V DC voltage.

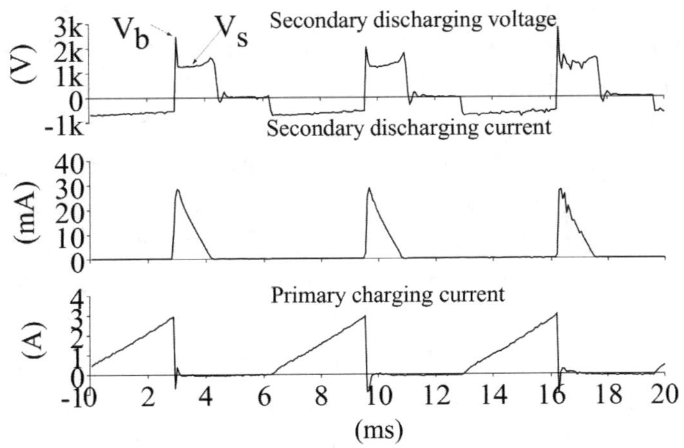

Fig.2 The electric characteristics of spark generating system (f=150Hz, α=50%, Q_{DME}=20mL/min).

2.3 Relative calculations

Eq. (1) is the actual reaction equation of DME partial oxidation. To assess the reforming effectiveness during plasma reforming of DME, the hydrogen yield, the conversion rate of DME (*CR*), the specific energy consumption (*SEC*) and the thermal efficiency (η) were calculated using equations (2)~(5), respectively.

$$C_2H_6O+z(O_2+3.76N_2)\rightarrow x(aCO+bCO_2+dCH_4+eH_2+3.76zN_2+fO_2)+gH_2O+jC \quad (1)$$

$$H_2yield(\%)=\frac{V_{H_2}^{out}(mL/\min)}{3\times V_{DME}^{in}(mL/\min)} \quad (2)$$

$$CR(\%)=\frac{V_{DME}^{in}-V_{DME}^{out}}{V_{DME}^{in}}\times 100\% \quad (3)$$

$$SEC(kJ/L-H_2)=\frac{P(W)\times 60}{V_{H_2}^{out}(mL/\min)} \quad (4)$$

$$\eta(\%)=\frac{V_{H_2}^{out}\bullet LHV_{H_2}}{V_{DME}\bullet LHV_{DME}}\times 100\% \quad (5)$$

where $V_{H_2}^{out}$ is the effluent velocity of H_2. V_{DME}^{in} and V_{DME}^{out} are the influent and effluent velocity of DME. V_{DME} is the velocity of converted DME. *LHV* is the low heat value of fuel.

3. Results and discussion

3.1 Effect of pulse duty ratio

Fig.3 shows the effect of pulse duty ratio on gas concentration in product for a DME supply rate of 20mL/min and a discharge frequency of 150 Hz. It indicates that as the pulse duty ratio increases, the H_2 and CO concentration increase first and then decrease. The maximum value of H_2

concentration is 28.2% at 80% pulse duty ratio. The concentration of CH_4 and DME present an opposite trend against the H_2 concentration. The CO_2 concentration is tiny and does not change significantly as the average value is 0.5%. Fig.4 shows the effect of pulse duty ratio on H_2 yield, DME conversion and spark energy. The DME conversion and H_2 yield has the similar trend as H_2 concentration, with maximal value of 87.6% and 39.4% at 80% pulse duty ratio, respectively. As the increase of pulse duty ratio from 40% to 80%, the charging time t_c of primary coil increases from 2.7ms to 5.3ms. The energy of a spark in discharge, which depends on charging time, was enhanced from 17mJ to 53mJ. Therefore, the ionization in plasma chamber is intensified. Moreover, the resident time of a spark in plasma chamber increases from 1.2ms to 1.8 ms, when pulse duty ratio increases from 40% to 70%. Therefore, ionization in plasma chamber is intensified, and H_2 yield and DME conversion are improved. However, by analysis of the electric characteristics at 80% pulse duty ratio, a spark can only resident for 1.3ms in plasma chamber. Clearly, a longer charging time results in a shorter residence time of a spark when the pulse duty ratio is beyond 70%. As a result, although the spark energy of 53mJ at 80% pulse duty ratio is much higher than that 42mJ at 70%, the H_2 yield and conversion only slightly increase when the pulse duty ratio is 80%. Especially, the situation at 90% pulse duty ratio is much serious, the resident time of a spark is much shorter and the hydrogen yield is depressed.

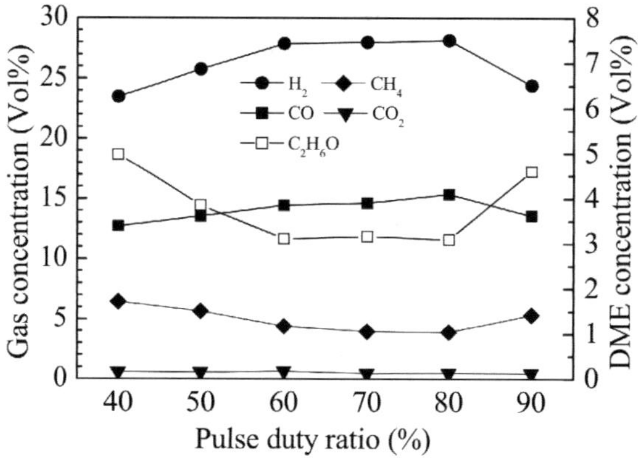

Fig.3 Effect of pulse duty ratio on gas concentration (f=150Hz, Q_{DME}=20mL/min, A/D=2, 3-stages).

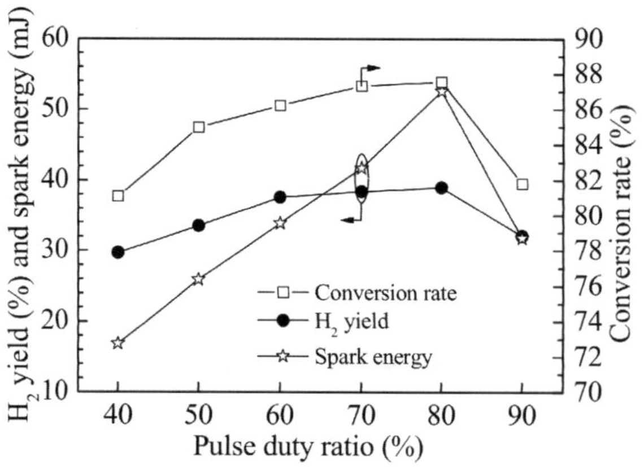

Fig.4 Effects of pulse duty ratio on DME conversion, H_2 yield and spark energy.

Fig.5 shows the influence of pulse duty ratio on energy consumption and thermal efficiency. As the increase of pulse duty ratio, the energy consumption increases smoothly when the pulse duty ratio less than 70%, and then increases sharply. It indicates that energy loss is increased with the increase of pulse duty ratio. The thermal efficiency increases first and then decreases, with the maximal value of 24.6% at 80% pulse duty ratio. Notably, 40% pulse duty ratio is desirable for low energy consumption and that 70% is desirable for high H_2 yield with moderate energy consumption. When pulse duty ratio is 70%, the charging time of primary coil was 4.7ms. However, when pulse duty ratio is less than 40%, the discharge of the coil is hardly started.

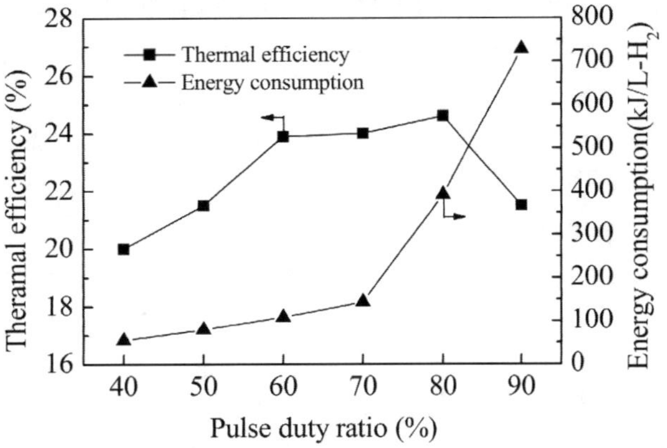

Fig.5 The influence of pulse duty ratio on energy consumption and thermal efficiency.

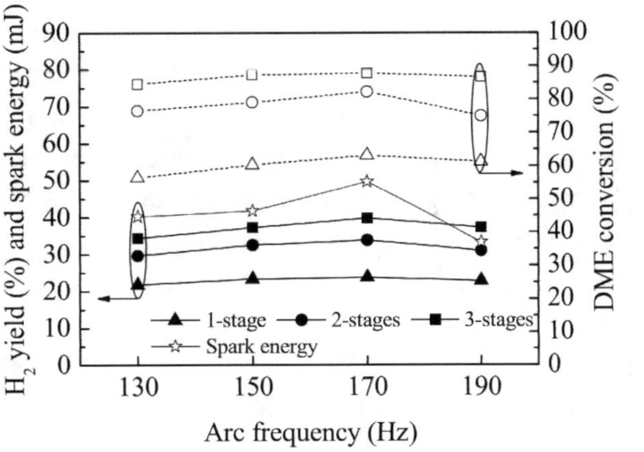

Fig.6 Effect of stage number in series on H_2 yield, spark energy and output power (t_c=4.7ms, A/D=2, Q_{DME}=20mL/min), (solid line: hydrogen yield, dotted line: DME conversion rate).

3.2 Effect of stage number in series

Fig.6 shows the effect of stage number on H_2 yield and DME conversion rate for a DME feed flow rate of 20mL/min and an A/D ratio of 2.0. Firstly, the experimental results reveal that as the increase of stage number the DME conversion and H_2 yield are improved. The maximal H_2 yield of 3-stages is 40% at 170Hz, which has 67% enhanced than that of single stage. That is because of more energy is added to the reactant to initiate the reactions. Secondly, the H_2 yield increases initially and then decreases with the increase of discharge frequency for every situation. When spark frequency is low, increasing arc frequency means reactants have more chance to contact with spark, so the reaction is more sufficient. Moreover, the spark energy for every stage increases from 40mJ to 50mJ, when the frequency increases from 130Hz to 170Hz as Fig.6 shown. The increase of discharge energy results in more energy is added to the reactant, and further intensifies the effect of spark frequency on hydrogen yield. But when frequency is high enough, the discharge energy and residence time of a spark in plasma chamber is decreased. That means the coil is not fully discharged, and hydrogen yield is depressed, especially at 190Hz, obviously.

Table.2 shows the influence of stage number on the thermal efficiency and the energy consumption. Notably, the energy consumption increases with the increase of stage number. It means that more discharge energy is lost in form of heat dissipation in the second and third stage, but the loss is not proportional to output power of DC source. Table.2 also indicates that the converter with 3-stages is desirable for higher thermal efficiency and H_2 yield, the single stage is desirable for low energy consumption.

Table 2 The influence of stage number on output power, energy consumption and thermal efficiency (f=170Hz, α= 79%, A/D=2, Q_{DME}=20mL/min).

Stage number	Output power (W)	Energy consumption (kJ/L-H_2)	Thermal efficiency (%)
1	45	189	20.6
2	94	272	22.5
3	136	339	24.8

3.3 Effect of heat preservation

The heat preservation configuration (HP) consists of double-deck asbestos and aluminium-foil papers. The temperature of outlet gas under heat preservation has a noticeable increase at each experimental condition as shown in table 3. It proves that the temperature of reactants in converter is enhanced. Fig.7 compares the effects of employing heat preservation configuration on H_2 yield and DME conversion with the natural heat dissipation situation (NHD). It is clear from the fig.7 that higher temperature results in higher DME conversion rate, and therefore better production of hydrogen. The reforming temperature plays a positive effect in reforming process. However, the improvement of H_2 yield is within 3%, the effect of heat preservation is not evident. The main reason being the heat preservation configuration does not fully prevent the heat dissipation and much of heat is taken away by exhaust gas. The energy consumption under HP are much lower than that under NHD at each experimental condition as table 3 shown, while the thermal efficiency are enhanced. It is indicates that employing heat preservation configuration is desirable for higher thermal efficiency and the reduction of energy consumption.

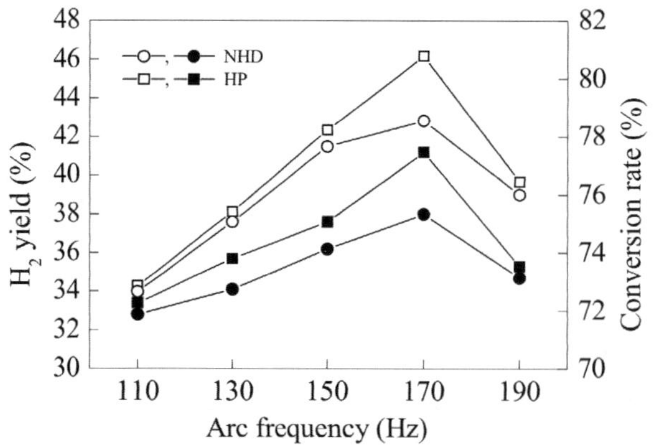

Fig.7 Effect of HP on DME conversion and H_2 yield (t_c=4.7ms, A/D=2.5, Q_{DME}=30mL/min) (solid line: hydrogen yield, dotted line: DME conversion rate)

Table 3 The effect of HP on outlet temperature, energy consumption and thermal efficiency.

f (Hz)	Outlet gas temperature (℃)		Energy consumption (kJ/L-H_2)		Thermal efficiency (%)	
	NHD	HP	NHD	HP	NHD	HP
110	52	67	82.3	77.4	24.7	25.1
130	57	80	92.6	84.1	24.8	25.9
150	61	91	96.7	92.6	25.4	26.3
170	67	108	204.7	186.2	26.4	27.9
190	69	112	369.7	341.3	24.9	25.2

3.4 Carbon deposit on electrodes

After the plasma oxidation reforming of DME, large amounts of fine carbon deposit is produced and deposited on the surface of anode and cathode as shown by the SEM photos in Fig.8. By element analysis, the dominant element is carbon with the content of carbon atoms being greater than 93%. Experiments indicate that the carbon deposit is mainly related to the air-DME ratio. When A/D ratio is beyond 2.5, with the increase of A/D ratio, the carbon deposit on both electrodes is clearly reduced. When A/D ratio is 4.0, almost no carbon deposit can be observed. That means the carbon deposit is due to insufficient oxygen. Fig.9 shows the effect of operating time on H_2 yield under different A/D ratio for a single stage system. It also reveals the influence of carbon deposit on H_2 yield. When A/D is 1.5, with the increase of operating time, the H_2 yield decreases gradually from 19.5% to 13.6%. The discharge was depressed and run down due to the increase of carbon deposit on electrodes surface. When A/D is 2.5, the H_2 yield does not change significantly as the average value is 23.1%. Experiment proves that the converter can work steadily for at least 24 hours when A/D is 2.5.

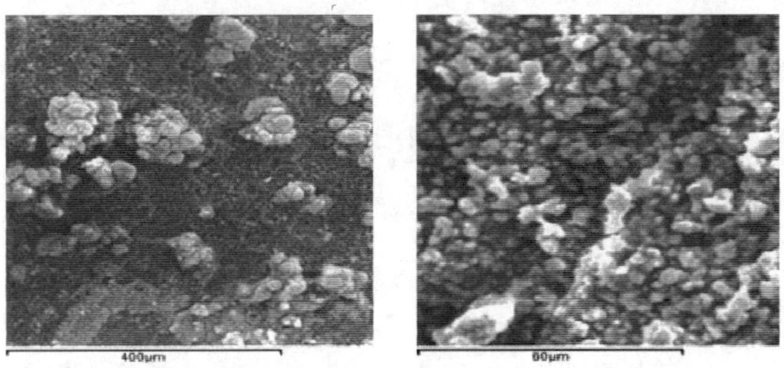

Fig.8 The SEM photographs of carbon deposit on the electrodes for an operating time of 60 minutes.

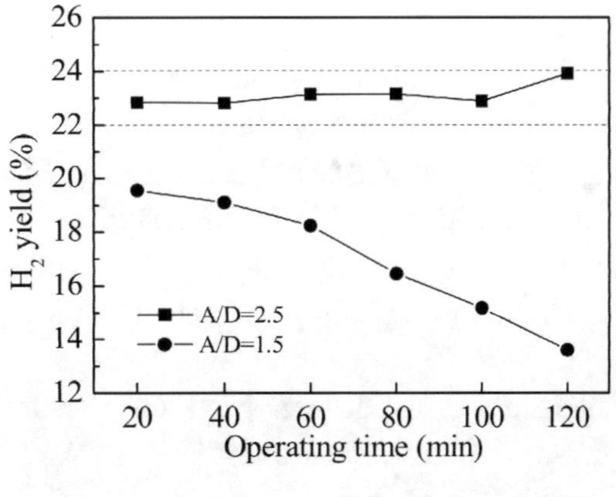

Fig.9 The effect of operating time on H_2 yield under different A/D ratio for a single stage system (f=150Hz, α=70%, t_c=4.7ms, Q_{DME}=30mL/min).

4. Conclusion

This paper successfully achieved hydrogen production by the multistage spark plasma system. The reforming characteristics of hydrogen production were investigated. It was shown that the pulse duty ratio of controlling signal played a pivotal role, which determined the spark energy directly. Although shorter pulse duty ratio (40%~50%) resulted in lower hydrogen production, but the energy consumptions could be kept less than 100 kJ/L-H_2 correspondingly, which were acceptable. The deficiency of shorter pulse duty ratio could be covered by increasing spark frequency. 3-stages configuration was desirable for higher H_2 yield and conversion, and that with single stage was desirable for low energy consumption. The carbon deposit forming on the electrode was clearly reduced when the A/D beyond 2.5 for the oxygen was sufficient. It was clear that to employ heat preservation was desirable for the decrease of energy consumption. So we may consider combining the converter with automotive exhaust system for the fully utilizing of exhaust heat of traditional engine, which added fossil fuels with hydrogen to reduce emissions.

References

[1] Changwei J, Shuofeng W. Effect of hydrogen addition on combustion and emissions performance of a spark ignition gasoline engine at lean conditions. Int J Hydrogen Energy 2009; 34:7823-34.

[2] Stanislaw S, Karol GR. Hydrogen combustion in a compression ignition diesel engine. Int J Hydrogen Energy 2009; 34:4413-21.

[3] JD H, J H, DL K, Y W. An overview of hydrogen production technologies. Catalysis Today 2009; 139:244-60.

[4] Nazim ZM, T.Nejat V. "Green" path from fossil-based to hydrogen economy: An overview of carbon-neutral technologies. Int J Hydrogen Energy 2008; 33:6804-39.

[5] Bromberg L, Cohn DR, Rabinovich A, Alexeev N. Plasma catalytic reforming of methane. Int J Hydrogen Energy 1999; 24:1131-37.

[6] Bromberg L, Cohn DR., Rabinovich A, Alexeev N, et al. System optimization and cost analysis of plasma catalytic reforming of natural gas. Int J Hydrogen Energy 2000; 25:1157-61.

[7] Thammanoon S, Piyaphon T, Sumaeth C. Partial oxidation of methane with air for synthesis gas production in a multistage gliding arc discharge system. Int J Hydrogen Energy 2007; 32:1067-79.

[8] Rong-Fang H, Hui-Hui H, Ming-Pin L, Chih-Sheng W, Wei-Cheng C. Characteristics of hydrogen production by a plasma–catalyst hybrid converter with energy saving schemes under atmospheric pressure. Int. J. Hydrogen Energy 2008, 33: 3719 – 3727.

[9] Rong-Fang H, Ming-Pin L, Yuh-Ping C, et al. Plasma-assisted catalytic reforming of propane and an assessment of its applicability on vehicles. Int J Hydrogen Energy 2009, 34: 6280–6289.

[10] Belen S, J. Javier B, Inmaculada G.V, Agustin R.GE, Jose C, Victor J.R. Hydrogen production by reforming of hydrocarbons and alcohols in a dielectric barrier discharge. Journal of Power Sources 2007; 169:140-143.

[11] JiJun Z, Yueping Z, ChangJun L. Hydrogen production from dimethyl ether using corona discharge plasma. J Power Sources 2007; 163:653-57.

[12] Lingjun S, Xinghu L, Tianlei Z. Onboard hydrogen production from partial oxidation of dimethyl ether by spark discharge plasma reforming. Int J Hydrogen Energy 2008; 33:5060-65.

(69h) One-Step Deposition of Photocatalytically Active TiO_2 Films Via Wire-to-Plate Dielectric Barrier Discharge Plasma at Atmospheric Pressure

Lanbo Di, Xiaosong Li and Aimin Zhu, School of Physics and Optoelectronic Engineering, Laboratory of Plasma Physical Chemistry, Dalian University of Technology, Dalian, China

One-step deposition of photocatalytically active TiO_2 films via wire-to-plate dielectric barrier discharge plasma at atmospheric pressure

Lan-Bo Di, Xiao-Song Li and Ai-Min Zhu*

Laboratory of Plasma Physical Chemistry, Dalian University of Technology, Dalian 116024, China

*Corresponding author. E-mail: amzhu@dlut.edu.cn

TiO_2 films play an important role in extensive applications such as photocatalysis, photovoltaics, biocompatibility, and sanitary disinfection [1, 2]. Low-temperature film fabrication processes are essential for all thermally sensitive or unstable substrate materials such as organic polymers and textiles. Plasma CVD is highly suitable for low-temperature fabrication, but it has mainly been low pressure incorporated with sophisticated discharge and vacuum systems that have been adopted. [3®C5] Thus, in recent years, considerable attention has been paid to the development of atmospheric-pressure, non-thermal plasma CVD technologies [6].

Dielectric barrier discharge (DBD) was widely used for generating atmospheric pressure, non-thermal plasmas and recently explored for preparing TiO_2 nanopowders or films [7-18]. In this paper, wire-to-plate DBD induced plasma CVD device, as illustrated in figure 1, was employed to fabricate photocatalytically active TiO_2 films on moving glass or plastic substrates at atmospheric pressure. The discharge gap has an adjustable distance from 0.3 mm to 0.6 mm.

1®CAr + $TiCl_4$, 2®Cdry air, 3®Calumina tube, 4®Cdischarge electrode, 5®C moving substrate, 6®Cdielectric, 7®Cground electrode.

$TiCl_4$ vapour, as the Ti precursor, was carried by Ar gas from a bubbler placed in an ice-water bath, mixed with Ar diluting gas and then flowed into the discharge area. Dry air was used as the oxygen precursor.

The photocatalytic reaction test for TiO_2 films was performed by complete oxidation of HCHO in simulated air with a continuous flow reactor under UV irradiation of a 254 nm lamp (8 W) [11].

The as-deposited TiO_2 films on cover glass or plastic substrates, prepared at 1 L∞§min^{-1} total flow rate, 27 W input power, 0.3 mm discharge gap, and 29 Pa $TiCl_4$ partial pressure, exhibited high photocatalytic activity. The apparent rate constants of the TiO_2 films on cover glass and plastic substrates are 0.99 s^{-1} and 0.52 s^{-1}, respectively.

The influence of input power, discharge gap and the partial pressure of $TiCl_4$ on the photocatalytic activity of the as-deposited TiO_2 films were also investigated.

References

[1] R Schaub, E Wahlström, A Ronnau, E Laesgaard, I Stensgaard, F Besenbacher 2003 *Science* **299** 377

[2] E W McFarland, J Tang 2003 *Nature* **421** 616

[3] K L Choy 2003 *Prog. Mater. Sci.* **48** 57

[4] T Busani, R A B Devine 2005 *Semicond. Sci. Technol.* **20** 870

[5] H Szymanowski, A Sobczyk, M Gazicki-Lipman, W Jakubowski, L Klimek 2005 *Surf. Coat. Technol.* **200** 1036

[6] J R Roth 2001 Industrial Plasma Engineering, Vol 2: Applications to Nonthermal Plasma Processing (Bristol: Institute of Physics Publishing)

[7] A-M Zhu, L-H Nie, X-L Zhang, C Shi, Z-M Song and Y Xu 2004 *Plasma Sci. Technol.* **6** 2546

[8] L-H Nie, C Shi, Y Xu, Q-H Wu and A-M Zhu 2007 *Plasma Process. Polym.* **4** 574

[9] X-L Zhang, L-H Nie, Y Xu, C Shi, X-F Yang and A-M Zhu 2007 *J. Phys. D: Appl. Phys.* **40** 1763

[10] A-M Zhu, L-H Nie, Q-H Wu, X-L Zhang, X-F Yang, Y Xu and C Shi 2007 *Chem. Vapor Deposition* **13** 141

[11] L-B Di, X-S Li, C Shi, Y Xu, D-Z Zhao and A-M Zhu 2009 *J. Phys. D: Appl. Phys.* **42** 032001

[12] Bai H, Chen C, Lin C, Den W and Chang C 2004 *Ind. Eng. Chem. Res.* **43** 7200

[13] C Chen, H Bai, H M Chein and T M Chen 2007 *Aerosol Sci. Tech.* **41** 1018

[14] C Chen, H Bai, S-M Chang, C Chang and W Den 2007 *J. Nanopart. Res.* **9** 365

[15] C Chen, H Bai and C Chang 2007 *J. Phys. Chem. C* **111** 15228

[16] S K Xu, J Z Xu and X B Peng 2006 *Plasma Sci. Technol.* **8** 293

[17] X W Zhang, Y Guo and G R Han 2007 *Plasma Sci. Technol.* **9** 674

[18] J L Hodgkinson, H M Yates, D W Sheel 2009 *Plasma Process. Polym.* **6** 575

See more of this Session: Plasma for Energy and Environmental Applications (II)
See more of this Group/Topical: Topical 2: Advanced Fossil Energy Utilization

(70a) Technical and Economic Assessment of Membrane-Based Systems for Capturing CO_2 From Coal-Fired Power Plants

Haibo Zhai and Edward S. Rubin, Department of Engineering and Public Policy, Carnegie Mellon University, Pittsburgh, PA

Coal-fired power plants contribute nearly 50% of U.S. electricity supply and account for about a third of national emissions of carbon dioxide (CO2), the major greenhouse gas associated with global climate change. Post-combustion carbon capture and storage (CCS) has been considered a key technology for deeply reducing CO2 emissions from existing and new coal-fired power plants. Membrane systems are among the advanced technologies being developed for more cost-effective capture. Membranes have been used commercially to separate CO2 from natural gas and other industrial gas streams, and have the potential for application to power plant flue gases. The objective of this study is to evaluate the technical feasibility and cost of membrane systems for CO2 capture at coal-fired power plants using newly developed cost and performance models.

Performance and cost models of polymeric membrane capture systems have been developed for use in this evaluation. We investigate the performance and cost of different capture system configurations including single- and multiple-stage modules to identify feasible membrane systems that are able to simultaneously achieve 90% CO2 capture and above 95% purity of CO2 product. We also examine a range of key factors affecting the capture system performance and cost. These factors considered include the inlet CO2 concentration, gas flow patterns, feed-side versus permeate-side pressure ratio, membrane properties, and other parameters. In particular, this study systematically evaluates the effects of membrane driving force designs for CO2/N2 separation. This is important since power plants flue gases are at atmospheric pressure, and typically have 10% to 15% CO2 by volume, which results in a low CO2 partial pressure. Furthermore, to increase the inlet CO2 partial pressure and decrease the energy required to generate a pressure difference across the membrane, boiler combustion air is used as a sweep gas for a two-stage, two-step membrane capture system.

The preliminary results show that multiple-stage membrane systems are capable of meeting the CO2 capture performance targets (90% capture and above 95% product purity) for given membranes. A hybrid design using both compressors and vacuum pumps for producing the separation driving force between feed and permeate sides is effective in reducing the energy requirements and cost of CO2 capture. A combination of combustion air sweep with the two-stage, two-step membrane capture system would further reduce the energy use and cost of CO2 capture. Plans for further model development also are discussed.

This work is supported under a contract from the U.S. Department of Energy's National Energy Technology Laboratory (DOE/NETL).

See more of this Session: Post-Combustion Techniques for Capture/Separation of Carbon Dioxide
See more of this Group/Topical: Topical 2: Advanced Fossil Energy Utilization

(70b) Development of Asymmetric Membrane Gas Permeation Simulation Modules for Generalized Plant-Wide Optimization of Power Plants with CCS: Effect of Detailed Porous Support Transport Sub-Model

Juan E. Morinelly and David C. Miller, U.S. Department of Energy, National Energy Technology Laboratory, Morgantown, WV

Development of asymmetric membrane gas permeation simulation modules for generalized plant-wide optimization of power plants with CCS: Effect of detailed porous support transport sub-model

NETL is currently developing a modular framework for the analysis and optimization of power generation systems, particularly coal-fired power plants with carbon capture and sequestration (CCS) [1]. The framework consists of a collection of simulation modules for various parts of a power plant and a structure that allows their integration into a specified configuration. Unlike most assessment studies for CCS technologies in literature, the integration of CCS modules to the modular framework provides plant-wide performance indicators and a common basis for the evaluation of new and existing CCS technologies. The modular approach is particularly advantageous given that beyond capital investment and the obvious compromises in operating cost and plant efficiency, CCS processes usually involve parasitic demands and modifications to other parts of the power generation system. This approach also permits the easy integration of new processes and the capability to carry out constrained multi-objective optimization routines by varying key system parameters for a given modular plant design. Rigorous models for gas permeation modules were developed in order to evaluate membrane carbon capture systems under the modular framework.

Gas permeation is a well established technology for many industrial applications [2]. As a carbon capture technology, a membrane system is advantageous because it does not require large amounts of water or solvents, and its non-moving, modular components are easy to operate and maintain. Recent studies conclude that multi-staged gas permeation processes can compete with amine absorption as a viable carbon capture technology [3-6]. Developments in membrane properties and process design are required in order to achieve the desired separation while minimizing energy and membrane area demands. Merkel et al. recently proposed a promising two-stage system configuration that incorporates different strategies to maximize the separation driving force [5]. The most prominent feature of the design is a counter-current sweep module that uses a portion of the air feed to the boiler as a sweep. The recirculation of carbon dioxide through the boiler increases its partial pressure in the flue gas stream to be treated, while the sweep flow decreases its permeate side partial pressure. Both effects increase the trans-membrane carbon dioxide partial pressure difference, which in turn reduce the process energy and area demands. A rigorous numerical model that captures the complexity of these effects is necessary when evaluating this type of processes.

There are several numerical models in the literature that describe the behavior of gas permeation in asymmetric membranes. Kaldis et al. provide a good summary of these efforts [7]. The asymmetric membrane architecture was developed in order to address conflicting requirements. In order to achieve high gas permeances the membrane must be sufficiently thin yet mechanically stable in order to endure the imposed pressure gradient. By coating a thin (0.5-1 μ m) selective layer on a porous support two to three orders of magnitude thicker both requirements are fulfilled. The transport across the thin selective layer is usually modeled according to the solution-diffusion model, while simplifying assumptions about the gas transport in the porous support are usually made. Experimental results [8, 9] suggest that the simplifying assumptions about the porous support are not capable of predicting observed behavior for gas permeation systems with sweep streams.

One-dimensional, multicomponent, hollow-fiber gas permeation models were developed according to the shell and fiber lumen flow equations of Pan et al. [10] and the detailed model for the porous support gas transport similar to Chan et al. [11]. Equivalent models, according to the limiting assumptions about the porous support were also developed. The models were used to simulate the performance of a module with and without sweep for the carbon capture from a flue gas stream. The predicted performance for no-sweep operation was almost identical regardless of the porous support gas transport sub-model. On the other hand, significant discrepancies were predicted for modules with sweep streams. Since sweep operated modules are a vital feature of novel processes, simulation models must be developed with enough detail to predict their performance accurately. The analysis of the different models under the modular framework will determine the sensitivity of the overall performance to the assumed transport mechanism for the porous support. This will in turn justify or reject the inclusion of the detailed sub-model, and potentially enable the optimization of porous support properties.

References

1. Miller, DC, Eslick, JC, Lee, A, Morinelly, JE. A modular framework for the analysis and optimization of power generation systems with CCS. Energy Procedia, 2010; In Press

2. Coker, DT, Freeman, BD, Fleming, GK. Modeling multicomponent gas separation using hollow-fiber membrane contactors. AIChE J., 1998; 44:1289-1302

3. Favre, E. Carbon dioxide recovery from post-combustion processes: Can gas permeation membranes compete with absorption? J. Membr. Sci., 2007;294:50-59

4. Hussain, A, H%ogg, M-B. A feasibility study of CO_2 capture from flue gas by a facilitated transport membrane. J. Membr. Sci., 2010;359:140-148

5. Merkel, TC, Lin, H, Wei, X, Baker, R. Power plant post-combustion carbon dioxide capture: An opportunity for membranes. J. Membr. Sci., 2010;359:126-139

6. Zhao, L, Riensche, E, Blum, L, Stolten, D. Multi-stage gas separation membrane processes used in post-combustion capture: Energetic and economic analyses. J. Membr. Sci., 2010;359:160-172

7. Kaldis, SP, Kapantaidakis, GC, Sakellaropoulos, GP. Simulation of multicomponent gas separation in a hollow fiber membrane by orthogonal collocation - hydrogen recovery from refinery gases. J. Membr. Sci., 2000;173:61-71

8. Sandru, M, Haukeb¯, SV, H‰gg, M-B. Composite hollow fiber membranes for CO_2 capture. J. Membr. Sci., 2010;346:172-186

9. Dittmeyer, R, H^llein, V, Daub, K. Membrane Reactors for hydrogenation and dehydrogenation processes based on supported palladium. J. Mol. Catal. A: Chem., 2001;173:135-184

10. Pan, CY. Gas separation by high-flux, asymmetric hollow-fiber membrane. AIChE J. 1986;32:2020-2027

11. Chan, CH, Khor, KA, Xia, ZT. A complete polarization model of a solid oxide fuel cell and its sensitivity to the change of cell component thickness. J Power Sources, 2001;93:130-140

See more of this Session: Post-Combustion Techniques for Capture/Separation of Carbon Dioxide
See more of this Group/Topical: Topical 2: Advanced Fossil Energy Utilization

(70c) CO_2-Selective Membranes Composed of Cross-Linked Polyethers

Katie Barillas, Department of Chemical Engineering, National Energy Technology Laboratory, University of Pittsburgh, Pittsburgh, PA, **Robert M. Enick**, Department of Chemical and petroleum Engineering, University of Pittsburgh, Pittsburgh, PA, David Luebke, US DOE/NETL, Pittsburgh, PA and B. D. Morreale, Office of Research and Development, U.S. Department of Energy, National Energy Technology Laboratory, Pittsburgh, PA

Membranes could be an integral part of the post-combustion capture of CO_2 from the effluent gas stream of coal-fired power plants. The objective of this work is to design polymeric membranes that have very high CO_2 permeability and high selectivity toward CO_2 (i.e. very low H_2 and N_2 permeability). It is our hypothesis that the favorable thermodynamic interactions that enable certain polymers to dissolve in dense CO_2 at extremely high pressure will also allow membranes composed of these polymers to exhibit high CO_2 permeability and CO_2 selectivity at low pressure. (These polymeric membranes have recently been assessed for the pre-combustion capture of CO_2 from a CO_2- and H_2-rich shifted gasifier stream).

The crosslinked polymeric membranes used in this study of post-combustion capture of CO_2 from a N_2-rich stream have been previously assessed and shown promise for the pre-combustion separation of CO_2 from a H_2-rich shifted fuel gas stream. Candidates include either (a) CO_2-philic polymers with linear repeat units such as poly(ethylene glycol) (PEG), polytrimethylene ether glycol (i.e. linear polypropyleneglycol), and polytetramethylene glycol ether (i.e. linear polybutylene glycol), or (b) CO_2-philic polymers with branched repeat units such as polypropyleneglycol and polybutylene glycol. Results of mixed gas permeability and selectivity values for CO_2/N_2 gas pairs will be discussed, and the performance of membranes with linear and branched repeat units will be contrasted.

See more of this Session: Post-Combustion Techniques for Capture/Separation of Carbon Dioxide
See more of this Group/Topical: Topical 2: Advanced Fossil Energy Utilization

(70e) Implications of Flue Gas Components on the Performance of Sorbents

James S. Hoffman, US DOE NETL, U.S. Department of Energy, National Energy Technology Laboratory, Pittsburgh, PA

Carbon sequestration is a viable scheme to mitigate carbon dioxide (CO_2) emissions from large point sources, such as power generators that burn fossil fuels. Although both the capture and storage steps are important, the most costly step is the capture one. The Fossil Energy Program in the U.S. Department of Energy has prompted the investigation of various techniques to capture carbon dioxide from coal-burning power generation point sources. A solid sorbent method, a category of post-combustion carbon capture, is a regenerable scheme where the CO_2 is removed from the flue gas stream by a sorbent and then regenerated. The technique can use an amine-based absorbent where the amine can be immobilized onto a support or encapsulated within a porous substrate. The sorbent becomes fully loaded once the CO_2 capacity of the sorbent is achieved. In a transport type of system, the spent sorbent is thermally regenerated in another vessel and recycled back to the absorption scrubbing unit.

With respect to CO_2 capture, loading capacity and regenerative heat duty are important parameters that define the technical/economic relevance of the sorbent. However, other components within the flue gas could have a direct impact on these parameters. The moisture of the flue gas can be important not only as it may be a reactant in certain CO_2–amine reactions, but by itself, it could possibly absorb onto the sorbent material, and thus cause an additional need for energy in the regeneration step. Depending on the requirement of upstream scrubbing of the flue gas in a flue gas desulfurization unit, the sorbent can be exposed to upwards of 15 vol% moisture in the flue gas. Additionally, if the flue gas emanates from the combustion of coal, it is eminent that the sorbent will need to contend with the trace gaseous components of sulfur dioxide and nitric oxides. Results of packed bed studies with an immobilized amine sorbent will be discussed. Gas mixtures simulated those of flue gas and the conditions for regeneration. The role of moisture in the absorption and regeneration steps will be described and its impact on CO_2 capacity and the regenerative heat duty will be addressed.

See more of this Session: Post-Combustion Techniques for Capture/Separation of Carbon Dioxide
See more of this Group/Topical: Topical 2: Advanced Fossil Energy Utilization

(70f) Evaluation of CO_2 Capture and Sequestration Using Oxyfuels with AMIGA Economic Modeling

Richard D. Doctor[1], Donald Hanson[2] and John C. Molburg[2], (1)Energy System Division, Argonne National Laboratory, Argonne, IL, (2)Decsion and Information Sciences, Argonne National Laboratory, Argonne, IL

Process Design studies for Oxyfuel retrofit to PC boilers of various sizes for several types of coal were linked with an ongoing effort employing the AMIGA (All-Modular Industry Growth Assessment) macroeconomic model to survey the domestic coal-fired power fleet. This project assessed the applicability of Oxyfuel CO_2 recovery for retrofit of the existing fleet. The coal preparation, power generation, existing environmental regulations, facility water use, pipeline CO_2 conditioning, and pipeline transport of CO_2 to sequestration were all considered and costs were estimated. This will provide input to a macroeconomic analysis employing the AMIGA model, a climate and energy-security model with a great deal of economic and technology detail. AMIGA steam power plant unit inventory (boilers, generators, control equipment) contains about 2000 boiler-level records and the questions examined will be where Oxyfuel systems fall on the dispatch against load curve, how to manage capacity expansion to meet load growth and to replace derated capacity, calculate the fuel demands, costs, and investment requirements. Cost and performance estimates for all major generating types may be viewed as technologies that are competing for investors. Hence, in terms of dynamic paths, the effort will show how transitions from Oxyfuels to IGCC could be managed. However, the technology options will not be limited to gasification, but will include coal-to-liquids, advanced refinery processes, CHP (combined heat and power), nuclear power, wind, PV, biomass, and end-use efficiency. These are embedded within a multi-sector, multi-region global model. Businesses and industries choose technologies and investment. Macroeconomic investment is build up from sector and technology investments. Household consumers by type and income choose energy-using durable goods. Changes in real income are calculated endogenously. While recognizing the importance of the CCS retrofits in the dispatch order as embodied in the AMIGA analysis, phased construction strategies for Air-fired PC to Oxyfuels and then to IGCC are a means of distributing the capital investment over time while realizing some level of benefits in the near term through building "capture ready" facilities. Oxyfuel firing requires the addition of air separation on the front end and CO_2 compression and liquefaction on the back end all of which are retrofitable in a transition to IGCC. Preliminary analysis supports the importance of investigating the benefits of Oxyfuels as a transitional strategy.

See more of this Session: Post-Combustion Tcchniques for Capture/Separation of Carbon Dioxide
See more of this Group/Topical: Topical 2: Advanced Fossil Energy Utilization

The role of heat exchangers in post-combustion CO_2 capture

Warren Ziegler

Abstract: The overall energy penalty for post-combustion carbon capture by aqueous amines as well as solid sorbent systems can be reduced significantly by recovering heat from the lean/hot solvents coming out of the regenerator and using the energy to pre-heat the rich/cool solvent absorber column. This paper will look at the potential of heat recovery via computer modelling of the cross exchanger in aqueous amine capture as well as heat recovery in solid from solid CO2 sorbent systems.

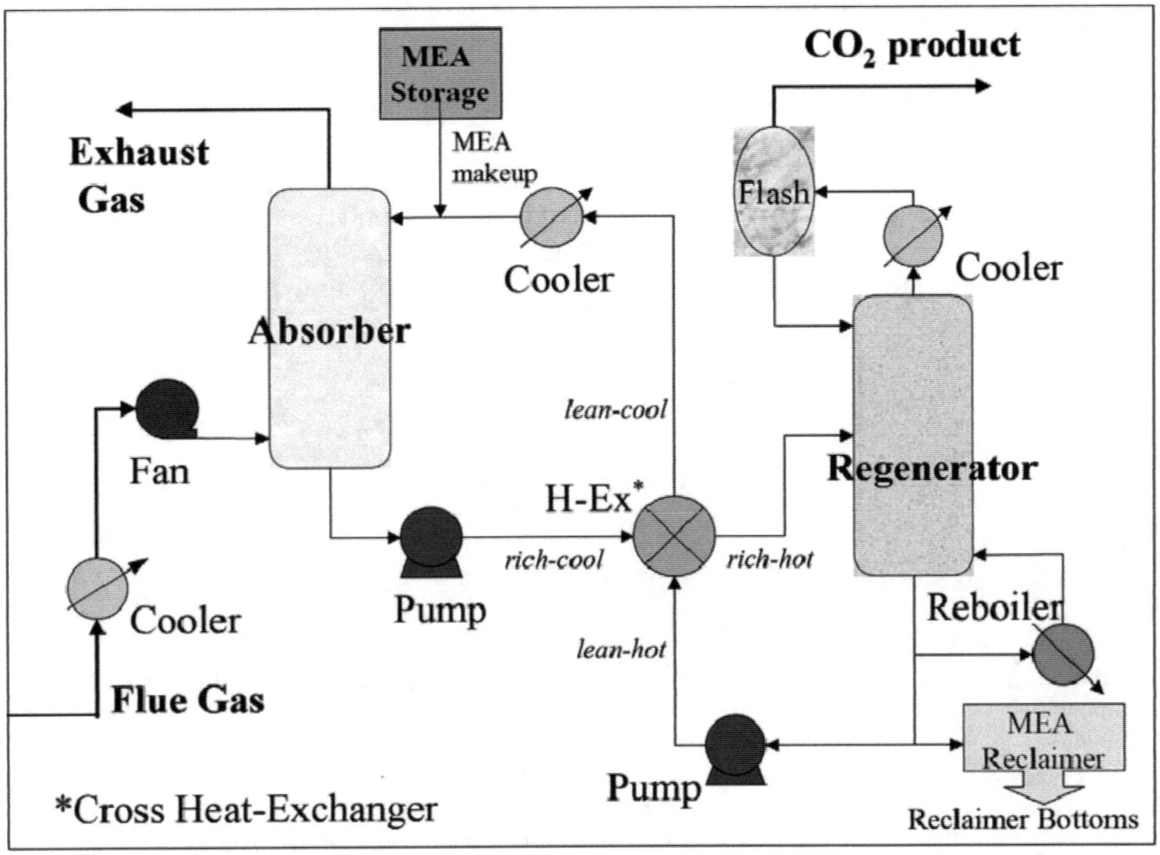

Figure 1. The Cross Exchanger in a post-combustion amine CO2 capture scenario

Heat Exchangers for Aqueous Amine CO_2 Scrubbers

For the post-combustion capture by aqueous amines having a heat exchanger recover heat from the hot CO_2 lean amines exiting the regenerator column and using that energy to heat the cool CO_2 rich stream from the absorber column will save energy by reducing what the reboiler needs to provide to the regenerator column (Figure 1). We will look at the energy recovered by the cross-exchanger under scenarios of different approach temperatures. Modeling shows that reducing the approach temperature reduces the heat required by the stripper column. Oyenekan[1]'s Aspen modeling demonstrated that for the baseline 7m concentration monoethanolamine (MEA) solvent with a rich loading= 0.563mol CO2/molAlk, and optimized lean loading,

P=160kPa. that lowering the approach temperature from 10°C to 5°C reduces the reboiler duty from 151 kJ/gmol CO_2 at 10°C to 127 kJ/gmol CO_2 at 5°C. (Table 1). This represents a 16% reduction in reboiler duty by reducing the approach temperature from 10 °C to 5°C.

Approach temp	Q_{sen} Sensible	$Q_{vap.H2O}$ Vaporization	Q_{des} Desorbtion	$\Sigma Q = Q_{sen} + Q_{vap} + Q_{des}$ Reboiler
5°C	29.8 KJ/g-molCO_2	32.2	65.7	127
10°C	59.4	27	65.0	151

Q in units of KJ/g-molCO_2

Table 1. Results of Oyenekan model for reboiler duty as a function of cross-exchanger approach temperature

The heat exchanger vendor Alfa-Laval[2] modeled one of their plate and frame heat exchangers in a hypothetical amine post-combustion scenario using a proprietary model similar to the package Xphe from the Heat Transfer Research Inc. with the following assumptions:

- 800MW plant
- No flashing of rich feed
- Counter-flow
- Energy cost $5/MMBTU (used to calculate the value of energy saved)

The Alfa-Laval model was run for approach temperatures from 11°F to 4°F (6.1°C to 2.2°C). The results are shown below in Table 2 and Figure 2.

Approach Temperature F	LMDT	Duty MMBTU/Hr	HT Area FT^2	Additional Energy Recovered MMBTU/hr	**Yearly Energy Benefit $	Cumulative yearly Energy Benefit $
11	18.6	760.8	96120			
10	17.6	767.3	104460	6.5	$334,941	$334,941
9	16.6	773.9	111720	6.6	$340,094	$675,035
8	15.5	780.4	119580	6.5	$334,941	$1,009,976
7	14.4	786.9	127380	6.5	$334,941	$1,344,918
6	13.3	793.5	134260	6.6	$340,094	$1,685,012
5	12.2	800	146300	6.5	$334,941	$2,019,953
4	11	806.6	195480	6.6	$289,080	$2,309,033

**Assumes 85% burner efficiency

Table 2. Results of Alfa-Laval model of a Cross-Exchanger for hypothetical 800MW plant

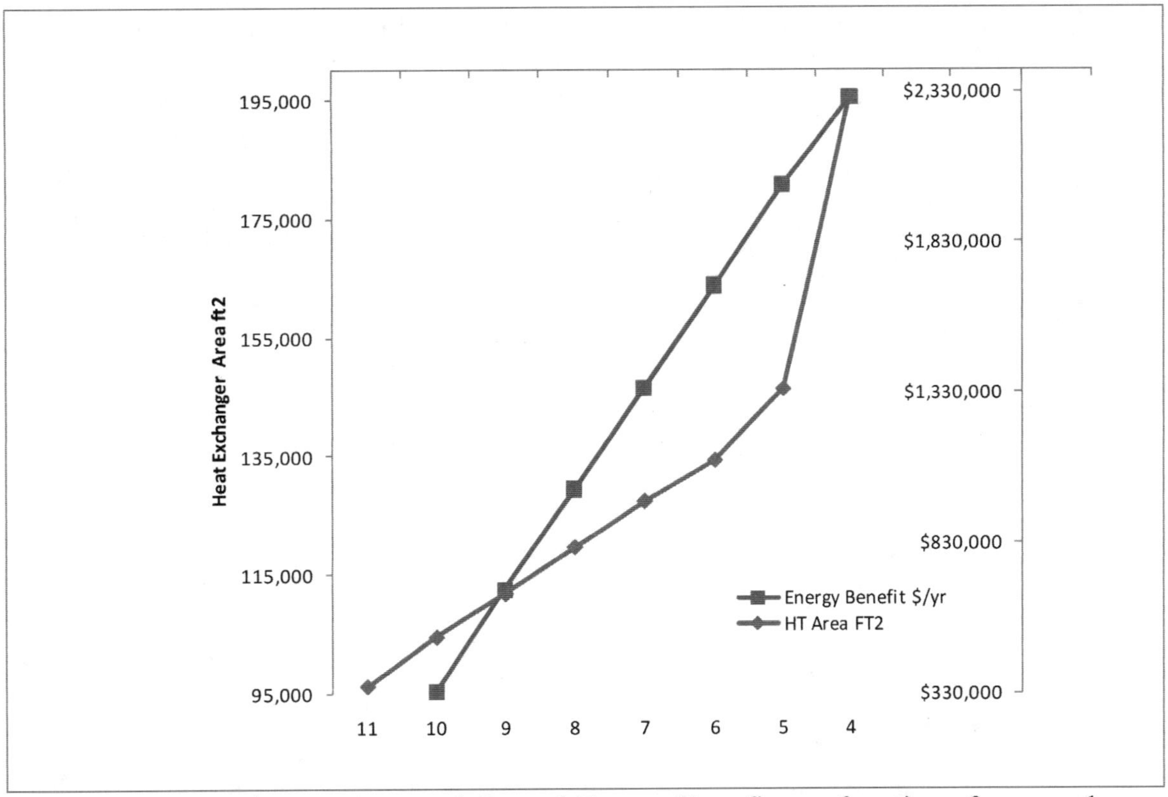

Figure 2. Heat Exchanger Area and Annual Energy Benefit as a function of approach temperature for a cross exchanger in amine CO2 scrubbing application

As expected from thermodynamics, the model results illustrate the difficulty in extaracting heat as the approach temperature decreases, however it may be possible to ecomically justify fairly large heat exchangers based on the value of the energy saved. In the hypothetical 800MW case above, going from a 10°F approach temperature to 5°F requires an increase of heat exchanger area of 41,840 ft^2, assuming the monetary value of the energy to be equivalent to $5/MMBTU gas and a burner with 85% efficiency then the equivalent energy savings is over $1.3M on an annual basis, each square foot of heat exchanger contributes an annual savings of $40/$ft^2$/yr. Going further, from an approach temperature of 5°F to 4°F increases the area by an additional 49,180 ft^2 but saves $289,000 per year, or in other words the marginal benefit per square foot of heat exchanger area is $6/$ft^2$/yr. Clearly a point of diminishing returns set in but if capital equipment is amortized over a multi-year period then even a 4°F approach temperature may be economically justified based on the value of the energy recovered.

Heat Exchangers for Solid CO2 Sorbents

Solid CO_2 sorbents have been proposed as an alternative to aqueous amine sorbents for post-combustion CO2 capture.Solid sorbents can potentially eliminate/decrease the environmental impact that liquid sorbents can inherently carry. The rich sorbent at 60°C enters a heat exchanger which brings the sorbent temperature up to 90°C before going into the regenerator, the regenerator strips the CO_2 off the sorbent by the addition of steam/heat and the solids reach a temperature of 110°C leaving the regenerator. The

110°C lean sorbent from the regenerator must be cooled before entering the absorber column, if the lean sorbent is cooled by a heat exchanger, then the waste heat can be used to pre-heat the rich sorbent before it enters the regenerator (see diagram below).

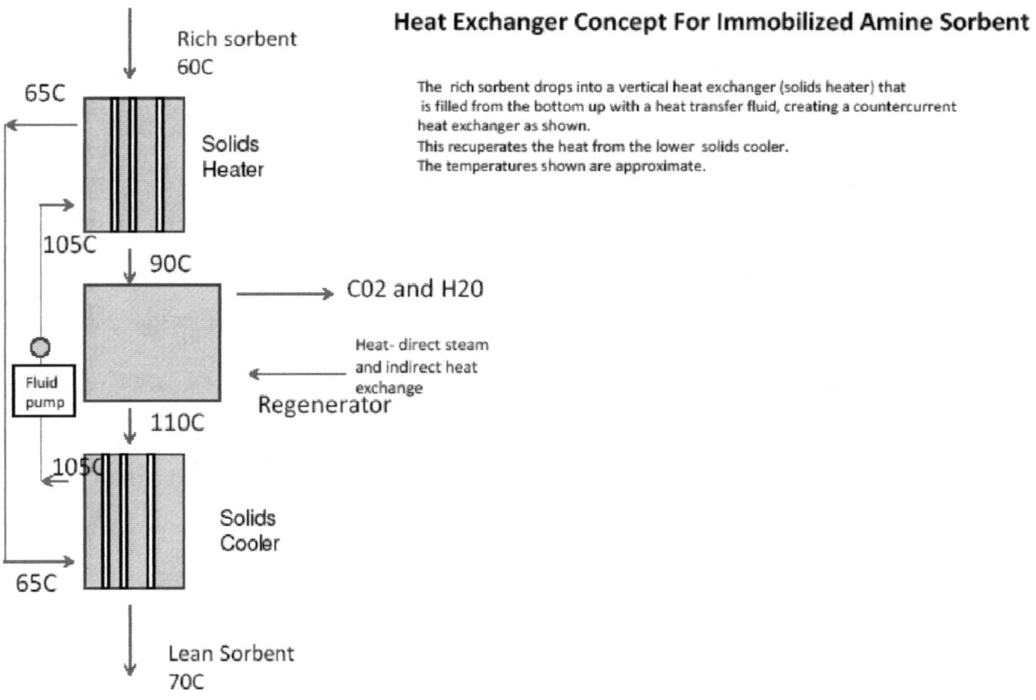

Figure 3. Recovering heat in a post-combustion solid CO2 sorbent scenario

Heat transfer in bulk solid system pose different challenges and thermal models are not easily developed as in liquid systems. The commonly used empirical formulae for fluid s do not hold valid for heat transfer in a bed of solids.

.

One commercially available solid phase heat exchanger is made by Solex Thermal Science Inc.. Solex makes solids heat exchangers for a variety of industries from chemical to fertilizers and even for food processing (see Figure 4).
Solex uses proprietary thermal modeling software to accurately design for the unique requirements of indirect heat transfer with bulk solids flowing on one side of the exchanger and liquid or gas on the other side of the exchanger.

As expected and similar to liquid systems, the greater the surface area , the more the energy that can be recovered, reducing boiler loads. The preliminary work and calculations on the solid heat exchangers have shown promising results, indicating considerable savings on energy. However, additional modelling work is required to establish the optimal point of balance in solid systems.Compared to liquid system the solids system operates at lower temperatures and hence sees a lower driving force for the heat transfer to occur.

Figure 4. Solex thermal solids heat exchanger

Solex Thermal modeled heat recovery for a solid sorbent CO2 post combustion capture scenario under the following model assumptions:

1) The sorbent will be an immobilized amine sorbent. Polyethylenimine (PEI) is deposited on a silica substrate.
2) Bulk density is about 44-lb/ ft^3.
3) Average particle size is projected to be 600 micron.
4) Specific heat is 0.45-Btu/lb-^{o}F as experimentally determined.
5) Moisture content is assumed to be zero (for simplicity at this time)
6) Angle of repose is approximated to be 37°.
7) Derived from the Hoffman et al.[4] where a 500-MW plant was used as a basis.
8) Solid flow (lean sorbent from the regenerator) into the absorber is 568 klb/hr or 4.7 ton per minute.
9) The solid flow into the regenerator (rich sorbent from the absorber) is 643 klb/hr or 5.36 ton per minute.

Results of the Solex model:

1) Multiple coolers/heaters needed
2) Each cooler/heater is 35' h x 6' sq
3) Each heater/cooler has 3000 ft^2 plate surface area
4) One cooler processes 35,000 lb/hr lean sorbent
5) Sixteen coolers needed to process 568,000lb/hr for hypothetical 500mw plant

6) Seven (6.5) heaters needed to process 643,000 lb/hr rich sorbent*

Discussion
Since we are preheating 643,000 lb/hr of sorbent from 65 to 90°C (140 to 194°F) and the specific heat of the sorbent is 0.45-Btu/lb-°F, we are recovering 15.62MMBTU/hr of heat energy. In monetary terms if the value of the energy is taken to be $5/MMBTU and we assume a burner with 85% efficiency then on an annual basis we are saving over $805,142 for the hypothetical 500MW plant.

We can now compare the relative performance of heat exchangers for liquid amine and solid sorbent applications:

1) For liquid/liquid cross heat exchanger for amine scrubbing get a heat transfer coefficient of ~425BTU/ft^2/hr/F
2) For the amine scrubbing case a cross-exchanger can yield a benefit of $2.3M/yr for the 800MW example or $2886/MW/yr
3) For solid/liquid heat exchanger for solid sorbents we get a heat transfer coefficient of ~11BTU/ft^2/hr/F
4) For the solid sorbent case the heat exchanger can yield a benefit of $805,142/yr for a hypothetical 500MW plant or $1610/MW/yr

Conclusions
For post combustion CO2 capture utilizing aqueous amines or solid sorbents recovering heat energy from the lean-hot absorber medium can yield valuable energy savings which can justify the additional cost of the exchanger.

*Solex has indicated that they have the flexibility of increasing the area and throughput of these exchangers by 20% so that perhaps fewer heat exchangers can be used for a given plant.

References:

1 Oyenekan, B. A. (2007). Modeling Strippers for CO2 Capture by Aqueous Amines, Ph.D. Dissertation, The University of Texas at Austin, May 2007

2 Chris Wajciechowski, Alfa-Laval private communication

3 Andy Nix, Solex Thermal private communication

4 Hoffman, J, Richards, G. , Pennline H., Fischer, D. , and Keller, G. "Factors in Reactor Design for Carbon Dioxide Capture with Solid, Regenerable Sorbents" Clearwater Conference paper

(70h) Use of Reversible Ionic Liquids as Solvents for Post-Combustion Recovery of CO_2 From Coal-Fired Power Plants

Amy Rohan[1], Ryan Hart[1], Kyle Flack[2], Swetha Sivaswamy[1], Jackson Switzer[1], Elizabeth Biddinger[1], Manish Talreja[1], Pamela Pollet[2], Charles Liotta[2] and Charles Eckert[1], (1)School of Chemical & Biomolecular Engineering, Georgia Institute of Technology, Atlanta, GA, (2)School of Chemistry and Biochemistry, Georgia Institute of Technology, Atlanta, GA

CO_2 emissions are likely to be regulated in the near future, so we need to develop energy and cost efficient strategies for CO_2 capture. Coal-fired power plants, essential for electricity generation, are one of the largest emitters of CO_2. Our unique reversible ionic liquids (RevILs) take advantage of a dual chemical and physical absorption capture mechanism. Unlike traditional ionic liquids, these silylamine-based solvents react with CO_2, forming an ionic liquid that may further capture CO_2 through a physical absorption process. Through thermal reversion, the CO_2-swollen ionic liquids return to their molecular components, releasing chemically and physically absorbed CO_2. The precursors thus recovered are then recycled for additional capture. Reversible ionic liquids can effectively capture CO_2 without the requirement of large quantities of a co-solvent such as water, contrasting with the current monoethanolamine (MEA) technology used for CO_2 capture. This allows for significant energy savings during the regeneration step.

We leverage an extensive understanding of structure-property relationships, and present the optimization of RevIL-based systems for CO_2 capture through viscosity, heats of generation and physical capacity. Further, we discuss the use of a unique *in-situ* technique to measure physical absorption, and hence values for Henry's Law constants, as a guide for identifying molecules with optimal capture properties.

See more of this Session: Post-Combustion Techniques for Capture/Separation of Carbon Dioxide
See more of this Group/Topical: Topical 2: Advanced Fossil Energy Utilization

Resin Wafer Electrodeionization for Flue Gas Carbon Dioxide Capture

Rebecca L. Stiles,[*] Jitendra Shah,[*] Jianwei Yuan,[*] Lisa Wesoloski,[*] Robert W. Dorner,[*] Wayne M. Carlson,[*] Yupo J. Lin,[+] Saurav Datta,[+] Michael P. Henry,[+] Cynthia S. Millard,[+] and Seth W. Snyder[+]
**Nalco Company, Naperville, IL*
[+]Argonne National Laboratory, Argonne, IL

Abstract

Energy efficient capture of CO_2 from coal-fired power plants is one of the grand challenges for the U.S. both to end dependence on imported energy and to reduce greenhouse gas emissions. Nalco Company and Argonne National Laboratory received a Department of Energy ARPA-E award to develop Resin Wafer Electrodeionization (RW-EDI) for the energy-efficient capture of CO_2 from coal flue gas. RW-EDI is being used as an electrochemically-driven platform that switches pH in the process stream to efficiently capture CO_2 from flue gas and then desorb it at atmospheric pressure without requiring heating, vacuum, or consumptive chemical usage. The acid/base pH shift drives the CO_2 capture/desorption and enzymes and/or inorganic catalysts drive the targeted kinetics. The goal of the project is to demonstrate 90% capture of CO_2 from synthetic flue gas (15% CO_2) and 90% purity of the released stream of CO_2. If successful, the CO_2 capture process could significantly reduce the ~30% parasitic energy loss associated with other capture technologies. Presented here will be recent results from a bench-scale system showing significant progress toward these goals.

Introduction

Advanced Research Projects Agency – Energy (ARPA-E) is a relatively new agency within the U.S. Department of Energy (DOE) created under the 2007 America Competes Act. It was formed to promote and enhance the economic and energy security of the U.S. through the development of breakthrough energy technologies. In October of 2009, Nalco and Argonne National Laboratory jointly received a $2.25 million award from the ARPA-E to develop the technology, Resin Wafer Electrodeionization (RW-EDI) for the capture of CO_2 from flue gas of coal-fired power plants. Argonne has been developing the RW-EDI technology for multiple applications over the past fifteen years and has numerous patents covering the technology.[1-5] To be competitive with other CO_2 capture technologies, the RW-EDI system must be cost-effective in reducing the parasitic energy loss of a power plant by a minimum of 30%. ARPA-E lays out specific goals that must be met in a two year time period for the project. The RW-EDI must be able to:

- Efficiently capture 90% of the CO_2 at 90% purity from the flue gas
- Minimize the parasitic load on the power plant by less than 30% to outperform competitive technologies
- Enable the capture & sequestration process for more than 300 kilotons of CO_2 per year

RW-EDI is based on the established commercial technology electrodionization (EDI), which combines the platforms of ion exchange (IX) and electrodialysis (ED) to electrically separate ions.[6-9] One of the designs of EDI is the plate-and-frame (also known as multiple cell pair) stack. The stack is comprised end-to-end with an endplate, anode electrode, [cell pairs]x, cathode electrode, and endplate. The cell pairs consist of an anion membrane, dilute spacer, cation membrane, and concentrate spacer. The dilute and concentrate spacers refer to channels where fluid is purified and concentrated, respectively. By alternating the anion-selective and cation-selective ion-exchange membranes between the dilute and concentrate spacer within a plate-and-frame module, a "stack" of parallel purifying and concentrating compartments is created. One or both of the dilute and concentrated spacers can be filled with ion exchange resin beads to allow the transport of ions in the channel across the membranes. Under an electric field, water-splitting reactions occur at both the resin/membrane and resin/resin interfaces. Water splitting in the presence of other counter-ions enables in-situ pH control of the solution by properly managing hydrogen and hydroxyl ions generated and transported. Figure 1 shows how water-splitting, ion-exchange resin beads, and ion-exchange membranes can be used together in an EDI system to control the pH in-situ.

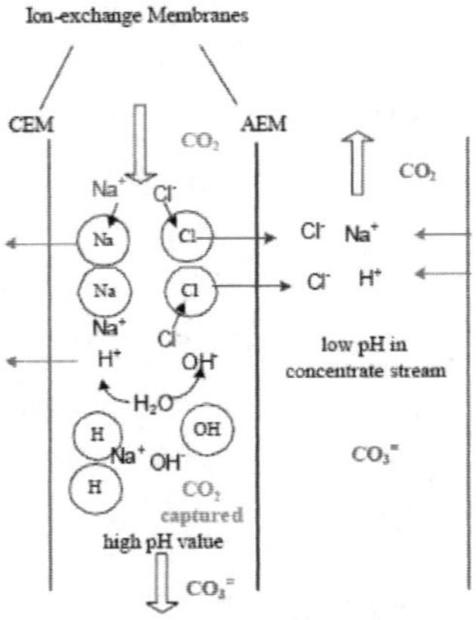

Figure 1. In-situ pH control in the EDI stack

RW-EDI differs from EDI in that the loose ion exchange resins beads of EDI are molded into a solid, porous, resin wafer. Figure 2 shows a depiction of a typical RW-EDI stack. The inlet is the feed water and the two outlets are the product water (deionized) exiting the dilute channels and the outlet waste (ion concentrated) leaving from the concentrated channels.

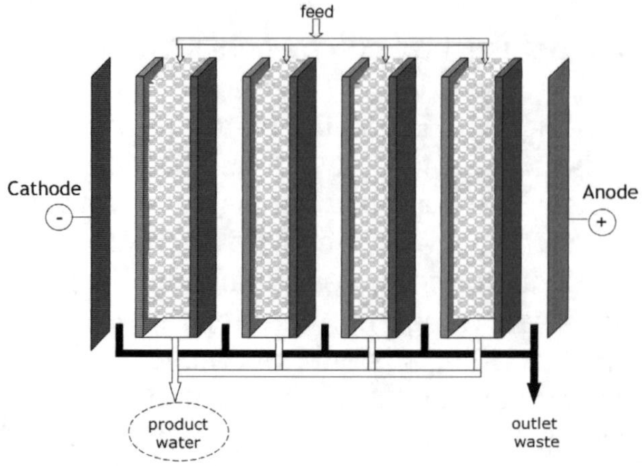

Figure 2. A depiction of an RW-EDI stack for water treatment.

In RW-EDI, the ion exchange resin beads are molded into solid, porous resin wafers, which lead to distinct advantages over the conventional EDI. Resin wafers ensure a more homogeneous bead distribution and avoid the problem of movement of loose beads within the stack. In addition, it is easier to replace solid wafers than flexible bags of beads, which can decrease the amount of time and labor spent assembling and servicing an EDI unit. The solid wafer also leads to lowered risk of internal leakages within the EDI stack. More significantly, the resin wafers are highly customizable in that conductivity agents, catalysts, protein binding agents, etc. can be embedded within the ion exchange resin matrix. For example, in the case of RW-EDI for CO_2 capture, the resin wafers provide a solid support to hold the inorganic catalyst and/or enzyme, which may be essential for enhancing the kinetics of CO_2 adsorption and desorption and overall performance of the CO_2 capture system.

Theoretical and Experimental Details

When CO_2 dissolves in water, carbonic acid (H_2CO_3) is formed. The reaction occurs as:

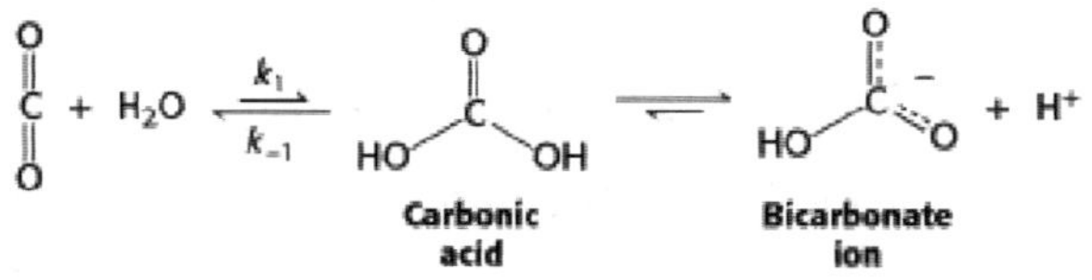

The equilibrium constant to form H_2CO_3 is $K = 1.7 \times 10^{-3}$, which indicates that only a small fraction of the acid exists and the primary form is CO_2.

Figure 3 shows the speciation of the CO_2-carbonate equilibrium as a function of pH.[10] The first acid equilibrium (pKa) of H_2CO_3 ($CO_2 + H_2O$) is 6.35, which indicates that the pH of the solution is 6.35 when equal concentrations of H_2CO_3 and bicarbonate ion (HCO_3^-) exist in solution. When the pH is greater than 6.35, there exists more HCO_3^- ions than H_2CO_3 and when the pH is below, H_2CO_3 is the more predominant species.

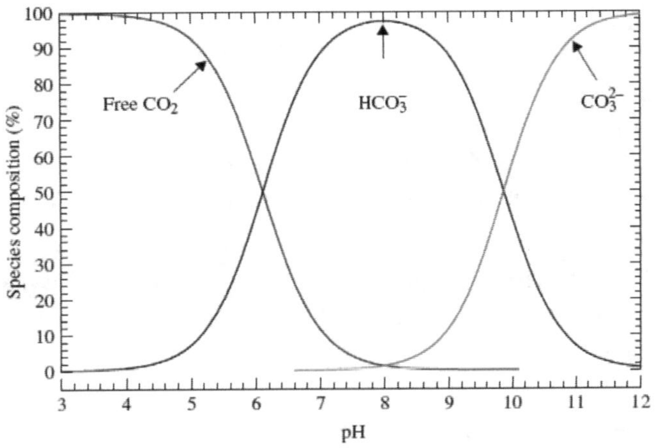

Figure 3. Carbonate equilibrium as a function of pH.8. Free CO_2 is formed from the dissociation of H_2CO_3 into H_2O and CO_2.

RW-EDI is a novel technology for the capture of CO_2 from flue gas emitted from coal-burning power plants. The proposed technology is to take advantage of EDI's electrochemical in-situ pH control to manage the CO_2/HCO_3^- equilibrium. The concept is that the RW-EDI cell pairs are separated into "capture" and "release" cells. The CO_2 is captured from the flue gas in a basic environment of the capture cells and then released as a pure CO_2 stream in an acidic environment from the release cells.

The capture cell is to contain a resin wafer between two membranes so that the solution within the cell is electrochemically shifted towards a slightly basic pH (8 – 8.5). This causes a cascade of dissolved CO_2 to convert into HCO_3^-. All other components of the flue gas are expected to pass through the system and separate from the process stream as CO_2-depleted gas. The release cell is to contain a resin wafer between two membranes so that the solution within the cell is electrochemically shifted towards a slightly acidic pH (4.5 – 5.0), which would release the HCO_3^- from the solution as a purified CO_2 gas stream at atmospheric pressure. The pH shift provides the thermodynamic driver for both adsorption and desorption of CO_2.

Figure 4 shows a depiction of a one cell pair RW-EDI stack being used for CO_2 capture. The stack is flanked by two electrodes (a negatively charged cathode and a positively charged anode) housed in end plates. Next to the cathode is a bipolar membrane with the cation-exchange membrane facing the electrode and the anion-exchange membrane facing the capture cell. Water splitting occurs at the bipolar membrane and the protons are attracted to the negatively charged cathode, while the hydroxyl ions are repelled by the cathode and migrate into the capture cell. The capture cell contains a resin wafer made up of a blend of cation- and anion-exchange beads. The solution running through the cells is a buffer solution. The flue gas is bubbled into the solution prior to entering into the capture cell. The migration of the hydroxyl ions into the capture cell leads to the slightly basic pH environment of the cell. This allows the CO_2 to be captured via the following net reaction:

$$(1) \quad CO_2 + OH^- \rightarrow HCO_3^-$$

The CO_2-depleted flue gas exits the system as a waste gas stream. The cation-exchange membrane allows the sodium ions to associate with the HCO_3^-, which is then pumped into the release cell.

The release cell is surrounded by a cation-exchange membrane that separates it from the capture cell as well as a bipolar membrane that separates it from the anode. The bipolar membrane is arranged so that the cation-exchange membrane faces the release cell and the anion-exchange membrane faces the anode. Again water splitting occurs at the bipolar membrane. The configuration shown in Figure 4 allows the hydroxyl ions to migrate toward the positively-charged anode and the protons to migrate toward the release cell. The release cell contains a resin wafer made up of a different blend of cation- and anion-exchange beads. The migration of the protons into the release cell leads to the slightly acidic pH environment of the cell. This allows the captured CO_2, now in the HCO_3^- form, to be released as high purity CO_2:

$$(2) \qquad HCO_3^- + H+ \rightarrow H_2CO_3 \rightarrow CO_2 + H_2O$$

The cation-exchange membrane allows the sodium ions to migrate back to the capture cell after the release of the CO_2. The CO_2 is released as a high purity gas stream, which can be available for beneficial reuse or storage. The buffer solution is pumped back into the capture cell and the whole series of reactions is repeated. In reality, a commercial RW-EDI unit could have more than 100 cell pairs separated by bipolar membranes to allow for water-splitting and the in-situ pH control.

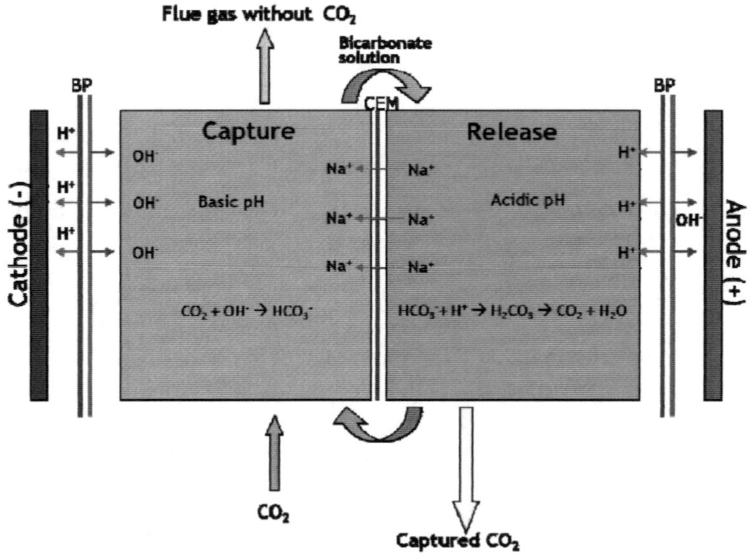

Figure 4. A depiction of one cell pair of the RW-EDI system being used for CO_2 capture.

Recent Results and Discussion

Resin Wafer Fabrication and Characterization

Resin wafers consist of dried ion exchange resins beads, a binder, and possibly other components, such as catalysts.[2, 5] In RW-EDI, several characteristics of resin wafers are of high importance, including ionic conductivity, porosity, compressibility, and distribution of components.

Ionic conductivity refers to the electric mobility of ionic species within a media. In RW-EDI, the ion exchange resin beads enhance the transport of ions across the wafer when a DC current is applied to the stack. Ionic conductivity of resin wafers is an important measurement as it can reveal how much power will be required by the system to carry the current across the stack. Ionic conductivity of the resin wafer is determined by saturating and equilibrating the wafer in various solutions of sodium chloride (NaCl) of known ionic concentrations and measuring the resistance of the wafer in a custom built conductivity cell using a Keithley® LCZ meter. An example of the results of ionic conductivity measurements of a resin wafer can be seen in Figure 5. It shows two conductivity curves for the ionic conductivity measurements performed using several different NaCl concentration solutions. The dashed curve refers to the 'blank' measurement, which is the conductivity cell without a resin wafer sample. The solid curve represents the same measurement repeated with a resin wafer sample inserted into the conductivity cell. The solid curve shows that at high salt concentrations past 3500 ppm of NaCl, the resin wafer has a lower conductivity than the 'blank' solution, possibly owing to the resistive nature of the wafer binder. Meanwhile at very low salt concentrations, the resin wafer has a higher conductivity than the 'blank' solution because of the enhanced ionic conductivity provided by the ion-exchange resin beads. The low salt concentration regime is highlighted in the red box in Figure 5 and is the region of focus when comparing resin wafers and their conductivities.

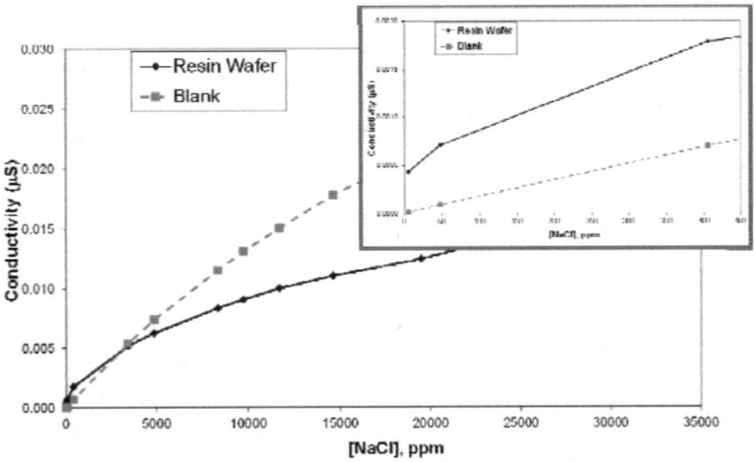

Figure 5. Ionic conductivity measurements of the flow-through cell (blank, pink dashed line) and of the flow-through cell containing a resin wafer (blue solid line).

Porosity is a measure of the void spaces in a material and is a fraction of the volume of voids over the total volume. In mercury porosimetry, mercury (a non-wetting liquid) is forced through the wafer sample pores at high pressure. The pore sizes are determined based on the external pressure needed to force the mercury into the pores against the opposing force of the mercury's surface tension. Figure 6 shows the pore size distribution of two resin wafers fabricated with the same recipe.

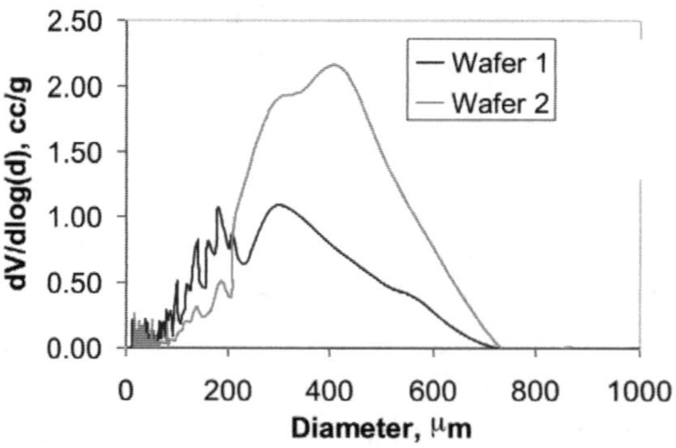

Figure 6. Mercury porosimeter measurements of two resin wafers showing pore size distribution centered around 300 – 500 μm.

CO_2 Capture with RW-EDI

Figure 7 shows a schematic of the RW-EDI system which contains a power supply attached to two electrode-containing endplates holding a stack of resin wafers separated by ion-exchange membranes. On either side of the stack is the release and capture tanks. The synthetic flue gas (15% volume CO_2) is flowed into the endplate holding the cathode electrode where it is mixed with the buffered solution at the capture inlet. The liquid-gas mixture then flows through the basic capture cells where the CO_2 is converted to bicarbonate and eventually flows out the opposite side of the stack at the capture outlet, where it then flows into the capture tank. The CO_2-depleted gas stream is released as the waste gas, and the liquid flows back into the RW-EDI stack through the release inlet. The liquid then flows through the acidic release cells where the bicarbonate is converted back into CO_2 gas. The liquid-gas mixture flows out of the stack through the release outlet and into the release tank where the pure CO_2 gas stream is released as product gas. The liquid is then pumped back into the capture inlet where it mixes with the incoming flue gas, and the process is repeated.

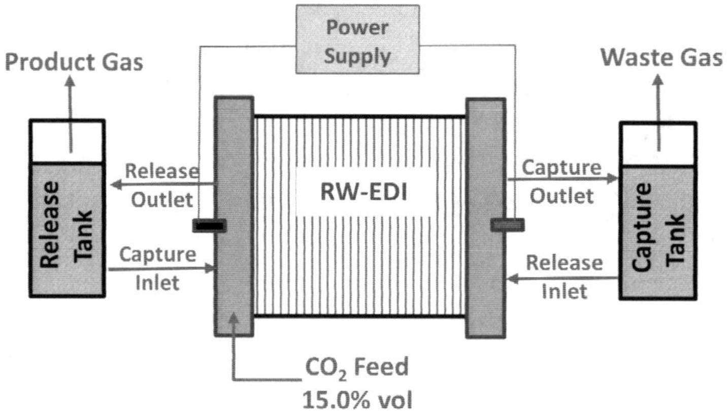

Figure 7. Schematic of RW-EDI system

Three methods have been developed to measure the amount of CO_2 captured in the RW-EDI stack. The first method is using an inorganic carbon analyzer to measure the amount of dissolved carbon in the solution at the inlets and outlets of the RW-EDI stack. The capture and release ratios as well as the overall capture efficiency can be calculated as follows:

$$(1) \qquad Capture\ Ratio = \frac{Capture\ Outlet - Capture\ Inlet}{CO_2\ Feed}$$

$$(2) \qquad Release\ Ratio = \frac{Release\ Inlet - Release\ Outlet}{Capture\ Outlet - Capture\ Inlet}$$

$$(3) \qquad Overall\ Capture\ Efficiency = \frac{Release\ Inlet - Release\ Outlet}{CO_2\ Feed} = Capture\ Ratio \times Release\ Ratio$$

The other two methods are gas-phase measurements using nondispersive infrared (NDIR) CO_2 analyzers. One method is to monitor the flow and measure the amount of CO_2 present in the waste gas that is released from the capture tank after the solution flows through the basic capture cells. From the amount of CO_2 left in the waste gas, it can be calculated how much CO_2 was captured in the RW-EDI stack. Ideally, if there is 100% CO_2 capture, there should be no CO_2 in the waste gas. The other gas-phase analysis method is to monitor the flow and measure the CO_2 in the product gas that is released from the release tank after the solution flows through the acidic release cells. Figure 8 compares the three methods for measuring the percent of CO_2 captured with the RW-EDI system over the course of eight hours under specific conditions.

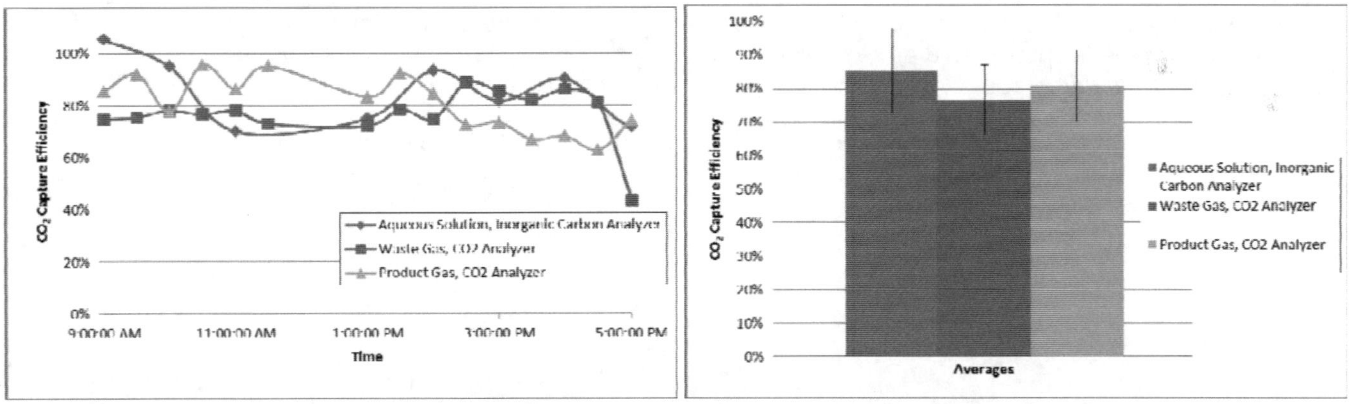

Figure 8. Left: CO_2 capture efficiency with the RW-EDI system measured using three different methods. Right: Comparison of the average CO_2 capture efficiencies with each method.

References

1. Arora, M.; Hestekin, J.; Lin, Y.; St. Martin, E.; Snyder, S. Porous solid ion exchange wafer for immobilizing biomolecules. US patent #7,306,934, 2007.
2. Datta, R.; Lin, Y.; Burke, D.; Shih-Perng, T. Electrodeionization substrate, and device for electrodeionization treatment. US patent #6,495,014, 2002.
3. Lin, Y.; Snyder, S.; Trachtenberg, M.; Cowan, R.; Datta, S. Carbon dioxide capture using resin-wafer electrodeionization. US patent application 20100300894, 2010.
4. Lin, Y. H., Jamie; Arora, M.; St. Martin, E. Electrodeionization method. US patent #6,797,140, 2004.
5. Lin, Y. H., Michael; Snyder, S. Electronically and ionically conductive porous material and method for manufacture of resin wafers therefrom. US patent #7,452,920, 2008.
6. Ervan, Y.; Wenten, I. G., Study on the influence of applied voltage and feed concentration on the performance of electrodeionization. *Songklanakarin J. Sci. Technol* **2002,** *24*, 955 - 963.
7. Kataoka, T.; Yoshida, H., Kinetics of Ion Exchange Accompanied by Neutralization Reaction. *AIChE Journal* **1988,** *34* (6), 1026.
8. Meng, H.; Peng, C.; Song, S.; Deng, D., Electro-regeneration Mechanism of Ion-Exchange Resins in Electrodeionization. *Surface Review and Letters* **2004,** *11* (6), 599.
9. Song, J.-H.; Yeon, K.-H.; Moon, S.-H., Effect of current density on ionic transport and water dissociation phenomena in a continuous electrodeionization (CEDI). *Journal of Membrane Science* **2007,** *291*, 165 - 171.
10. Kemmer, F. N., *Nalco Water Handbook Third Edition*. McGraw-Hill, Inc.: New York, NY, 1988.

(70j) Secondary Effects of Reducing Carbon Dioxide Emissions

Eloy Flores III[1], Michael Hartmann[2], Thomas Andrews[2] and Maoqi Feng[2], (1)Chemical Engineering Department, Southwest Research Institute, San Antonio, TX, (2)Chemical Engineering, Southwest Research Institute, San Antonio, TX

Secondary Effects of Reducing Carbon Dioxide Emissions

While climate change has lost some interest to the global economy, the United States and the World continue to move forward on advancing research and technologies to capture, sequester, or reuse the green house gas carbon dioxide. Capturing carbon dioxide from point sources such as power plants and cement kilns provides the secondary effect of reducing other regulated pollutants such as heavy metals, sulfur and nitrogen oxides, and particulates. This paper evaluates several types of carbon capture technologies (pre and post combustion, and carbon recycling) currently under development and the secondary effect of reducing other regulated emissions. Evaluations are done by chemical and energy balances along with simulation models. It also evaluates how the other regulated emissions common found with carbon dioxide gas affect the overall carbon capture, sequestration, or reuse processes and what engineering process or process integration considerations are required to handle these affects. The secondary net effects of carbon capture, sequestration, or reuse will likely improving overall air quality and reduce healthcare cost. The new cost imposed by implementing carbon capture, sequestration, or reuse would have likely been required to meet new air emission regulations in the future.

See more of this Session: Post-Combustion Techniques for Capture/Separation of Carbon Dioxide
See more of this Group/Topical: Topical 2: Advanced Fossil Energy Utilization

(87a) The Dry Gasification Oxy-Combustion (DGOC) Power Production Cycle

Michael E. Walker[1], Javad Abbasian[1], Donald J. Chmielewski[1] and Marco J. Castaldi[2], (1)Chemical and Biological Engineering, Illinois Institute of Technology, Chicago, IL, (2)Department of Earth and Environmental Engineering (HKSM), Columbia University, New York, NY

Proposed is a novel coal conversion process, dubbed the "Dry Gasification Oxy Combustion Cycle" (DGOC), because of its unique two stage conversion process. Feed coal will be partially oxidized at very high pressures (60 bar) in an oxygen-blown gasification unit, using recycled flue gas as a gasification agent (≈62% CO_2, 32% H_2O b/v). This is in contrast to typical gasification systems that use steam as a gasification agent. In addition, the reducing environment of the gasifier will provide an opportunity to perform *in situ* sulfur removal through sorbent based capture.

The second stage oxy-combustion unit will also utilize recycled flue gas; for flame temperature moderation. The oxy-combustion stage will provide energy to raise steam for power generation, while maintaining a process effluent that is very concentrated in CO_2. This effluent stream, already at high pressure, will utilize ambient cooling to 1) flash out the water from the effluent stream and 2) condense remaining CO_2 into a high purity liquid stream, ready for further compression and sequestration.

Analysis of a preliminary design was carried out using process simulation models developed with Aspen Plus®. Results suggest that DGOC can achieve carbon capture and sequestration (CCS) goals with a 5.2% benefit to plant efficiency, when compared to oxy-combustion systems configured for this purpose. Results also suggest that DGOC will remain competitive against the IGCC process, in terms of efficiency and fresh water consumption.

See more of this Session: Carbon Capture/Separation From Gasification-Based Systems
See more of this Group/Topical: Topical 2: Advanced Fossil Energy Utilization

(87b) Investigation of the Adsorbed Species of CO_2 on $Mg(OH)_2$

James C. Fisher II, National Energy Technology Laboratory, Department of Energy, Morgantown, WV and Ranjani V. Siriwardane, National Energy Technology Laboratory, U.S. Department of Energy, Morgantown, WV

Introduction

Fossil fuels supply a majority of the world's energy needs. However, the combustion of fossil fuels is one of the major sources of the greenhouse gas CO_2. Integrated gasification combined cycle (IGCC) is one the most efficient power generation systems that use coal. However, because the IGCC gas streams are at high pressure and elevated temperatures, there is a thermal efficiency loss if the fuel gas stream is cooled for CO_2 removal. The current commercial technologies, such as the Selexol process, require cooling of the gases to near ambient temperature for water and CO_2 removal. There are tremendous advantages of developing technologies for CO_2 capture at moderate/hot gas temperatures. One advantage is that the moderate/hot fuel gas after CO_2 removal can be directly introduced to the turbine systems for power generation were as the Selexol process requires the fuel gas to be reheated after low temperature capture. IGCC systems with CO_2 removal would be more energy efficient if the sorbents were operational at warm gas temperatures (i.e. 150 to 350 °C). Further, successful development of such a CO_2 capture sorbent would enable the use of other warm gas clean up technologies such as HCl and H_2S. Although warm gas clean up removal processes are expected to increase the overall efficiency of the IGCC process significantly, only a few studies related to regenerable sorbents with sufficient CO_2 removal capacity at 150 to 350 °C are reported in the literature.

A novel magnesium hydroxide-based sorbent that can capture CO_2 from 200 to 315 °C was developed and patented by NETL researchers.[1-2] The sorbent is also regenerable at 375-400 °C. The capture process with this novel warm gas CO_2 removal sorbent involves a chemical reaction, as shown in Reaction 1 below. $Mg(OH)_2$ has a very high CO_2 sorption capacity at 200ñ315 °C which is considerably higher than that of the commercial Selexol process. The capture reaction/process will be referred to as "sorption."

$$Mg(OH)_2\ (s) + CO_2\ (g) \rightarrow MgCO_3\ (s) + H_2O \qquad (1)$$

The carbonate formed during the reaction will thermally decompose to release carbon dioxide and regenerate according to Reaction 2 at 375ñ400 °C.

$$MgCO_3\ (s) + H_2O \rightarrow Mg(OH)_2\ (s) + CO_2 \qquad (2)$$

Reaction 2 may also occur in two reaction steps as shown in Reactions 3 and 4:

$$MgCO_3\ (s) \rightarrow MgO\ (s) + CO_2\ (g) \qquad (3)$$

$$MgO\ (s) + H_2O\ (g) \rightarrow Mg(OH)_2\ (s) \qquad (4)$$

Experimental

The $Mg(OH)_2$-based sorbent contains a mixture of $Mg(OH)_2$, bentonite binder, and a promoter, the detailed synthesis has been described elsewhere.[2] This sorbent powder was formed into 1-2 mm pellets by combining with water followed by drying at 150 °C for 4 hours. The pellets were then ground into a powder and placed in a high pressure, high temperature FTIR-DRIFTS reactor (Thermo Fisher - Nicolet). The reactor was purged with N_2 that flowed through a saturator at 50 sccm. The saturator provided approximately 4% H_2O water to the gas flow. After purging for 30 minutes, the pressure was set to 280 psig and the reactor was heated to 425 °C to clean the sorbent for 2 hours. The clean sorbent was cooled to 200 °C for 2 hours while flowing hydrated N_2 to rehydrate the sorbent per Reaction 4. The hydrated sorbent was then exposed to 5% in CO_2 in hydrated N_2 for 12 hours. The FTIR was set to series mode to scan the sorbent through the entire capture cycle.

Pellets of 1-2 mm were placed in a bench-scale, packed-bed reactor (0.5 inch inner diameter) to form a 6 inch bed. The sorbent bed was then heated to 200 °C and exposed to the capture gas consisting of 28% CO_2, 28% Ar, 10% H_2O, and 34% N_2 gas mixture flowed downward with gas hourly space velocity (GHSV) of 250 hr^{-1}. The capture was complete when the CO_2 concentration of the outlet matched the inlet. After the capture was complete, the sorbent was regenerated to remove the captured CO_2. Regeneration occurred by flowing 10% H_2O with 90% N_2 at 500 hr^{-1} GHSV through the reactor maintained at 375 °C until CO_2 was no longer present in the effluent as determined by a mass spectrometer. Upon finishing the regeneration, the reactor was cooled to 200 °C and a mixture of 20% H_2O in N_2 was introduced for 2 hours to hydroxylate the remaining MgO into $Mg(OH)_2$. After hydroxylation, the CO_2 capture gas stream was introduced and subsequent capture cycles were conducted.

Results and Discussion

Figures 1 and 2 show FTIR spectra during CO_2 sorption over the $Mg(OH)_2$ sorbent. The spectra are in absorbance form where the single beam spectra collected immediately before the CO_2 introduction was used as the background. In absorbance form, species with a positive rate of formation are shown with a positive IR band while species with negative rates of formation are shown with negative IR bands. As Reaction 1 proceeds, the carbonates in Figure 1 are shown increasing while the hydroxyl groups shown in Figure 2 are decreasing.

Figure 1 shows the 1300 to 1750 cm^{-1} range which contains the carbonate formations. The three peaks shown correlate to two carbonate species: 1597 cm^{-1} correlates to the C=O stretch of bidentate carbonate, 1780 cm^{-1} correlates to the C=O stretch from bridged carbonate, and 1260 cm^{-1} correlates to the asymmetric stretch from bridged carbonate. [3] After 10 minutes of CO_2 exposure the bridged carbonate reached a maximum while the bidentate carbonate concentration continued to increase. To establish the nature of the continued bidentate carbonate growth, the hydroxyl species associated to Mg $(OH)_2$ are shown in Figure 2. The hydroxyl groups show continued decreasing through the 12-hour CO_2 exposure confirming the sorbent captured CO_2 throughout the 12-hour capture period. It is interesting to note that the rate of growth of the carbonate peaks does not correspond to the rate of disappearance of the hydroxyl groups. This could be the result of equilibrium where as the initial hydroxyl groups are consumed, new hydroxyl groups are formed. The new hydroxyl groups are formed from the hydrated gas and MgO that was not hydroxylized during the hydroxylation step. The new hydroxyl groups replace the consumed hydroxyl groups maintaining the surface concentration and equilibrium.

THE GRAPHIC ON THIS PAGE
WAS NOT AVAILABLE ON
THE E-MEDIA VERSION OF
THIS PAPER.

Figure 1. FTIR spectra of the carbonate formation during CO_2 sorption.

THE GRAPHIC ON THIS PAGE
WAS NOT AVAILABLE ON
THE E-MEDIA VERSION OF
THIS PAPER.

Figure 2. FTIR spectra of the hydroxylate consumption during CO_2 sorption.

Figure 3 shows the MS profile of CO_2 during a capture over $Mg(OH)_2$ sorbent. The initial capture shows that nearly 90% of the CO_2 was captured. After 15 minutes of CO_2 exposure, the rate of CO_2 capture decreased to about 15% of the CO_2 flow. Comparing this profile to the IR data suggests the initial capture was the result of the formation of both bridged and bidentate carbonates. The lower rate of capture in the bench scale reactor may correlate to the formation of the bidentate formation. In addition, powders were used in FTIR while pellets were used in reactor tests. Diffusional resistance through the product layer and pores in the pellets may also contribute to the lower rates with increasing time.

THE GRAPHIC ON THIS PAGE WAS NOT AVAILABLE ON THE E-MEDIA VERSION OF THIS PAPER.

Figure 3. The MS CO_2 profile from the packed bed bench scale reactor during CO_2 sorption.

This paper will discuss these results in greater detail and further investigation on how to promote the rate of formation of bidentate carbonate. Increasing the rate of formation of bidentate carbonate would effectively increase the usable capture capacity (e.g. capture capacity above 90%).

Conclusions

Two carbonate species formed during CO_2 capture over $Mg(OH)_2$ were identified using an FTIR.

Bridged carbonate reached a maximum concentration after 10 minutes where as the bidentate carbonate intensity continued to increase. The paper is going to investigate the correlation between the rate of formation of bidentate carbonate and the slow secondary capture observed in the bench scale reactor. If a correlation exists, promoting the rate of formation of bidentate carbonate to shift this capacity to a useable capacity will also be discussed.

References

1.Siriwardane, Ranjani V. and Robert W. Stevens Jr., *Novel Regenerable Magnesium Hydroxide Sorbents for CO2 Capture at Warm Gas Temperatures.* Ind. Eng. Chem. Res., 2009. **48**: p. 2135-2141.

2.Siriwardane, R. V., *Regenerable Sorbents for CO2 Capture From Moderate and High Temperature Gas Streams.* 2008, The United States of America as represented by the United States Department of Energy.

3.Little, L. H., *Infrared Spectra of Adsorbed Species.* 1966, New York: Academic Press Inc.

See more of this Session: Carbon Capture/Separation From Gasification-Based Systems
See more of this Group/Topical: Topical 2: Advanced Fossil Energy Utilization

(87c) Physical Solvents That Are Alternatives for PEGDME In CO_2 Absorption

Matthew B. Miller, Dept.of Chemical and Petroleum Engineering, University of Pittsburgh, Pittsburgh, PA, David Luebke, US DOE/NETL, Pittsburgh, PA and Robert M. Enick, Department of Chemical and petroleum Engineering, University of Pittsburgh, Pittsburgh, PA

Advanced gasification power plants will employ the water-gas shift reaction producing a high pressure gas-phase mixture containing CO_2, H_2 and water. This gas mixture at elevated pressures provides ample driving force in order to use a physical solvent that will selectively absorb, not chemically bind, to the CO_2. Potential physical absorbers include (a) liquid mixtures of CO_2-philic polydisperse oligomers, (b) small, volatile, liquid CO_2-philic solvents, and (c) CO_2-philic solids capable of melting in the presence of CO_2. The objective of this study was to identify alternative physical solvents that would selectively dissolve only CO_2 from this mixture.

The first phase of our research involved the comparison of the CO_2 solvent strength of PEGDME with that of other low volatility oligomers that are known to be "CO_2-philic". These oligomeric solvent candidates include poly(propylene glycol) di-methyl ether (PPGDME), poly(propylene glycol) di-acetate (PPGDAc), poly(butylene glycol) di-acetate (PBGDAc) with linear or branched C4 monomers, poly(dimethyl siloxane) (PDMS), perfluoropolyether (PFPE), and glycerol tri-acetate (GTA). Pressure-composition phase diagrams are presented for the pseudo-binary systems of CO_2 with PEGDME n=6, PPGDME n=6, PDMS n=6, PBGDAc n=3.2, PPGDAc n=6.7, and GTA (which is analogous to a trimer of polyvinyl acetate), each at 25 ^oC and 40 ^oC. Although the performance of PPGDME, PEGDME and PDMS are comparable on a weight basis, PPGDME and PDMS appear to be the best CO_2 solvents based on the ability to absorb CO_2. Further, the viscosity, an property of interest upon commercialization, of these compounds at 22 ^oC and 40 ^oC indicates that PDMS is significantly less viscous than all of the others, including PEGDME. Further, PDMS and PFPE are essentially immiscible with water, and water is only slightly soluble in PPG- and PBG-based oligomers and GTA, whereas water and PPGDME are completely miscible.

Small volatile CO_2 solvents were also examined in the second phase of this work. Phase behavior results in the form of pressure-composition diagrams are presented for the binary systems of CO_2 and acetone, methyl acetate, 1,4-dioxane, 2-methoxyethyl acetate, methanol, 2-nitropropane, n,n-dimethylacetamide, acetylacetone, 1-nitropropane, isooctane, 2-(2-butoxyethoxy)ethyl acetate, nformylmorpholine, propylene carbonate, 2-butoxyethyl acetate, and n-tert-butylformamide. Acetone, methyl acetate and 1,4-dioxane are the most CO_2-philic compounds on a weight basis, however, their boiling points are relatively low. 2-methoxyethyl acetate, the next best solvent, has a significantly higher boiling point, however. (The best performing solvents on a molar basis are 2-(2-butoxyethoxy) ethyl acetate, methyl acetate, and 2-methoxyethyl acetate.) Hydrocarbon solvents that are highly oxygenated tend to be CO_2-philic so long as the oxygens are contained in carbonyl group, ether, or acetate groups. The hydroxyl group, however, is a CO_2-phobic moiety.

Lastly, solid CO_2-philes have been investigated as a potential CO_2 solvent. These solids are typically sugar acetates, or benzene rings having tert-butyl groups and acetate groups attached to the aromatic ring the benzene, or ether oxygens within the ring. The solid solvents that were investigated are β-D-ribofuranose 1,2,3,5-tetraacetate, 2,6-di-tert-butyl-4-methylphenol, 1,2,4-triacetoxybenzene, 2,4-di-tert-butylphenol, sucrose octaacetate, and 1,3,5-trioxane. The unique property of solid solvents is that the CO_2 can be desorbed at an elevated pressure (the three-phase SLV pressure), as opposed to a typical pressure swing desorption pressure that is usually around or a little above 1 atmosphere. The ability of these solvents to melt in the presence of substantial amounts of hydrogen in the gas mixture and selectively absorb CO_2 will be presented.

See more of this Session: Carbon Capture/Separation From Gasification-Based Systems
See more of this Group/Topical: Topical 2: Advanced Fossil Energy Utilization

(87d) Complimentary Theoretical and Experimental Studies: CO_2 Capture with Supported Ionic Liquid Membrane

Wei Shi, National Energy Technology Laboratory, U.S. Department of Energy, URS Corporation, Pittsburgh, PA, **David Luebke**, US DOE/NETL, Pittsburgh, PA, Christina Myers, NETL/US DOE, Pittsburgh, PA and Dan C. Sorescu, Computational Chemistry, National Energy Technology Laboratory, Pittsburgh, PA

Researchers at NETL have examined techniques for the capture of CO_2 using supported ionic liquids membranes. In this talk, we will show recent molecular simulation and experimental results involving separation of mixtures of CO_2 and H_2 using ionic liquids membranes. Specifically, we will show that

1) The molar volume of the ionic liquid is very important in determining its H_2 solubility. Based on that result, we are focusing on identifying and synthesizing novel ionic liquids which have small molar volume and interact with CO_2 through complexation or chemical reaction. One example ionic liquid will be shown which exhibits very high CO_2/H_2 selectivity and very high CO_2 permeability.

2) Confinements in pores significantly affects ionic liquid properties in separation of CO_2 and H_2. We have found that CO_2 and H_2 permeability and CO_2/H_2 selectivity are improved when the ionic liquid is confined in carbon nanotube.

Although we have made significant progress in finding promising ionic liquids which exhibit high CO_2 permeability and high CO_2/H_2 selectivity, there are still some issues, such as hydrothermal stability and viscosity, to be addressed for this separation technology. These issues will also be discussed in the presentation.

See more of this Session: Carbon Capture/Separation From Gasification-Based Systems
See more of this Group/Topical: Topical 2: Advanced Fossil Energy Utilization

(87e) Evaluation of Advanced Solvents for CO_2 Capture From Syngas

Tony Wu[1], Alex Bonsu[1], Robert S. Dahlin[2], E. Carl Landham[2] and John Wheeldon[1], (1)National Carbon Capture Center, Southern Company, Wilsonville, AL, (2)National Carbon Capture Center, Southern Research Institute, Wilsonville, AL

Integrated gasification combined cycle for power generation has been shown to have the potential to achieve high electric efficiency and low emissions. It also offers process advantages to remove CO2 from synthesis gas due to its significantly higher CO2 partial pressure, comparing to that in flue gas from a pulverized coal power plant. Currently, physical solvents (e.g. Selexol) are the primary choices to capture CO2 from the synthesis gas stream.

Power System Development Facility in Wilsonville, Alabama is a large-scale gasification facility that was established to develop and improve advanced coal-based power generation technologies. In 2009, a new National Carbon Capture Center was created at the site with the mission to provide slipstreams of syngas for testing and demonstration of various advanced CO2 capture technologies at multiple scales. Performances data gathered from the testing with coal-derived synthesis gas could help technology development efforts for real world operation and accelerate the commercialization process. Technologies tested so far include chemical solvents, physical solvents, H2 membranes and CO2 membrane as well as water gas shift catalysts and trace metal removal. In this work, characterization of advanced solvents in a high-pressure, stirred batch reactor is described. The absorption and regeneration behaviors of several solvents in both simulated and coal-derived synthesis gases are discussed.

See more of this Session: Carbon Capture/Separation From Gasification-Based Systems
See more of this Group/Topical: Topical 2: Advanced Fossil Energy Utilization

MOLECULAR SIEVE CARBONS FOR CO_2 CAPTURE

J. Donald Carruthers, Melissa A. Petruska, Edward A. Sturm, Shaun Wilson

ATMI, 7 Commerce Dr., Danbury, CT 06810

Abstract

With the current environmental interest in controlling greenhouse gases, there is a new focus on materials that can discriminate between gas molecules by capturing those gases which must be curtailed and releasing those which pose no adverse effects to the atmosphere. Although zeolites can be applied to such tasks, molecular sieve (MS) carbons have an additional advantage in their ability to both adsorb and desorb molecules cyclically with comparatively little energy demand.

In addition to characterizing pore sizes and pore size distributions of MS carbons, we recognize the importance of characterizing pore 'portals', i.e., those pores which orchestrate movement of molecules from the gas phase into the inner pore structure of the carbon. This paper describes studies of ATMI commercial carbons obtained from PVDC pyrolysis that have experienced different levels of activation and additives to modify the sizes of these pore 'portals'. The structures are probed using molecules in the 0.3–0.62 nm size range. One significant feature of these materials is that total pore volume for smaller molecules is maintained during the 'tailoring' of the portal pore size. Because these carbons have additional benefits such as high strength and high density, they become ideal candidates for capturing greenhouse gases. Data for certain carbons indicate capacities at ambient conditions for CO_2 of >140 v/v (~ 20 wt.%) and, in simulated power plant exhaust, Henry's Law Separation Factors > 6. Through collaboration with SRI International, ATMI has developed carbons that exhibit >97% capture of CO_2 in such streams with subsequent desorption of 98% purity CO_2 in the stripper section of a pilot-scale unit. In this application, these carbons outperform the claims made for MOFs and other similar materials.

Introduction

Eighty percent of worldwide energy needs are currently met through combustion of fossil fuels which is the primary reason for the increasing level of CO_2 in the atmosphere and its influence on climate change. There is now intense interest in controling CO_2 and other greenhouse gas emissions. The use of electric-powered vehicles suggests a way of limiting CO_2 emissions and yet recent reports indicate that the use of such vehicles in the US and UK will have minimal overall impact because most electricity generation comes from coal and oil burning power plants compared to their use in Europe (France and Germany), where nuclear power contributes as much as 60% of the electric power (Holdway et al. 2010). There is therefore an urgent need to control CO_2 emissions from fossil-fuel burning electricity-generating power plants.

Currently there are three primary methods under evaluation for controlling CO_2 emissions from electric power plants: cryogenic technology; amine-based absorbents; and, dilute alkali (Weiss, 2010) or chilled dilute ammonia absorbents (Koss, 2010). These approaches are both energy-intensive and expensive (House et al. 2009). Selective adsorption of CO_2 by a porous material is one possible path to Carbon Capture and Storage (CCS), with porous metal-organic framework (MOF) materials suddenly becoming the focus of much attention among the research community. Zeolites generally fail in this task because of their preferred adsorption of H_2O over CO_2. Certainly porous materials can be used to reversibly capture and release CO_2 from a flue gas stream using temperature swing (TSA) or pressure swing adsorption (PSA) procedures but attention must be given to the US Department of Energy (DOE) targets for a system achieving 90% CO_2 capture at a 35% maximum increase in the cost of electricity by 2020. To achieve such a target via a TSA or PSA system will require new sets of criteria for the porous medium: high reversible CO_2 adsorption capacity at ambient pressures and temperatures; rapid kinetics; low H_2O capacity at low partial pressures of H_2O, high selectivity for CO_2 over H_2O and N_2, completely regenerable adsorbent with a low heat of adsorption for easy regeneration at moderate temperatures; stability over a large number of cycles; good hydrothermal stability with high attrition resistance; high bulk density; and low cost.

The development of MOFs (MOFs, MILs, COFs and ZIFs) over the past 15 years has generated more than 1000 interesting materials. These ordered, crystalline materials, varying in metal atom and ligand, have provided porosity researchers with almost limitless opportunities to advance the science and a large number of papars focus on the ability of these materials to adsorb CO_2. For example, Millward and Yaghi (2005) report 'exceptionally high CO_2 capacities' for ten MOFs including their best, MOF-177, but only at high pressures (up to 45 bar); Barbaro & Jiang (2008) show similarly high capacities (~27 mmol/g) for IRMOF-1, but only at pressures above 10 bar. Walton et. al (2008) report high capacities for IRMOF-1 but at low temperatures. Yang Q. et al. (2008) report studies of IRMOF-11 and MOF-177 but showing significant capacities only at pressures above10 bar. Liu et al. (2010) were more realistic when they emphasized the need to evaluate the CO_2 capacities at pressures lower than 1 bar and at 25°C. For HKUST-1 and Ni/DOBCD they measured capacities of up to 6 mmol/g. However their work points to the need for high hydrothermal stability, showing evidence for some degradation of the Ni/DOBCD after a mere 10 cycles. Low et al. (2009) has furthered this evaluation by conducting steam-stability studies on a series of MOFs which demonstrate high steam stability in ZIF-8, Al-MIL-110, and MOF-74 and very low stabilities for MOF-69C, MOF-5, and MOF-508.

Banerjee et al. (2008, 2009) in two publications report CO_2 capacities for a series of ZIFs (-2 to -81), with the best achieving ~ 3.1 mmol/g. An & Rosi (2010a) report capacities for bio-MOF-1 at atmospheric pressure - 3.41 to 4.46 mmol/g - but their isosteric heat calculations at low coverage are quite high - at 35kJ/mol and 55 kJ/mol. Similar capacities were reported by them for bio-MOF-11 (An et al. 2010b) but again with quite high heats of adsorption (~33 kJ/mol average). Bao et al. (2011) report data for Mg-MOF-74 that show excellent capacities (~ 8 mmol/g) at 1 bar, but the isosteric heats are very high at 73 kJ/mol, comparable to MOF-100 at 63kJ/mole which was reported by Llewellyn et al. (2008). Caskey et al. (2008) report capacities for DOBDC MOFs with different metal centers – (Mg, Zn, Ni and Co). Although the capacities are quite respectable at ~ 5.5 mmol/g, adsorption occurs with high heats of adsorption (~ 37–47 kJ/mol). Low heats of adsorption were reported by Farrusseng et al. (2009) for HKUST-1 and IRMOF-3 (~ 25 kJ/mol), but the capacities at atmospheric pressures and ambient temperatures were low (~1.5 mmol/g). Mu et al. (2010) considers the performance of MOF, UMCM-1 for CO_2 adsorption compared to MOF-177, Maxsorb carbon, two zeolites (MCM-41 and SBA-15), and two mesoporous carbons (FCMK-3, and CMK-3). The capacities at 5 bar pressure are all comparable at ~ 4 mmol/g, but their densities are low.

One material type that has not been evaluated effectively for CO_2 capture is the MS carbon. ATMI commercialized several MS carbons from the pyrolysis and activation of poly(vinylidene chloride), (PVDC) polymers and recently collaborated with SRI International to develop a MS carbon and process for flue gas CO_2 capture that meet the criteria outlined above. This paper describes the characterization and application of MS carbons selected for CCS.

Experimental

The porosity characterization involved organic vapor adsorption measurements, obtained up to 0.1MPa pressure and at 294.3K using a modified Micromeritics ASAP 2405 porosimeter, replacing the supplied vacuum pump with an Adixen Molecular Drag Pump and an Adixen Rotary Vane roughing pump vented to a Novapure® S520 dry scrubber. N_2 adsorption measurements at 77.4K and up to 0.1MPa and CO_2 adsorption at 273.2K and up to 0.1MPa pressure were obtained on a Micromeritics ASAP 2420.

Carbon dioxide adsorption and kinetic measurements up to 0.1MPa pressure and at 283.3, 293.3 and 303.3K were obtained using a VTI Corp. GHP-300 Gravimetric Magnetic Suspension Balance. High vacuum was achieved via an Adixen turbo-molecular drag pump with dry roughing pump. CCl_4 isotherms and single-point analyses were obtained using a gas manifold system and off-line weighing of gas cylinders containing the carbon samples.

Dynamic adsorption equilibrium breakthrough measurements were obtained using a laboratory apparatus similar to that described by ASTM Method D6160-95. A Bruker Tensor 27 FTIR spectrometer with 15cm path length , 2cm^{-1} resolution, and 4 scan set was used to quantify the continuous on-stream off-gas analysis. The CO_2 feed concentration was held at 20% in a nitrogen stream at atmospheric pressure and 293.2K . The total gas flow rate was 692 cm^3/min in a bed diameter of 5.08 cm.

Results & Discussion

I. Pore Size Distribution

The successful application of a MS carbon depends on a complete understanding of its porous structure. Whereas the measured values of 'equivalent surface areas' and micropore volumes, (MPV), are useful as a guide, micropore size distribution and pore-aperture size (or pore constriction size) are key elements. Unfortunately these values are not easily obtained. N_2 adsorption isotherms and pore size derived from isotherms can be very misleading. In Figure 1, N_2 isotherms at 77 K and CO_2 isotherms at 273 K. are shown for two molecular sieve carbons. Density Functional Theory (DFT) analyses are shown in Figure 2.

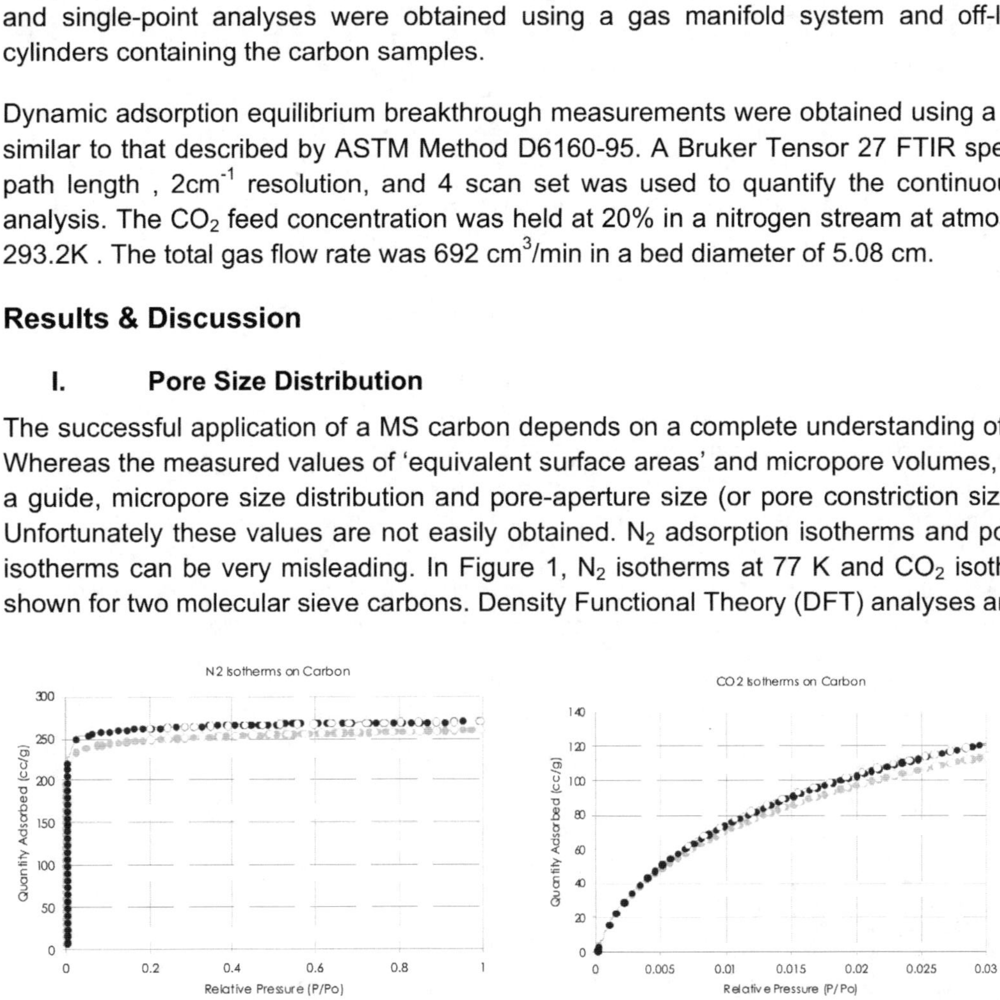

Figure 1 N_2 and CO_2 adsorption isotherms on two MS carbons: yellow – Carbon 1, blue, - Carbon 2.

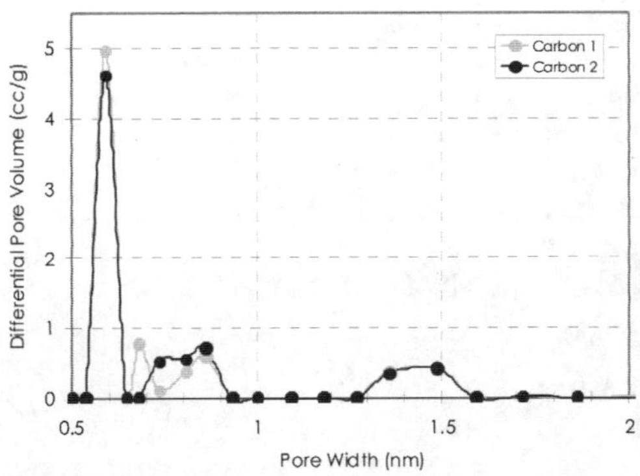

Figure 2 DFT (N_2) analysis of the two carbons.

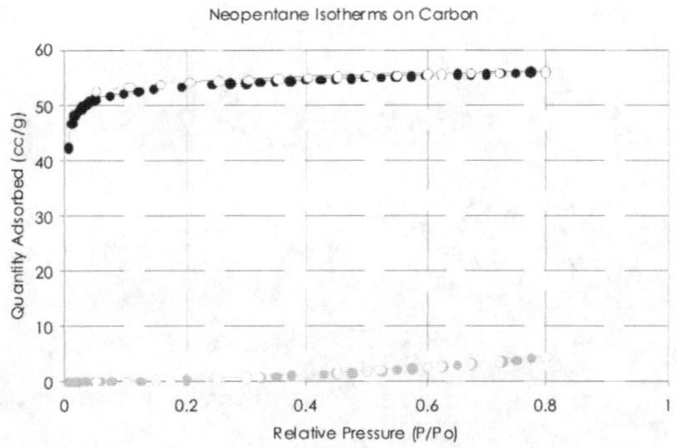

Figure 3 Neopentane isotherms on the two carbons.

The conclusion one might initially draw is that both samples exhibit similar porosities and should perform similarly in gas capture. In Figure 3 isotherms are shown for the same two carbons with the molecule neopentane, (0.62 nm kinetic diameter) adsorbed at 273 K. Not only is there a major difference in pore volume between the two samples, but the volume for Carbon 2, when computed as a liquid volume, is a full 70% of the micropore volume reported for nitrogen adsorption, implying that the porosity in Carbon 2 is mainly in slits of width greater than 0.62 nm while the porosity of Carbon 1 is all within slits much narrower than 0.62 nm or is only accessed through (apertures) slits of smaller width.

Similar to MS carbons, many MOFs exhibit restricted gas movement because of such constrictions. Wang et al. (2008) describe two ZIFs (ZIF-95, ZIF-100) with 'colossal' cages of 3.01 nm x 2.00 nm (inner diameter) which exhibit slow gas diffusion into the porous materials as a result of small pore apertures that act to restrict the flow of gas into the structures. Such constrictions, the authors suggest, are 0.365 nm in diameter. Based on these observations, Yaghi et al. suggest a role for these MOFs in CCS with data showing 'high' CO_2 capacities of ~1.7 mmol/g at 273K and 0.1 MPa corresponding to a 28.2 v/v CO_2.

To better define the pore structure of the MS carbons, adsorption isotherms were obtained on a series of post-treated carbons using the following probe molecules: N_2(0.30nm), CO_2(0.32nm), AsH_3(0.38 nm), i-C_4H_{10} (0.46 nm), SF_6(0.49 nm), CCl_4(0.57 nm) and neopentane (0.62 nm). The micropore volumes were measured using Dubinin-Ashtakov and α-S procedures and are reported in Figure 4. The significance of the results is that the micropore volumes are high and similar for molecule access sized between 0.30 and 0.46 nm but, depending upon the degree of treatment, the access could be restricted for molecules in the 0.46 – 0.62+ nm size range. For our CO_2 capture materials, ACS-1, -2 and -3, high capacities are achieved with these carbons that show a pore size distribution providing access to molecules of >0.6 nm in diameter. The data for ACS-3 is shown alongside the other carbon properties in Figure 4.

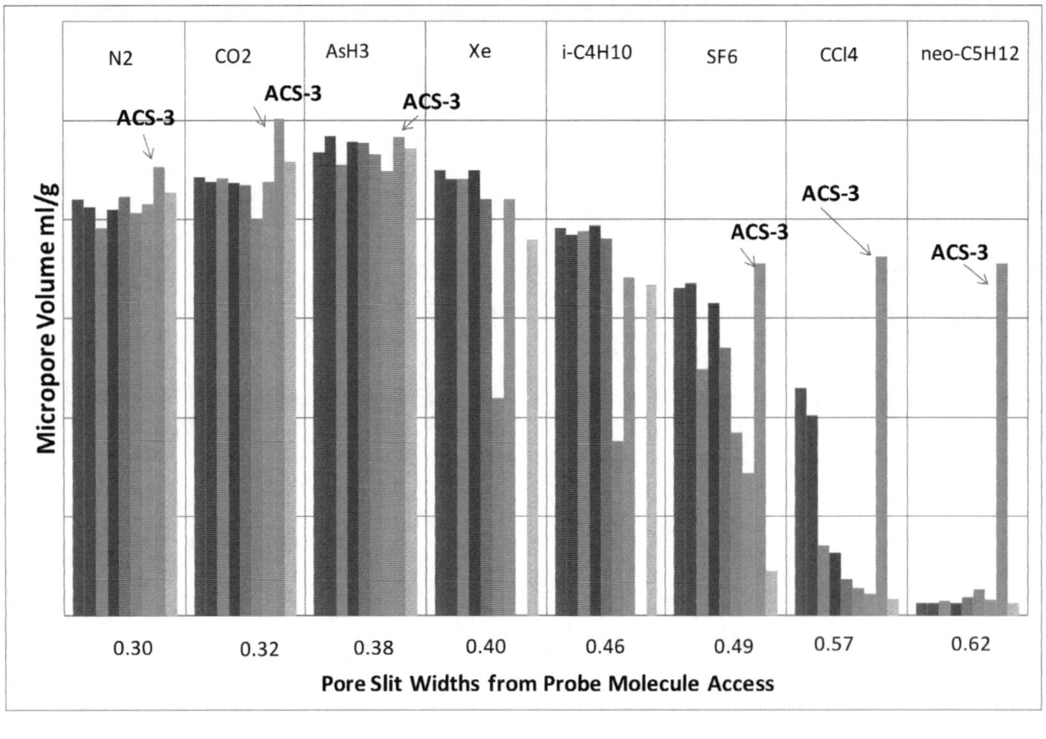

Fig. 4 Measured micropore volumes for eight probe molecules varying in size from 0.30 to 0.62 nm.

II. Properties of the MS Carbons Selected for CO_2 Capture.

Sample ID	Method		ACS-1	ACS-2	ACS-3
Apparent Surface Area N2	BET method	m2/g	1030	1059	1135
Micropore Volume N2	t-method	ml/g	0.373	0.383	0.408
Micropore Volume N2	α-S method	ml/g	0.406	0.413	0.426
Micropore Volume N2	Dubinin-Rad.	ml/g	0.409	0.422	0.452
Micropore Volume N2	Dub. Astakhov	ml/g	0.41	0.42	0.453
Median Pore Width N2	Horvath-Kawaz.	nm	0.534	0.532	0.541
Median Pore Width N2	Dubinin-Stoeckli	nm	0.548	0.555	0.581
Exponent	Dub.Astakhov		1.959	2.0865	1.9747
Apparent Surface Area CO2	BET method	m2/g	768	773	832
Micropore Volume CO2	Dubinin. Rad.	ml/g	0.449	0.449	0.48
Micropore Volume CO2	Dub. Astakhov	ml/g	0.442	0.449	0.502
Particle Size Median (No.)	Number	microns	225	225	295
Bulk Density	ASTM 4164	g/cc	0.68	0.68	0.68
Powder Attrition Test	ASTM D5757	Wt.%/hr			<0.01

Table 1. Physical Properties of three MS carbons

III. CO_2 Adsorption Isotherms, Isosteric Heats, and Henry's Law Separation Factors.

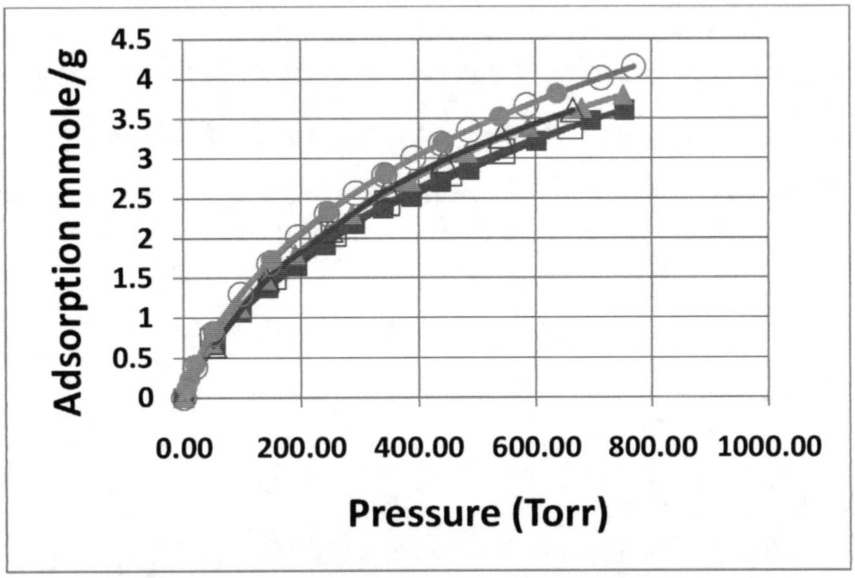

Figure 5 Adsorption (filled) and desorption (open) CO_2 isotherms for MS carbons: squares ACS-1, triangles ACS-2, and circles ACS-3 at 294 K

CO_2 adsorption isotherms for ACS-1, -2 and -3 are shown in Fig. 5. The highest CO_2 capacities are measured for ACS-3 and adsorption isotherms were obtained for this sample at three different temperatures (8.1 °C, 20.07 °C and 29.69 °C) using the VTI gravimetric porosimeter. The isosteric heat of adsorption was calculated from the slope of plots of ln(Pressure) vs. 1/T for constant values of the amount adsorbed using a modified Clausius-Clapeyron equation: $[\delta(\ln P)/\delta(1/T)]_n = q_{st}/RT$. The isosteric heats ranged from 27.16 to 25.52 kJ/mol, averaging 26.52 kJ/mol. These values are similar to isosteric heats reported by Krishnan et al (2010) for ACS-1 of 24.2 – 28.7 kJ/mol.

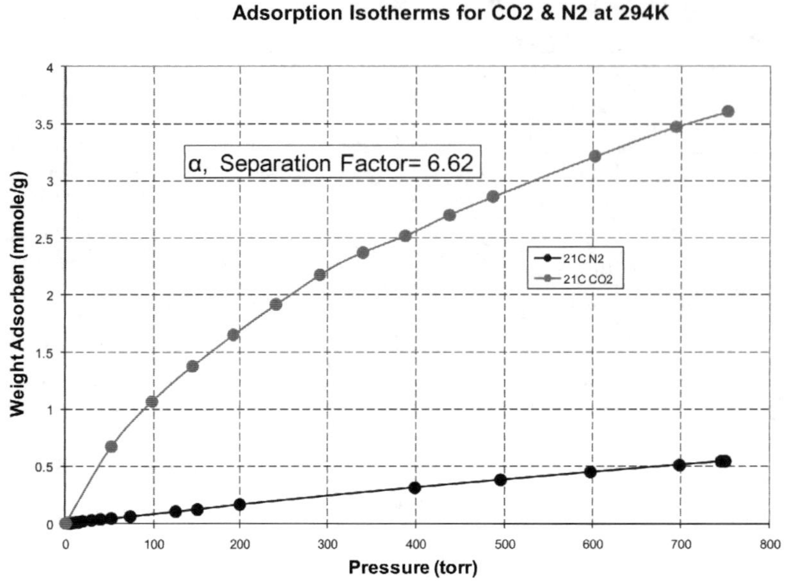

Figure 6 CO_2 and N_2 isotherms for MS carbon ACS-1 at 294°K

Extrapolation of the adsorption isotherms for CO_2 and N_2 on ACS-1 into the Henry's Law region allows calculation of a separation factor, 'α' between the two gases of 6.62, which is substantially higher than the minimum required for successful gas separation (Yang R.T., 2008) assuming no kinetic issues (Fig. 6).

IV CO_2 Adsorption Kinetics and Dynamic CO_2 Adsorption Breakthrough Curves

The rate of adsorption of CO_2 by ACS-3 in the pressure region of 100 to 200 Torr can be assessed from the curves shown in Figure 7. Relatively rapid equilibrium is achieved within 10 minutes with 90-95% achieved within 5 minutes.

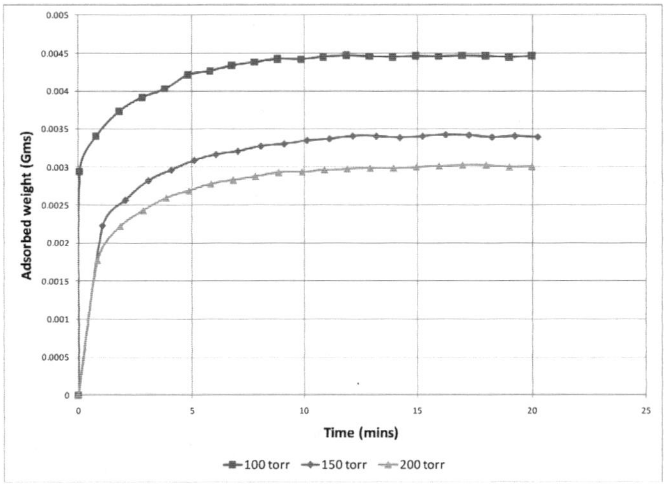

Figure 7 CO_2 adsorption vs. time curves at 293 °K

Dynamic CO_2 adsorption and breakthrough in a 20% CO_2/80%N_2 stream is plotted as a function of time in Figure 8. The first two points were collected with only N_2 gas flowing, while points 3 and 4 were obtained with the 20% CO_2 stream flowing through a bypass loop The bypass valve was then closed and the stream was allowed to flow through the carbon bed immediately following point # 4

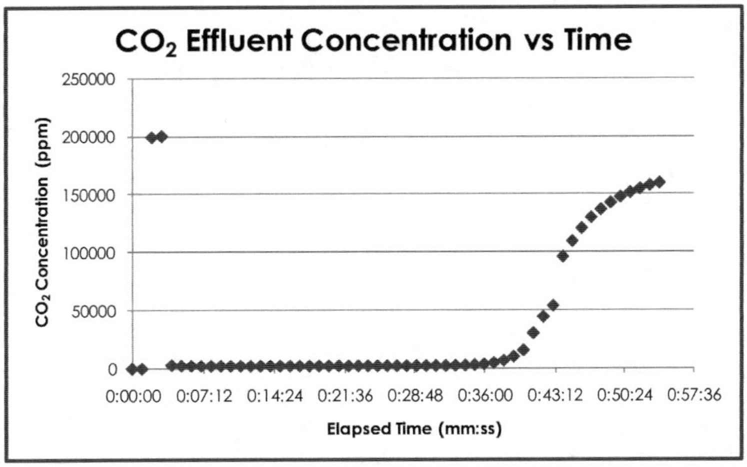

Fig. 8 Dynamic CO_2 Adsorption in a 20% concentration in N_2 on ACS-3

From the breakthrough data, the dynamic adsorption capacity is 0.026 g CO_2/cm^3 carbon with a critical bed volume of 53.3 cm^3 and a critical bed depth of 2.6 cm.

V Pilot Scale CO_2 Capture in a 5cm x 183 cm Vertical Reactor.

Experiments have been carried out by collaborators at SRI International (Krishnan, 2010) with a pilot scale vertical reactor comprising a powder feed hopper, an adsorption section with static mixer, a stripper section fed with 1 bar steam and a receiving hopper as depicted in Figure 9. The flue gas enters at 15% CO_2 in N_2 and at a 150 sec cycle is reduced by >99%; the stripper section releases CO_2 at a 98% purity (Fig. 10). Water vapor in the entering flue gas has been shown to have a negligible effect on the CO_2 capacity. The next phase of the SRI project is the scale up to a 15.25 cm diameter x 610 cm column reactor and production of a 1000-cycle test with integrated adsorption-desorption conditions.

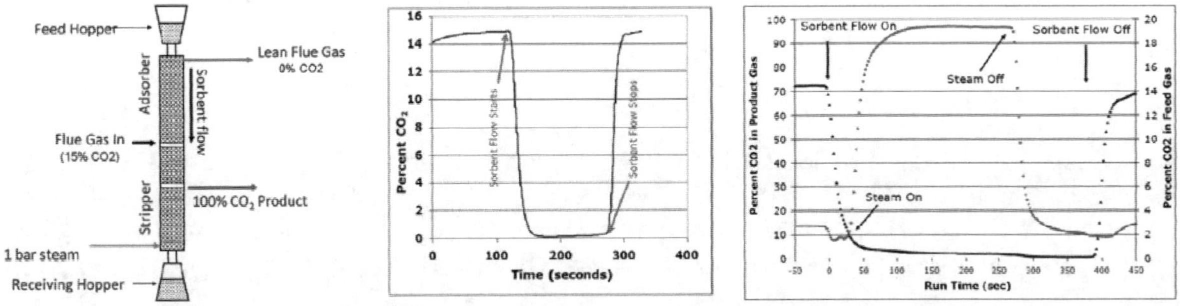

Figure 9 Schematic of the pilot scale vertical reactor and plots of the CO_2 capture & recovery cycle

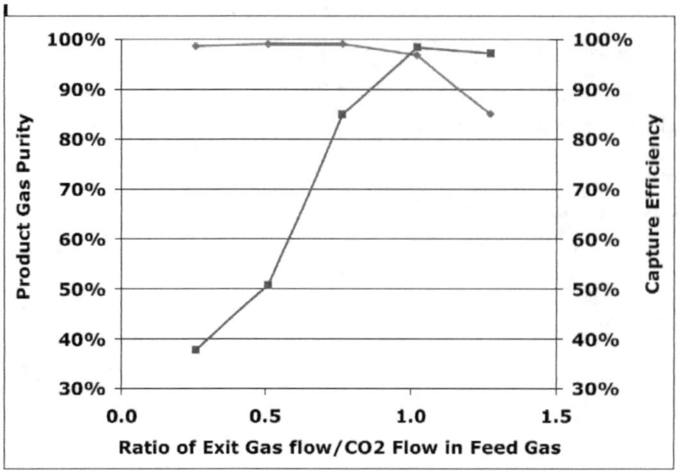

Figure 10 Plots of product gas purity (green) and capture efficiency (red)

The data reported here reveal a CO_2 capture process using an advanced MS carbon powder, ACS-3, at atmospheric pressure and ambient temperatures, with a capacity of 4.2 mmol/g, a significant separation factor of 6.62, an isosteric heat of adsorption of 26.5 kJ/mole, a density of 0.68 kg/L, an attrition resistance of <0.01%/hour, and excellent hydrothermal stability with negligible impact by water vapor. In contrast, much of the work reported for MOFs describe capacities either at much higher pressures or lower temperatures and often with high heats of adsorption.

Conclusions

Although a great deal of attention has been focused on developing a MOF for capturing flue-gas CO_2 , it is clear that simply targeting high CO_2 capacities is quite inadequate. When the other criteria are included in the specifications for such material, a MS carbon such as ACS-3 example described here as looks promising. Pilot scale test work at SRI has reinforced the belief that this carbon can achieve the required CO_2 capture under moderate energy demands.

Acknowledgements

The authors wish to acknowledge the important contribution from the SRI International collaborators, Gopala Krishnan, Marc Hornbostel, Anoop Nagar and Angel Sanjorjo (DOE# DE-NT0005878) and thank ATMI for permission to publish this work.

References

An, J., Rosi, N.L.,2010a, "Tuning MOF CO2 Adsorption Properties via Cation Exchange", *J. Am Chem. Soc. 132, 5578-5579.*

An, J., Geib, S.J., Rosi, N.L. 2010b, "High and Selective CO_2 Uptake in a Cobalt Adeninate Metal-Organic Framework Exhibiting Pyrimidine- and Amino-Decorated Pores", *J. A. Chem. Soc. 132 38-39.*

Banerjee, R., Phan, A., Wang, B., Knobler, C., Furukawa, H., O'Keefe, M., Yaghi, O.M., 2008, "High Throughput Synthesis of Zeolite Imidazolate Frameworks and Application to CO2 Capture", *Science, 319, 939-943.* Banerjee, R., Furukawa, H., Britt, D., Knobler, C., O'Keeffe, M., Yaghi, O.M., 2009, " Control of Pore Size and Functionality in Isoreticular Zeolitic Imidazolate Frameworks and their Carbon Dioxide Selective Capture Properties" , *J. Am. Chem. Soc. 131, 3875-3877.*

Bao, Z., Yu, L., Ren, Q., Lu., X., Deng, S., 2011, "Adsorption of CO_2 and CH_4 on a Magnesium-Based Metal Organic Framework", *J. Coll. Int. Sci., 353, 549-556*

Barbarao, R., Jiang, J. 2008, "Molecular Screening of Metal-Organic Frameworks for CO_2 Storage", *Langmuir, 24, 6270-6278.*

Barcia, P.S., Bastin, L., Hurtado, E.J., Silva, J.A.C., Rodrigues, A.E., Chen, B., 2008, " Single and Multicomponent Sorption of CO_2, CH_4, and N_2 in a Microporous Metal-Organic Framework", *Separation Sci. and Technol., 43 3494-3521.,*

Caskey, S.R., Wong-Foy, A.G., Matzger, A.J.,2008, " Dramatic Tuning of Carbon Dioxide Uptake via Metal Substitution in Coordination Polymer with Cylindrical Pores", *J. Am. Chem. Soc., 130, 10870-10871.*

Farrusseng, D., Daniel, C., Gaudillere, C., Ravon, U., Schuurman, Y., Mirodatos, C., Dubbeldam, D., Frost, H., Snurr, R.Q., 2009, "Heats of Adsorption for Seven Gases in Three Metal-Organic Frameworks: Systematic Comparison of Experiment and Simulation" *Langmuir, 25, 7383-7388.*

Holdway, A.R., Williams, A.R. Inderwildi, O.R., King, D.A., 2010, "Indirect Emissions for Electric Vehicles: Emissions from Electricity Generation", *Energy Environ. Sci, 3, 1825-1832.*

House, K.Z., Harvey, C.F., Aziz M.J. Schrag, D.P., 2009, "The Energy Penalty of Post-Combustion CO_2 Capture & Storage and its Implications for Retrofitting the US Installed Base." *Energy Environ. Sci., 2009, 2, 193-205.*

Krishnan, G., Hornbostel, M., Nagar, A., Sanjorjo, A., 2010, "Development of Novel Carbon Sorbents for Carbon Dioxide Capture", *2010, NETL CO_2 Capture Technology Meeting, September 13 – 17, 2010, Pittsburgh, PA.*

Koss, U.,2010, "An Update on Carbon Capture Using Chilled Ammonia", *Separations Technology VIII: Sustainable Separation Technology for Energy and Environmental Challenges, Dec. 5 -9, ECI, NY, NY.*

Liu, J., Wang, Y., Benin, A.I., Jakubczak, P., Willis, R.R., LeVan, M.D., 2010 " *CO_2/H_2O Adsorption Equilibrium and Rates on Metal-Organic Frameworks: HKUST-1 and Ni/DOBDC" Langmuir, 26, 14301-14307.*

Llewellyn, P.L., Bourrelly, S., Serre, C., Vimont, A., Daturi, M., Hamon, L., De Weireld, G., Chang, J-S, Hong, D-Y., Hwang, Y.K., Jhung, S.H., Ferey, G., 2008, " High Uptakes of CO_2 and CH_4 in Mesoporous Metal-Organic Frameworks MIL-100 and MIL-101", *Langmuir, 24, 7245-7250.*

Low, J.J., Benin, A.I., Jakubczak, P., Abrahamian, J.F., Faheem, S.A., Willis, R.R., 2009, " Virtual High Throughput Screening Confirmed Experimentally: Porous Coordination Polymer Hydration", *J. Am. Chem. Soc., 131, 15834-15842.*

Millward, A.R., Yaghi, O.M., 2005, "Metal-Organic Frameworks with Exceptionally High Capacity for Storage of Carbon Dioxide at Room Temperature" *J. Am. Chem. Soc 127, 17998-17999.*

Mu, B., Schoenecker, P.M., Walton, K.S., 2010, "Gas Adsorption Study on Mesoporous Metal-Organic Framework UMCM-1", *J.Phys. Chem. C, 114, 6464.*

Walton, K.S., Millward, A.R., Dubbeldam, D., Houston, F., Low, J.J., Yaghi, O.M., Snurr, R.Q., " Understanding Inflections and Steps in Carbon Dioxide Adsorption Isotherms in Metal-Organic Frameworks" *J. Am. Chem. Soc.130, 406-407.*

Wang, B., Cote, A.P., Furukawa, H., O'Keefe, M., Yaghi, O.M.,2008, "Colossal Cages in Zeolitic Imidazolate Frameworks as Selective Carbon Dioxide Reservoirs" *Nature 453, 207-212*

Weiss, M., "Carbon Dioxide Mineralization with Beneficial Use", *Separations Technology* 2010 *VIII: Sustainable Separation Technology for Energy and Environmental Challenges, Dec. 5 -9, ECI, NY, NY.*

Yang, Q., Zhong, C., Chen, J-F., 2008, "Computational Study of CO_2 Storage in Metal-Organic Frameworks" *J. Phys. Chem. C, 112, 1562.*

Yang, R.T., 2008 *Gas Separation by Adsorption Processes, Series on Chemical Engineeering Vol. 1, Imperial College Press.*

(87g) CO_2 Chemical Absorption Based on Self-Heat Recuperation Technology for Pre-Combustion Process

Akira Kishimoto, Yasuki Kansha and Atsushi Tsutsumi, Institute of Industrial Science, The University of Tokyo, Tokyo, Japan

Coal has performed an important role as the primary energy source since the industrial era in the eighteenth century. Recently, the importance of coal has been growing in the world. In fact, IEO2010 reports that energy consumption of coal is growing from 132 quadrillion Btu to 206 in 2035. The combustion of coal emits CO_2, which drives green house effects in the world. Therefore, CO_2 capture process has been paid more attention in the energy production sectors, especially coal combustion and gasification field.

In the energy production sectors, integrated coal gasification combined cycle power plant (IGCC) has been attracted more attention for improved electric generation efficiency compared to conventional power plants. In this plant, $CO/CO_2/H_2$ mixtures are produced by gasification/reforming and water gas shift reaction. Then, these gases are fed into gas turbines for power generation.

Although IGCC can achieve higher power generation efficiency than conventional power plants, CO_2 capture process reduces this efficiency. This is because CO_2 separation from gas mixtures requires huge amount of energy.

To overcome this problem, in this study, we analyzed the pre-combustion process combined CO_2 separation process based on self-heat recuperation technology (SHR). The electric generation efficiency can be improved by SHR compared with conventional IGCC.

See more of this Session: Carbon Capture/Separation From Gasification-Based Systems
See more of this Group/Topical: Topical 2: Advanced Fossil Energy Utilization

Stack Fueling Perspectives

Subir Roychoudhury
Precision Combustion, Inc. (PCI), 410 Sackett Point Rd., North Haven, CT 06473.

Abstract

Fuel cell power systems are an attractive solution for future portable and stationary power sources in many applications such as shipboard, military base, remote sites, and truck APU. These fuel cell stacks have primarily been operated with synthetic fuels or with stored H_2. It has become increasingly apparent that a lack of fueling solutions utilizing practical fuels has limited the migration of fuel cells to field applications. In developing practical fuel cell-based power generators, durability of reforming systems to safely supply hydrogen in-situ and to permit stable, coke-free fuel cell stack operation via catalytic reforming of commonly available fuels is crucial. The use of practical liquid fuels includes distillate fuels such as diesel or kerosene, as well as renewable fuels such as bio-fuels. Making these fuels suitable for use in the fuel cell stacks include reformation of the fuels such as to minimize higher hydrocarbon content in the reformate to an acceptable level (depending on the fuel cell type), sulfur cleanup to <<1 ppm in the reformate stream, and low parasitic losses. Additionally, water neutrality is an essential component of these systems. This requires sufficient water recovery and separation from the anode exhaust. The ability to achieve all of these objectives requires accurate flow management, thermal management, and robust control algorithms.

Efforts to develop and optimize practical fueling solutions for different fuel cell stacks will be discussed. These solutions include integrated reformer systems. Topics such as fuel flexibility (gaseous and liquid fuels, biofuels), sulfur tolerance (e.g., with distillate and logistics fuels), water neutrality, BOP development (e.g., pumps, sensors, heat exchangers) and control protocols will be briefly reviewed. Reformer and stack performance data will be shared. Recent advances in H_2 production, pre-reforming and partial oxidation approaches will also be briefly highlighted. The goal will be to provide insights into Fuel Cell fueling initiatives and corresponding system design approaches.

Introduction

Important aspects in the development of integrated reformer systems include fuel flexibility, sulfur tolerance, and water neutrality. These are required for efficient H_2 production, while operating at low S/C ratios to reduce or eliminate back-up water supply. Three catalytic reforming technologies, namely autothermal reforming (ATR), catalytic waterless partial oxidation (dry CPOX), and steam reforming (CSR) are being developed at Precision Combustion, Inc. (PCI) and optimized for various applications. This paper will focus on the development and maturation of the integrated ATR system, including sulfur fuel tests and water recovery approaches. The reformer utilizes a limited amount of water that can be recovered from the integrated system exhaust (e.g., anode gas exhaust), while continuously producing H_2 with complete fuel conversion and reforming efficiency of ≥80% (ratio of LHV of reformate to fuel) without coke formation.

Fuel Processor Background Work at PCI: The fuel-processing prototype demonstrated at PCI consists of an ATR reactor packaged with a fuel/air/steam injector, igniter, steam generator and desulfurizer. It has connections for controlled fuel, air and water inlet. The ATR is operated at a steam-to-carbon (S/C) ratio of 0.90 and an oxygen-to-carbon (O/C) ratio of 0.95, to minimize water need while achieving complete conversion of the fuel to C1 products (i.e., CO, CO_2, and CH_4) and fuel-sulfur to H_2S (and COS) at sustainable temperatures. Since the ATR is operated at very low S/C ratios, H_2S (and COS) in the reformate stream is readily adsorbed in downstream desulfurizer bed. A schematic and photo of the reforming approach being developed at PCI is shown in Figures 1 and 2. In Figure 2, a 2 L soda bottle is shown next to the prototype to reference the size. The flow through the ATR is indicated by arrows. Water, fuel and air (all at ambient temperatures at startup) are fed to the fuel processor. The ATR is ignited via glow plug and started with ambient fuel and air in CPOX mode. The heat from the CPOX operation is used to raise steam from the ambient temperature water. It is then transitioned to an ATR mode. These designs were developed at PCI via ASPEN system modeling software. When integrated thermally with a stack, the air and/or steam may be preheated by heat exchange with the stack before entering the ATR. This fuel processor has been used to feed an SOFC stack and found to permit satisfactory stack operation.

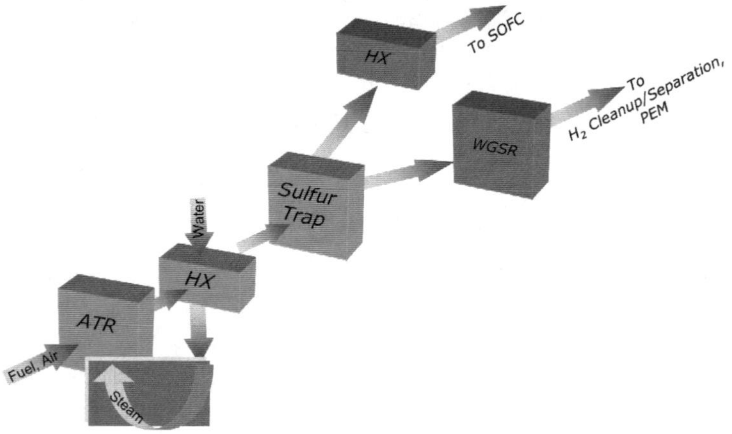

Figure 1: Design approach for the ATR system to generate syngas or H_2 for fuel cell application.

In the conceptual layout shown in Figures 1 and 2, note that after the ATR, the temperature is dropped to 350 – 400°C for sulfur removal. PCI's ATR catalysts have shown tolerance up to 400 ppm_w sulfur in fuels. We also observed that sulfur content affects the H_2/CO concentration ratios in the reformate stream and thus the overall reforming efficiency. Higher sulfur results in lower H_2 and higher CO concentrations as well as lower reforming efficiency and vice-versa with lower sulfur fuels. The desulfurizer beds, which can remove H_2S, were tested with actual JP-8/diesel reformer outlet. For a 5 kW_{th} system, a bed volume of ~80 in^3 (1.3 liter) was used. The desulfurizer bed lifetime will depend on the fuel-sulfur content; e.g. a 1 L bed could last for more than 2000 hours providing <1 ppm_v H_2S in the reformate with ~20 ppm_w fuel-sulfur as in Tier II diesel. The prototype desulfurizer tested in-line with JP-8 and diesel reformate exiting the ATR confirmed this result.

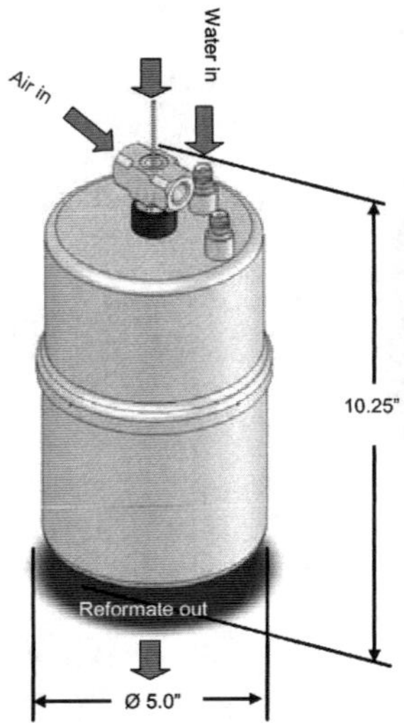

Figure 2: Rendering, photo of thermally integrated 5 kW_{th} ATR + Fuel Injector + Steam generator + Igniter + Sulfur trap. Ambient temperature water, fuel, air in → SOFC quality reformate out.

Overview of Microlith® Technology: The compact, lightweight, and modular catalytic fuel reformers discussed here are based on PCI's patented Microlith® substrates and catalyst technology.[1] The Microlith technology consists of a series of catalytically coated, microfabricated metal meshes. The meshes each have ultra short channel lengths, high cell density, and low thermal mass. The resulting reactor is very compact, has rapid transient response, and only requires small amounts of precious metal catalysts due to the high catalytic activity. The benefits of this design stem from the reduction in mass transfer limitations as compared to conventional monoliths having relatively thick boundary layers. The Microlith substrate also provides about three times higher geometric surface area over conventional reactors (e.g., monoliths) with equivalent volume and open frontal area. Additionally, it packs more active surface area into a small volume, and thus provides more effective fuel conversion for a given pressure drop as well as a higher catalyst utilization. The effectiveness of the Microlith technology and long-term durability of PCI's proprietary catalyst coatings have been systematically demonstrated in different applications, including exhaust aftertreatment, catalytic combustion, partial oxidation of methane,[2,3] as well as liquid fuel reforming and water gas shift reactors in fuel processors.[4,5,6]

Experimental results

In order to develop integrated reformer systems for H_2 (or syngas) production for fuel cell application, several possible system configurations were studied using ASPEN simulation software. Figure 3 illustrates one possible system configuration capable of operating with sulfur fuels under water neutral condition with a target reforming efficiency of ≥80% (LHV-based). In this particular configuration, the fuel processor system includes fuel/air/steam injector, ATR core reformer, steam generator, downstream sulfur trap, and water recovery unit. Additional downstream components, such as water gas shift reactors and membrane, may be required depending on the type of fuel cell stacks being used.

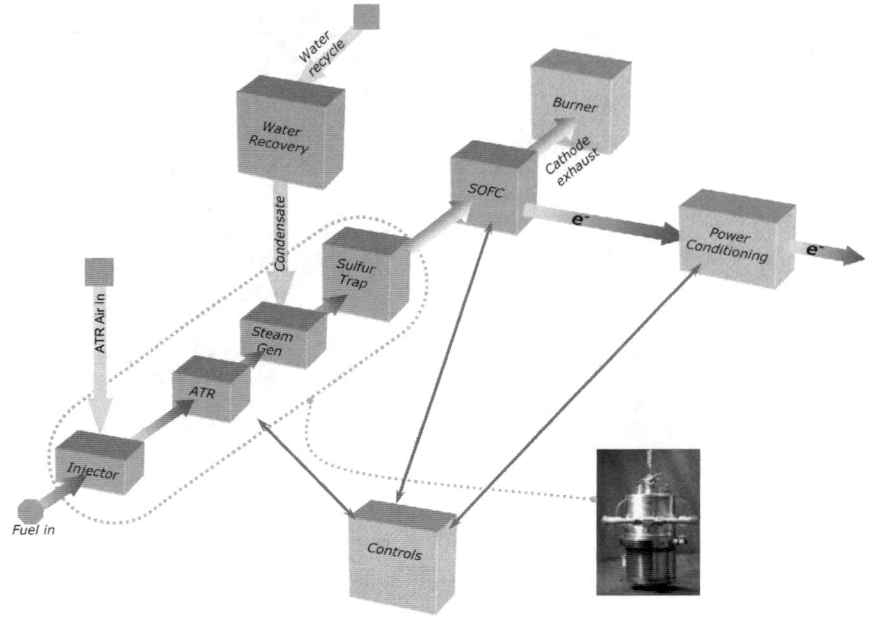

Figure 3: One possible configuration for the integrated ATR system that can operate with sulfur fuels under water neutral condition for efficient H_2 (or syngas) production for fuel cell application.

Recently, a 5 kW_{th} integrated ATR reformer system was designed, fabricated, and assembled at PCI. This fuel reformer system consisted of ATR core reformer, fuel injector, steam generator, and sulfur removal unit, and was fully instrumented with multiple thermocouples, pressure gauges, and gas sampling ports. The fuel reformer was then integrated with the necessary BOP components, including pumps, mass flow controllers, and control system, in a portable, compact test rig (Figure 4).

Figure 4: A 5 kW_{th} integrated ATR reformer unit with full system assembly in a stand-alone test cart.

In order to study the sulfur tolerance of the ATR reformer system and to demonstrate its ability to produce SOFC-grade reformate stream (i.e., H_2S <1 ppm_v and low hydrocarbons slippage), 50-hr test and 1000-hr long-term durability test were performed with JP-8 containing 400 ppm_w naturally occurring sulfur. During these tests, the catalyst temperature profile and the reformate composition were continuously monitored to evaluate catalyst activity over time. Here, the ATR reformer was operated by maintaining the catalyst bed peak temperature using a feedback closed-loop control system. The S/C ratio was maintained at ~0.90 while the O/C ratio was regulated as necessary by modulating the air flow rate. This control strategy allowed the reformer to maintain complete fuel conversion and ≥80% reforming efficiency throughout the duration of the test with minimal degradation/catalyst deactivation. The ability to operate the reformer at a low S/C ratio with sulfur fuels while avoiding coke formation and catalyst deactivation was also demonstrated. This allows for ATR operation under water neutral condition by recycling water from the exhaust of the integrated system (e.g., anode gas exhaust). Also, since minimal amount of water was present in the reformate stream, H_2S could be readily adsorbed in the downstream sulfur removal bed.

Sulfur Tests: The results obtained from the 60-hour test of the ATR with 400 ppm_w naturally occurring sulfur JP-8 indicated stable conversion (~100%) and high reforming efficiency (>80%). H_2 and CO concentrations in the reformate stream changed very little over the 60-hour test. Table 1 lists the average reformate composition (mole %, wet basis) obtained from the 400 ppm_w sulfur test. The composition obtained from the 4 ppm_w sulfur test was included for comparison. In general, the presence of higher sulfur level in JP-8 reduced the H_2 concentration and increased the CO concentration in the reformate due to the inhibition of WGS reaction. This also led to a lower CO_2/CO ratio. Additionally, the presence of C2s (i.e., ethane and ethylene) and C3s (i.e., propane and propylene) compounds in the reformate stream were observed. Nevertheless, total C2s concentration remained <50 ppm_v and the total C3s concentration <10 ppm_v over the 60-hr test.

Table 1: Typical reformate compositions with low and medium sulfur JP-8.

Fuel	O:C	S:C	H_2	CO	CO_2	N_2	H_2O	CH_4	C2	C3
4 ppm_w S JP-8	0.95	0.90	26.0	13.2	7.4	39.7	13.6	0.03	0	0
400ppm_w S JP8	1.02	0.90	24.7	14.1	6.8	40.8	13.6	0.07	0.0024	0.0008

Sulfur analysis of the ATR reformate stream during operation with 400 ppm_w sulfur JP-8 using GC/FPD (gas chromatography equipped with a flame photometric detector) showed the presence of COS in addition to H_2S. The COS concentration, however, was stable at ~2-3 ppm_v over 60 hours. At one point during the 60-hour ATR test, the ATR reformate stream was passed through a desulfurizer bed for 24 hours. The H_2S and COS concentrations in the desulfurizer bed *outlet* stream were monitored using GC/FPD in order to evaluate their concentrations as a function of time. During this 24-hr test, no H_2S and COS breakthroughs were observed downstream of the bed. Both the H_2S and COS outlet concentrations were below the detection limit of our GC/FPD instrument, which was ~0.4 ppm_v. This indicated that the desulfurizer bed was also capable of absorbing COS to some extent. A longer test to evaluate the H_2S and COS breakthrough, over hundreds of hours, has been performed to validate the result.

Water Neutrality: Two water recovery methods, namely anode gas exhaust recycle and condensation approach were evaluated and compared for water neutral operation. For the anode gas recycle (AGR) method, PCI's ATR reformer was operated with a simulated AGR mixture composition and the operating condition (i.e., S/C, O/C ratio, and inlet gas mixture temperature) was optimized. For the condensation approach, tests were performed by passing a simulated anode gas exhaust mixture through condensers, and the water recovery was calculated to evaluate their performance. It was observed that water neutrality for PCI's ATR reformer system could be achieved via both AGR and condensation approaches.

Conclusions

A novel sulfur tolerant and water neutral Microlith-based fuel processor system was designed, developed and demonstrated. The unit demonstrated durability during operation with up to 400 ppm_w sulfur fuels. Test results indicated stable, coke-free reformer operation at low S/C ratio with complete fuel conversion and ≥80% reforming efficiency.

References

1. U.S. Patent #5,051,241, September 24, 1991.
2. Lyubovsky, M.; Karim, H.; Menacherry, P.; Boorse, S.; LaPierre, R.; Pfefferle, W.C.; Roychoudhury, S., *Catal. Today*, **83**, 183 (2003).
3. Lyubovsky, M.; Roychoudhury, S.; LaPierre, R., *Catal. Lett.*, **99**, 113 (2005).
4. Roychoudhury, S.; Castaldi, M.J.; Lyubovsky, M.; LaPierre, R.; Ahmed, S., *J. Power Sources*, 152, 75 (2005).
5. Castaldi, M.J.; LaPierre, R.; Lyubovski, M.; Pfefferle, W.C.; Roychoudhury, S., *Catal. Today*, 99, 339 (2005).
6. Roychoudhury, S.; Junaedi, C.; Walsh, D.; Morgan, C.; Mastanduno, R., *Prepr. Pap. - Am. Chem. Soc., Div. Fuel Chem.*, **54** (2), 417 (2009).

Contact

Author for correspondence: Subir Roychoudhury (sroychoudhury@precision-combustion.com)

(98b) Micro-Structured Fuel Processors as Hydrogen Source for Fuel Cells In Future Distributed Energy Generation Systems

Gunther Kolb, S. Keller, M. O'Connell, J. Schuerer, D. Tiemann and M. Wichert, Energy Technology and Catalysis Department, Institut f¸r Mikrotechnik Mainz GmbH, Mainz, Germany

Future sustainable and distributed energy generation will rely in many cases on fuel cell technology. To address the critical issue of hydrogen supply, fuel processing of fossil and renewable fuels is a viable option to meet the limited space demands of auxiliary power units (APUs) due to the high energy density of liquid fuels [1]. Especially for stationary applications of the smaller scale, mobile and portable power generation systems require a compact hydrogen supply when fuel cell technology is applied. Microstructured reactors provide significant potential for the reduction of system size, system complexity and catalyst costs. Scaled-up micro-channel reactors containing thousands of channels create less scale-up issues compared to conventional (fixed bed) systems. They allow a careful heat management during start-up and normal operation, which increases the service life of the catalyst.

At IMM, the development of complete fuel processors for alcohol and hydrocarbon fuels is ongoing, applying microstructured plate heat-exchanger technology. Complete fuel processors are under development applying steam reforming, partial oxidation and partial de-hydrogenation reactions on a large variety of fuels such as methanol[2], ethanol [3,4], natural gas, LPG [5], gasoline, kerosene and diesel [6]. The systems size varies between 100 W and about 20 kW thermal power of the hydrogen produced. All development steps required for building a fuel processor have been addressed i.e. reactor and system simulation and design, catalyst development, catalyst stability against impurities in the fuel and set-up/testing of complete fuel processors for systems up to 13 kW thermal power output. The choice of fuel converted to hydrogen mainly depends on the application and existing infrastructure. While trucks have diesel fuel readily available, LPG is the right choice for caravans. Having demonstrated the feasibility of compact fuel processing, the aspects of future mass production for such devices for consumer applications have been addressed. Complete fuel processor systems (e.g. a 5 kW diesel fuel processor, a 250W fuel processor/ fuel cell system running on LPG as fuel which will be commercialized end of 2010 and a 100 W methanol fuel processor operated with a high temperature PEM fuel cell) and novel components such as a reactor for the partial de-hydrogenation of kerosene will be presented and discussed.

References

[1] Kolb, G.; *Fuel Processing for Fuel Cells,* 1 Ed.; Wiley-VCH, Weinheim (2008).

[2] Kolb, G., Schelhaas, K.-P., Wichert, M., Burfeind, J., Hesske, C., Bandlamudi, G.;

"Development of a Microstructured Methanol Fuel Processor coupled to a High Temperature PEM Fuel Cell", Chem. Eng. Technol. **32,** 11 (2009) pp. 1739-1747.

[3] Men, Y., Kolb, G., Zapf, R., Hessel, V., Löwe, H.; *"Ethanol steam reforming in a microchannel reactor"*, Process Saf. Environ. Prot. **85,** B5 (2007) pp. 413-418.

[4] Kolb, G., Men, Y., Schelhaas, K. P., Tiemann, D., Zapf, R., Wilhelm, J.; *"Development Work on a Microstructured 50 kW Ethanol Fuel Processor for a Small Scale Stationary Hydrogen Supply System "*, Ind. Eng. Chem. Res. **DOI: 10.1021/ie100602w,** (2010).

[5] Wichert, M., Men, Y., O'Connell, M., Tiemann, D., Zapf, R., Kadhum, A. A. H., Butschek, S., Frank, R., Schiegl, A.; *"Self-sustained operation and durability test of a 300 W-class micro-structured LPG fuel processor"*, Int. J. Hydrogen Energy **submitted for publ.,** (2010).

[6] O'Connell, M., Kolb, G., Schelhaas, K. P., Schürer, J., Tiemann, D., Ziogas, A., Hessel, V.; *"Development and evaluation of a microreactor for the reforming of diesel fuel in the kW range"*, Int. J. Hydrogen Energy **34,** (2009) pp. 6290-6303.

See more of this Session: Fuel Processing for Hydrogen Production (I)
See more of this Group/Topical: Topical 2: Advanced Fossil Energy Utilization

(98c) Ultra-Pure Hydrogen Production From Renewables for PEM Fuel Cells

Hazem Tawfik[1], Steven Lopez[1], Thomas Butcher[2] and Devinder Mahajan[2], (1)Mechanical Engineering Technology, Farmingdale State College, Farmingdale, NY, (2)Sustainable Energy Technologies, Brookhaven National Laboratory, Upton, NY

Recent concerns over the security and reliability of the world's energy supply have caused a shift towards the study of renewable sources. The generation of ultra-pure hydrogen (H_2) from biomass-derived synthesis gas and its integration with fuel cell systems has the potential for a high-energy efficiency, ultra-clean environmental performance, and near-zero greenhouse gas emissions. Biomass gasification is a process that produces synthesis gas (syngas) from organic matters such as wood chips, farm waste, and forest debris. In order to efficiently produce H_2 gas suitable for operating fuel cells, carbon monoxide (CO) must be reduced to less than 10 parts per million (ppm) to prevent the poisoning of the fuel cell's platinum catalyst. In this study, the Water Gas Shift (WGS) system was utilized to reduce CO in the syngas to an acceptable level for powering a PEM fuel cell. Initially, the WGS system consisted of two-stage commercial catalysts: a high temperature shift catalyst (HTSC), Fe_3O_4/Cr_2O_3, and a low temperature shift catalyst (LTSC), $Cu/Zn/Al_2O_3$ to establish baseline data. Runs are now underway to develop a more efficient catalyst for WGS by evaluating the effect of support on the efficiency of CO removal.

See more of this Session: Fuel Processing for Hydrogen Production (I)
See more of this Group/Topical: Topical 2: Advanced Fossil Energy Utilization

Decentralized production of hydrogen for residential PEM fuel cells from piped natural gas by low temperature steam-methane reforming using sorption enhanced reaction concept

Michael G. Beaver and Shivaji Sircar (Speaker)

Chemical Engineering Department
Lehigh University, Bethlehem, Pa 18015, U.S.A.

Extended Abstract

Decentralized generation of small-scale stationary power (< 250 KW) for residential or commercial use has been a subject of much interest during the last decade and many corporations around the world have engaged in research and development of fuel cell technology for this application. The driver for this technology is rapidly expanding worldwide demand for more heating, cooling and electrical supply by increasing populations and growing economics.

Proton exchange membrane or polymer electrolyte membrane (PEM) fuel cell technology which transforms the chemical energy liberated during the electrochemical reaction between hydrogen and oxygen to electric energy as opposed to the direct combustion of hydrogen and oxygen to produce thermal energy has attracted most attention.

The fuel cell needs pure hydrogen, oxygen (from air) and water (to moisten the membrane) for operation. The purities of H_2 and O_2 used in the PEM hydrogen fuel cell are critical issues. The catalytic activity of the platinum electrodes in a PEM fuel cell is poisoned by the presence of trace amounts of CO, NH_3, H_2S, HCN in H_2, as well as by the presence of trace amounts of SO_2 and H_2S in the O_2 (air). Presence of CH_4 in H_2 is regarded to be inert towards the performance of the electrodes. CO_2 itself is also regarded to be inert, but the formation of trace CO by reverse water gas shift reaction (RWGS) at the anode [$CO_2 + H_2 \leftrightarrow CO + H_2O$] due to the presence of CO_2 in H_2 can have the same detrimental effect as in the presence of trace CO in H_2. Consequently, essentially CO_x, sulfur, and NH_3 free H_2 and air stream are needed as feed gases for efficient and durable operation of a PEM fuel cell. A 2005 draft specification of the fuel cell grade hydrogen suggested by the U. S. Department of Energy is provided in Table 1.

Components	Levels
Hydrogen	>99,9 %
CO	~0.1 ppm
CO_2	~5.0 ppm
NH_3	~1 ppm
Non CH_4 hydrocarbons	~100 ppm

Table 1: Suggested specification of H_2 purity for PEM fuel cell.

The high purity H_2 required by a PEM fuel cell must be easily available at the point of location of residential use. One potential solution is to directly produce fuel-cell grade H_2 at the site of the fuel cell by steam reforming of methane. A network of natural gas pipe lines to supply the gas for domestic (heating or cooking), and commercial (heating) applications already exists in the infrastructure of many developed countries. The gas is typically available at a pressure of 2 – 60 psig, and is processed to remove impurities like H_2O, CO_2, S, He, and heavy hydrocarbons.

The conventional method for production of high purity H_2 (99.999+ %) for fuel cell use from natural gas is high temperature (~850 -950 C) catalytic steam-methane reforming (SMR), followed by catalytic water gas shifting (WGS) at ~ 300 – 400 C, and finally purification of the reactor effluent gas to produce pure H_2 at feed gas pressure by employing a multi-step, multi column pressure swing adsorption (PSA) process. The feed natural gas to the process is pre-treated to remove trace S and N impurities if needed. The

PSA process is operated at a near ambient temperature (20 -40 C) by employing physi-sorbents like zeolites, aluminas, and activated carbons for removal of the impurities (H_2O, CO_2, CO, CH_4, N_2) from the H_2 product gas. The feed gas to the PSA system typically contains 15 – 25 % CO_2 + 1 – 4 % CO + 1 – 5 % CH_4 + 0.2% N_2 in H_2 (dry basis). A waste gas containing all of the carbon impurities and un- recovered hydrogen is also produced by the PSA system which is used as fuel in the SMR furnace.

Although, the conventional route has become the state of the art technology and commercially being used for production of ~ 1 to 150 MMSCFD of 99.999+% H_2 from CH_4, there are several unattractive but unavoidable features for scaling down the process for residential fuel cell use (H_2 demand for a 250 KW PEM fuel cell is only ~0.15 MMSCFD). These include (a) operation of the SMR reactor at high temperature, (b) use of a part of the purified H_2 product (8 -25%) to regenerate the PSA adsorbents by purge, which reduces the over-all recovery of H_2 produced by the SMR and WGS reactions, (c) generation of export steam in order to recover the excess heat required for the operation of the relatively low efficiency, radiant SMR reactor, and (d) the fairly complex nature of the process using several unit operations as well as a large footprint.

An alternative process scheme has been developed for production of fuel cell grade H_2 from CH_4 by SMR. It replaces the PSA purification step of the conventional scheme of by a catalytic PROX /SELOX (Preferential/ Selective Oxidation) reactor which selectively oxidizes the residual CO (~ 1 - 4 %) from the WGS reactor effluent gas to CO_2 in presence of H_2 at a moderate temperature of 80 – 200 C. A small quantity of air is added to the PROX reactor feed to supply the oxygen needed for this purpose. The CO level can be reduced to ~ 10 ppm by the PROX concept.

Clearly, a practical PROX catalyst for producing an essentially CO free H_2 for fuel cell application must exhibit ~ 100 % CO conversion in presence of CO_2

and H_2O. Less than 100 % CO selectivity may be acceptable albeit with the loss of some H_2 produced by SMR. The presence of CO_2 in the effluent gas from a PROX reactor, however, can still be the cause of anode deactivation of a PEM fuel cell due to reformation of CO by RWGS reaction.

Recently, several novel adsorptive process concepts called 'sorption enhanced reaction (SER)' have been designed to substantially enhance the performance of SMR and WGS reactors for production of fuel cell grade H_2 from CH_4 by circumventing the thermodynamic limitations of these reactions. The key benefits include:

- Drastically increase the H_2 product purity and conversion in a single unit operation
- Significantly lower the SMR reaction temperature without sacrificing process performance
- Enhance the kinetics of the forward SMR reaction
- Increase the over-all H_2 recovery from the plant
- Reduce the plant foot print by integration of SMR, WGS reactors and the PSA unit of the conventional route in a single unit, and by lowering the temperature of reaction (easier heat management).

These advantages are achieved by applying the Le Chatelier's principle whereby one of the reaction products, CO_2, is selectively removed from the reaction zone at the reaction temperature. An admixture of a reversible CO_2 chemisorbent, which can selectively sorb CO_2 in presence of steam at the reaction temperature, and an SMR catalyst is used in a sorber-reactor for this purpose. The chemisorbent is periodically regenerated for re-use by desorbing the CO_2 using the principles of pressure swing adsorption (PSA) or thermal swing adsorption (TSA) processes.

A fixed bed cyclic SER process employing a CO_2 chemisorbent utilizing surface reaction may be preferred for the residential fuel cell application because of (a) direct production of high purity H_2 by SMR with high conversion of CH_4 to H_2, (b) relatively lower temperature of operation, (c) ease of CO_2 regeneration using conventional principles of pressure or thermal swing adsorption processes , (d) use of steam purge for CO_2 desorption, (e) process compactness, (f) faster chemisorption kinetics, and (g) absence of sorbent transportation. Figure 1 is a conceptual drawing of a generic fixed-bed SER process concept for directly producing fuel cell grade H_2 by low temperature SMR.

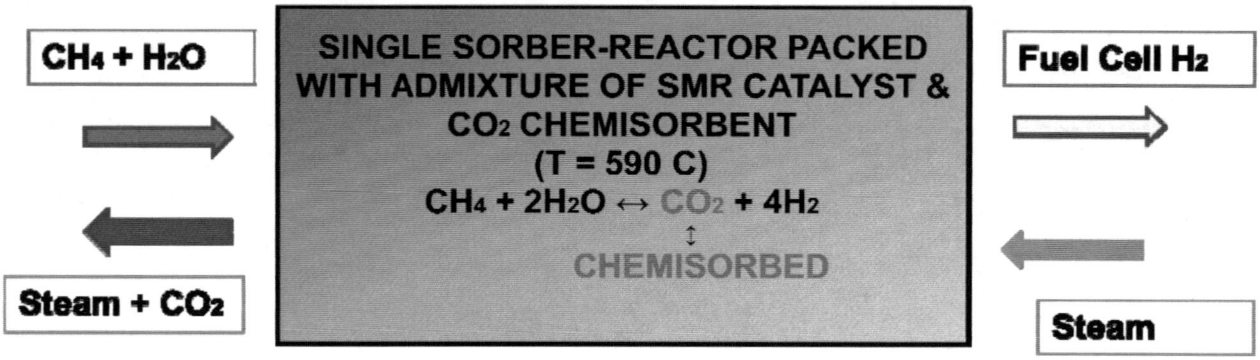

Figure 1: Conceptual sketch of SER SMR process concepts.

Rapid Thermal swing sorption enhanced reaction (RTSSER) process:

A rapid thermal swing sorption enhanced reaction (TSSER) process for low temperature (~ 520 - 590 C) steam methane reforming was recently designed by Sircar and coworkers . The process employed a pair of fixed bed sorber-reactors and it could directly produce a fuel-cell grade H_2 using K_2CO_3 promoted hydrotalcite as the CO_2 chemisorbent in the process. The process uses two cyclic steps:

(a) **sorption-reaction step** where a mixture of H_2O and CH_4 is fed at a pressure of ~ 1.5-2.0 bar and a temperature of ~ 490 C into a fixed-bed reactor, which is packed with an admixture of the SMR catalyst and the chemisorbent, and which is pre-heated to ~ 520 - 590 C. The effluent from the reactor is fuel-cell grade H_2 at feed pressure.

(b) **thermal regeneration step** where the reactor is simultaneously depressurized to near-ambient pressure and counter-currently purged with superheated steam at ambient pressure and at ~ 520 - 590 C, followed by counter-current pressurization of the reactor with steam at ~520 - 590 C to the feed pressure. The reactor effluent is a CO_2 rich waste gas.

The performance of a TSSER process design [sorber-reactor tubes packed with an admixture of a commercial SMR catalyst (10 %) and promoted hydrotalcite (90 %) as a reversible CO_2 chemisorbent]was estimated using a mathematical process model which simulated the operation of the individual steps (10 minutes each) of the process. The equilibrium and column dynamic characteristics of the CO_2 chemisorbent used in the design were experimentally measured in our laboratory and modeled using a novel mechanism of the chemisorptions process developed by us. Figure 2 is a heat integrated flow diagram of a TSSER concept designed for production of hydrogen for a 250 KW residential PEM fuel cell which requires ~ 3 kilo liters of H_2 per minute. The system contains two shell and tube sorber-reactors, heat exchangers, make-up heater and blowers. Each sorber-reactor contains 2665 tubes [2.54 cm ID x 250 cm long, intra tube void fraction = 0.25, each packed with ~ 1.1 kg of an admixture of the SMR catalyst (10%) and CO_2 chemisorbent].The feed (5:1 steam: methane) to the reactor was at 450 C and at a pressure of 1.5 bar. The reaction temperature was 590 C. The cycle time for each step was 10 minutes.

A first pass estimation of the capital and operating costs ($/kg of H_2) of the TSSER process for H_2 production for a 250 KW residential fuel cell indicates that the cost is competitive with the currently estimated cost of distributed production of fuel cell grade H_2 from natural gas.

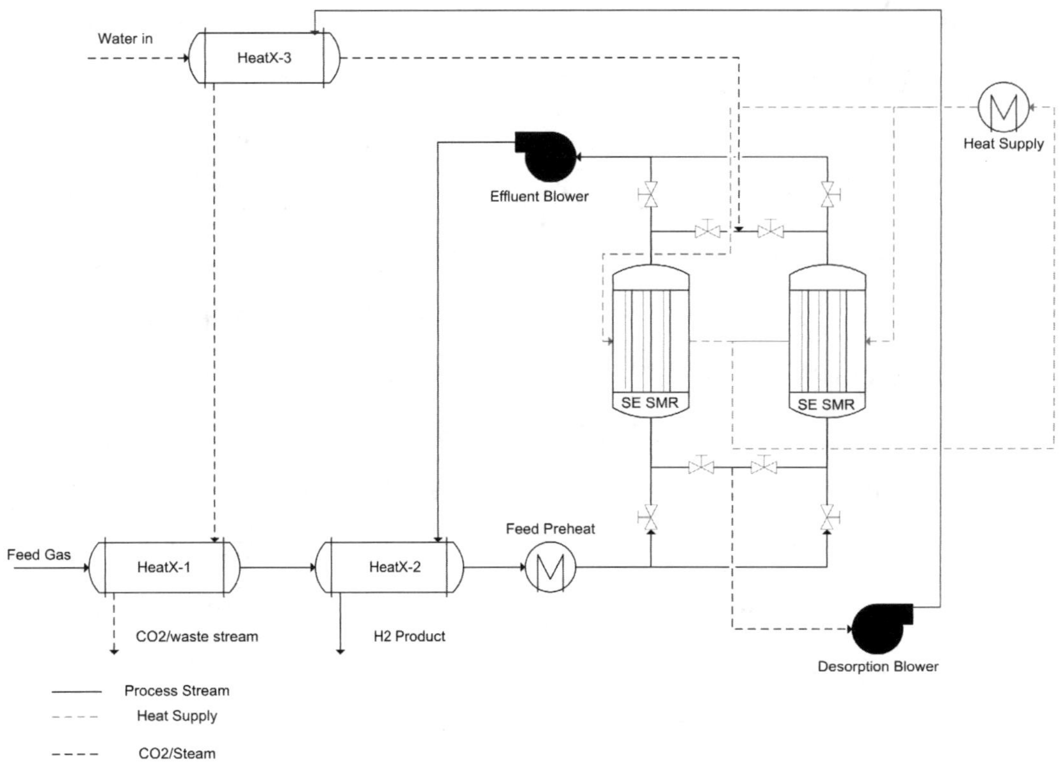

Figure 2: Tentative flow sheet for a RTSSER system supplying H_2 to a 250 KW residential PEM fuel cell.

The TSSER process will potentially provide a relatively simple and very compact alternative for direct production of fuel-cell grade hydrogen by low temperature SMR without producing export steam.

(98e) Preparation of Perovskite-Based Catalysts and Fuel Injection System for Diesel Reforming

Yong-Gun Shul[1], Junki Rhee[1], Jae Choon Yang[2], Sangsun Park[1] and Hyun Khil Shin[2], (1)Chemical and Biomolecular Engineering, Yonsei University, Seoul, South Korea, (2)Fuel Cell Team Value Creation Center, GS Caltex Corporation, Daejeon, South Korea

Autothermal reforming processes of hydrocarbon liquids such as diesel fuels are spotlighted as methods to produce hydrogen for Fuel cell. However, the use of higher hydrocarbons as feedstocks for hydrogen production causes some problems which increase the catalyst deactivation by the carbon deposition. Coking can be inhibited by increasing the water dissociation on the catalyst surface. This results in catastrophic failure of whole system. Performance degradation of diesel autothermal reforming (ATR) leads to increase of undesirable hydrocarbons at reformed gases and subsequently decrease the performance. And it is hard to feed diesel fuel homogeneously because of high evaporation point. Non-homogeneous injection of fuel cause poor performance and fail of system.

In this study, perovskites (ABO3)-based catalysts were investigated as alternatives to substitute the noble metal catalyst for the autothermal reforming (ATR) of diesel fuels. The investigated perovskite structure was based on LaCrO3. And some metals were added at the A-site to enhance oxygen ion mobility, transition metals were doped on the B-site to enhance the reformation.

And we compare peroveskite based only catalysts and physically mixed provskite & noble catalyst(Pt-CGO). Pt-CGO catalyst is novel metal catalyst and it was used widely in the industrial fields. But high cost of novel metal is critical weak point. If the amount of noble catalyst were reduced without the degradation of performance, it can be significant. Substituted Lanthanum chromium perovskites were made by aqueous combustion synthesis, which can produce high surface area conveniently. For LaCrO3, lanthanum nitrate and chromium nitrate were dissolved in water and Ruthenium nitrate solution was added to the solution. Glycine as a fuel was added in the 1:1 molar ratio to the sum of La and Cr. The solution was slowly boiled in a quartz beaker and heated to cause the combustion. The produced oxides were calcined in air at 1250°C for 4hrs.

The physically mixed catalyst were made by milling 24hr in 1:1 wt ratio of LaCrRu & Pt-CGO.

Furthermore, for the homogeneous fuel supply, we made ultrasonic injection system for reforming. The ultrasonic injector made form of fuel to fine particle, so it is possible to get homogeneous mixing condition. We compared durability of evaporation system and ultrasonic system for fuel injection.

See more of this Session: Fuel Processing for Hydrogen Production (I)
See more of this Group/Topical: Topical 2: Advanced Fossil Energy Utilization

(98f) Development of a Structured Graded Catalyst System for the Catalytic Partial Oxidation of Hydrocarbons

Mark Smith[1], David Berry[2], Dushyant Shekhawat[2], Daniel J. Haynes[2] and James J. Spivey[3], (1)U.S. DOE NETL, Morgantown, WV, (2)National Energy Technology Laboratory, Morgantown, WV, (3) Louisiana State University, Baton Rouge, LA

Reforming hydrocarbons into a H_2- and CO-rich gas stream for solid oxide fuel cell transportation applications requires a catalyst system with thermal and mechanical stability and resistance to deactivation in the presence of common contaminants. Liquid fuels considered appropriate for these applications often contain sulfur and aromatic compounds that poison and deactivate catalysts, specifically through carbon formation. Typically, precious metal catalysts like Rh have been used to achieve the desired performance for the CPOX of these hydrocarbons. However, these catalysts are very expensive. This work examines the use of a graded catalyst bed to reduce the amount of precious metal needed to produce the level of performance necessary for commercial viability. Further, this work examines the development of a commercially representative catalyst system by applying the graded catalyst approach to a structured monolith. The CPOX of a diesel surrogate and a biodiesel fuel have been demonstrated previously using a single-composition monolith. This work demonstrates the potential benefit of a multi-composition monolith comprised of Ni-, Ru-, and Rh-substituted oxide catalysts deposited onto alumina foams.

See more of this Session: Fuel Processing for Hydrogen Production (I)
See more of this Group/Topical: Topical 2: Advanced Fossil Energy Utilization

(98g) Diesel Reforming Catalytic Kinetics - Experimental Results and Model Development

Rajesh D. Parmar[1], Dushyant Shekhawat[2], Brant. A. Peppley[1] and Kunal Karan[1], (1)Queen's - RMC Fuel Cell Research Centre, Queen's University, Kingston, ON, Canada, (2)National Energy Technology Laboratory, Morgantown, WV

In regions where electricity is supplied using diesel generators (e.g., Northern territories of Canada) and regions where diesel engines are used as a standby system when there is no electric power from grid (e.g., India, Pakistan), diesel reforming coupled with a Solid Oxide Fuel Cell (SOFC) system is a promising technology to efficiently convert the chemical energy of the fuel into electricity with less pollution issues. The present work will discuss the technical issues related to diesel reforming and research undertaken by SOFC Canada and SECA to address these issues.

Experimental studies have been performed to characterize and screen two catalysts for diesel reforming. Fundamental surface science studies have been carried out using a Diffuse Reflectance Infrared Fourier Transform Spectroscopy (DRIFTS) cell to study surface reactions as well as gas phase species formed during the catalytic reforming of small molecules such as methane and carbon monoxide. A blank reactor study using a bed of inert material was done to characterize the temperature and flow profiles, and a gas phase microkinetic model was implemented in order to define the species reaching the catalyst bed in later experiments. Using the results of the inert run it was possible to deconvolute the surface reaction mechanism from catalytic kinetic studies. The first phase of the development of a Langmuir-Hinshelwood-Hougen-Watson (LHHW) type model based on elementary surface reactions from the previous literature on the catalytic kinetics of hydrocarbon steam reforming will be presented.

See more of this Session: Fuel Processing for Hydrogen Production (I)
See more of this Group/Topical: Topical 2: Advanced Fossil Energy Utilization

(98h) Effect of the Characteristic Dimension of Catalytic Wall Microchannels and Microslits on the Performance of Microreactors Coupling the Methane Steam Reforming and Combustion Reactions: A CFD Simulation Study

Gurutze Arzamendi[1], Irantzu Uriz[1], Pedro M. Dieguez[1], Mario Montes[2], Miguel Angel Centeno[3], José Antonio Odriozola[4] and **Luis M. Gandia**[1], (1)Quimica Aplicada, Universidad Publica de Navarra, Pamplona, Spain, (2)Department of Applied Cehmistry, Faculty of Chemistry, 20018 San Sebastian, Spain, (3)Inorganic Chemistry Department and Materials Science Institute, University of Sevilla-CSIC, Sevilla, Spain, (4)Instituto de Ciencias de Materiales de Sevilla, Centro mixto CSIC-Universidad de Sevilla, 41092 Sevilla, Spain

Effect of the Characteristic Dimension of Catalytic Wall Microchannels and Microslits on the Performance of Microreactors Coupling the Methane Steam Reforming and Combustion Reactions: A CFD Simulation Study

G. Arzamendi[1], I. Uriz[1], P.M. Diéguez[1], M. Montes[2], M.A. Centeno[2], J.A. Odriozola[3], and L.M. Gandía[1,*]

[1]Departamento de Química Aplicada. Universidad P˙ blica de Navarra, Campus de Arrosadía s/n, E-31006 Pamplona. Spain

[2]Grupo de Ingeniería Química, Departamento de Química Aplicada, Facultad de Ciencias Químicas de San Sebasti·n, UPV/EHU, Paseo Manuel de Lardiz·bal 3, 20018 San Sebasti·n, Spain

[3]Instituto de Ciencia de Materiales de Sevilla, Centro Mixto CSIC-Universidad de Sevilla, Avda. Américo Vespucio 49, 41092 Sevilla, Spain

[*] **Corresponding author.** *E-mail address:* **lgandia@unavarra.es (L.M. Gandía)**

Both H_2 and syngas (synthesis gas, a mixture of H_2 and CO) are expected to play an increasingly important role in our energetic system. H_2/syngas production technologies can be integrated in Natural Gas Combined Cycle plants for precombustion CO_2 capture, in advanced systems as the Integrated Gasification Combined Cycle with or without capture of CO_2 for future coal-based power plants and, of course, for synthetic liquid fuels production through the gas-to-liquids (GTL) and coal/biomass-to-liquids (CTL/BTL) processes [1]. Syngas can be used also as fuel for high-temperature solid oxide and molten carbonate fuel cells [2].

Steam reforming of natural gas in conventional packed-bed reactors is the preferred technology for commercial large-scale H_2 and syngas manufacture. However, for small-scale stationary or mobile/portable applications as well as production offshore or in remote areas, microreaction technology offers a convenient solution [3-5].

In this work, a Computational Fluid Dynamics (CFD) simulation study on the thermal integration of the endothermic steam reforming of methane (SRM) and exothermic methane combustion reactions in microstructured devices is presented. In a previous study on this system, the effects on the performance of catalytic wall microchannels of the gas streams space velocities, SRM catalyst loading and steam-to-carbon (S/C) ratio were investigated [6]. In this work, the effect of the characteristic dimension (d = 0.35, 0.70, 1.40 and 2.80 mm) of square microchannels of 20 mm of length on the microrreactor performance is shown. Moreover, microslides are also considered as a new geometry with very different aspect ratio compared with square microchannels. The characteristic dimension of microslides has been also varied between 0.35 and 2.80 mm whereas their length has been kept at 20 mm. CFD models of these geometries have been developed with commercial ANSYSÆ CFX software considering both parallel and cross flow arrangement. It has been assumed that thin layers of typical Ni and Pd catalysts for the SRM and methane combustion reactions, respectively, have been deposited onto the walls of the microchannels or microslides. The loadings were established at 2-4 mg/cm^2 and 1 mg/cm^2 for the Ni and Pd catalysts, respectively. Heterogeneous catalytic reactions (SRM, water-gas shift-WGS, and methane combustion in air) have been modeled considering the inner walls as sources of products and sinks of reactants. Kinetic expressions of the relevant catalytic reactions have been taken from the literature [6]. The feedstreams consisted of methane and steam with S/C = 2 and 2 wt.% methane in air for the SRM and combustion reactions; the inlet temperature was set at 600¬∫C for both streams.

Steady state simulations have evidenced a marked influence of the characteristic dimension of both channels and slides on the microrreactor performance. For example, at GHSV of 10000 h^{-1} the SRM methane outlet conversion decreases according to 99.9, 99, 96 and 87% as the microchannel characteristic dimension increases as 0.35. 0.70, 1.40 and 2.80 mm, respectively. In contrast, the selectivity was almost unaffected, resulting a reformate with a H_2/CO ratio close to 3.8. To take into account the very different surface-to-volume ratio associated to each dimension the performance was compared also on a WHSV basis. Although the differences decreased the performance still improved with the decrease of the characteristic dimension. For example, at WSHS of 13542 h^{-1} the microchannels with d = 0.35 mm yielded 93% methane conversion while those with d = 2.80 mm gave 87%. An analysis of the results has shown that mass transport limitations became an issue for the largest characteristic size as evidenced by concentration profiles as the ones that can be seen in Figure 1. Therefore, decreasing the characteristic dimension has the advantages of a higher surface-to-volume ratio and a lower influence of transport limitations.

When comparing the two geometries considered in this work it is found that the performance of the microchannels is better than that of the microslides; however, the differences are small, of the order of 3-6 percent units of methane conversion at the reactor exit. This is due to the improved heat transfer characteristics and surface-to-volume ratio of microchannels compared with the microslides. Nevertheless, the microslides can be manufactured more easily and at lower cost so they are an interesting option to develop steam methane microreformers.

SHAPE * MERGEFORMAT

Literature cited

1. Wei W, Kulkarni, P, Liu K. in: Hydrogen and Syngas Production and Purification Technologies (Eds.: Liu K, Song C, Subramani V). AIChE - John Wiley & Sons Inc., Hoboken, NJ. 2010, p. 451-485.

2. Song C. in: Hydrogen and Syngas Production and Purification Technologies (Eds.: Liu K, Song C, Subramani V). AIChE - John Wiley & Sons Inc., Hoboken, NJ. 2010, p. 1-13.

3. Tonkovich AY, Perry S, Wang Y, Qiu D, LaPlante T, Rogers WA. Microchannel process technology for compact methane steam reforming. Chem. Eng. Sci. 2004;59:4819-4824.

4. Tonkovich ALY, Yang B, Perry ST, Fitzgerald SP, Wang Y. From seconds to milliseconds to microseconds through tailored microchannel reactor design of a steam methane reformer. Catal. Today 2007;120:21-29.

5. Subramani V, Sharma P, Zhang L, Liu K. in: Hydrogen and Syngas Production and Purification Technologies (Eds.: Liu K, Song C, Subramani V). AIChE - John Wiley & Sons Inc., Hoboken, NJ. 2010, p. 14-126.

6. Arzamendi G, Diéguez PM, Montes M, Odriozola JA, Falabella Sousa-Aguiar E, Gandía LM. Methane steam reforming in a microchannel reactor for GTL intensification: A computational fluid dynamics simulation study. Chem. Eng. J. 2010;154:168-173.

See more of this Session: Fuel Processing for Hydrogen Production (I)
See more of this Group/Topical: Topical 2: Advanced Fossil Energy Utilization

Highly conductive metal honeycomb reactor for methanol autothermal reforming

Dong Hyun Kim, Hyun Chan Lee, Jitae Lee*
Department of Chemical Engineering
Kyungpook National University, Daegu 702-701
Korea

1. Introduction

Hydrogen is economically produced by steam reforming of fossil fuels. Among the fuels, methanol is easy to reform at low temperatures of 200 ~ 300 °C. With methanol, hydrogen can be produced by steam reforming or partial oxidation.

$$CH_3OH + H_2O \rightarrow CO_2 + 3\ H_2,\quad \Delta H_{298\ K} = 49.5\ kJ/mol \qquad (1)$$
$$CH_3OH + ½\ O_2 \rightarrow CO_2 + 2\ H_2,\quad \Delta H_{298\ K} = -192.3\ kJ/mol \qquad (2)$$

Depending on the reaction for hydrogen production, the reaction heat should be either supplied or removed. Instead, the two reactions can be combined to make a reaction having a zero enthalphy change (autothermal reforming).

$$CH_3OH + 0.795\ H_2O + 0.102\ O_2 \rightarrow CO_2 + 2.795\ H_2,\quad \Delta H_{298\ K}= 0\ J/mol \qquad (3)$$

In the reactor, however, the exothermic reaction rate and the endothermic reaction rate are not balanced as the stoichiometry of equation (3) indicates. In the front part of the reactor where oxygen is present, the exothermic partial oxidation dominates and after all the input oxygen is consumed, only the endothermic steam reforming occurs. With a conventional packed-bed reactor, the unbalance in heat generation and consumption gives rise to a hot spot as high as 570 °C due to poor thermal conductivity (~ 1 W/m/K) in the catalyst bed[1]. In order to make the hot spot disappear, it is essential to quickly transfer the reaction heat of partial oxidation to the latter part of the reactor where steam reforming consumes the heat.

For this purpose, we develop in this study a highly conductive metal honeycomb reactor for the autothermal reforming of methanol. The metal structure of the reactor facilitates the heat transfer from the front section to the latter section of the reactor, making the temperature profile smooth and suppressing hot-spot formation. We also develop a reactor model for the honeycomb reactor and analyze the experimental results obtained with an aluminum honeycomb reactor. The design parameters affecting the performance of the honeycomb reactor are identified and the effect of each parameter is described.

2. Experimental

The honeycomb is made of an aluminum block with 34 holes of 2 mm in diameter. The reactor is shown in Fig. 1. A 28 g of 20 at % $CuO\text{-}CeO_2$ (0.36 mm) was packed into the reactor. The reactor outside was covered with ceramic wool to reduce heat loss from the reactor. The premixed liquid reactant (methanol:water=1:1 in mole ratio) was vaporized, mixed with oxygen and fed into the reactor. The gas phase feed flow rates were 600 ml/min methanol, 600 ml/min water and 50- 110 ml/min oxygen.

Fig. 1. Reactor before catalyst packing and assembly.

3. Reactor model

The mass balance equations for methanol and oxygen in the reactor are

$$F_M^0 \frac{dX_M}{dZ} = (r_{POX} + r_{SR}) \quad (4)$$

$$F_O^0 \frac{dX_O}{dZ} = \frac{r_{POX}}{2} \quad (5)$$

where F_M^0 is the superficial molar feed flow rate of methanol, F_O^0 is the superficial molar feed flow rate of oxygen, X_M is the methanol conversion, and X_O is the oxygen conversion. The catalyst particles are packed in each hole. As the diameter of the hole is small (2 mm), the temperature difference between the catalyst particle and the inner wall of the hole would be negligible. Then the energy balance can be written as

$$\frac{A_w k_w}{A}\frac{d^2T}{dZ^2} - \sum_i F_i Cp_i \frac{dT}{dZ} + (r_{POX}(-\Delta H_{POX}) + r_{SR}(-\Delta H_{SR})) = 0 \qquad (6)$$

where T is the temperature of the wall and the flow, A_W is the cross sectional area of the wall, A is the cross sectional area of the honeycomb including the frontal open area and A_W, k_W is the thermal conductivity of the wall, F_i is the superficial molar flow rate of the component i and Cp_i is the heat capacity of the component i, normally a function temperature. At the entrance of the honeycomb, we assume that the heat transferred from the honeycomb by conduction is heating the feed, i.e.

$$-k_e \left.\frac{dT}{dZ}\right|_{Z=0} = \sum_i F_i^0 \int_{T_0}^{T(0)} Cp_i dT \qquad (7)$$

As a result, there is a jump discontinuity in the temperature at $Z = 0$. The condition is analogous to the Danckwerts boundary condition for the inlet of an axially dispersed plug-flow reactor. At the exit of the honeycomb, the condition is

$$\left.\frac{dT}{dZ}\right|_{Z=L} = 0 \qquad (8)$$

4. Results and discussion

4.1. Simulation

In simulation, the kinetics of steam reforming and partial oxidation are needed. The steam reforming kinetics over a $CuO\text{-}CeO_2$ catalyst has been reported[2] , but the kinetics of partial oxidation is still unclear. It is reported to be much faster than the steam reforming[3]. In preliminary simulation of the reactor, we have found that the temperature profile along the reactor as well as the exit methanol conversion is virtually independent of the kinetics of partial oxidation as long as it is faster than the steam reforming.

The honeycomb reactor we consider for simulation is 10 cm×10 cm square in cross section and 15 cm in length. The fraction of cell open area to the total cross sectional area is 0.5. Except when specified, the feed mole ratios are methanol/water=1 and oxygen/methanol=0.13, and the feed temperature is 230 ^{o}C. Fig. 2 shows the simulated temperature profiles along the honeycomb reactor and a conventional packed-bed reactor. For the packed bed, the axial thermal conductivity is set to zero in the simulation. A hot spot as high as 600 ^{o}C is seen to be formed near the inlet, when there is no axial conduction. But in the aluminum honeycomb reactor, the hot spot disappears, and the temperature varies smoothly within 15 ^{o}C along the reactor. This demonstrates the effectiveness of the heat conduction through the metal structure.

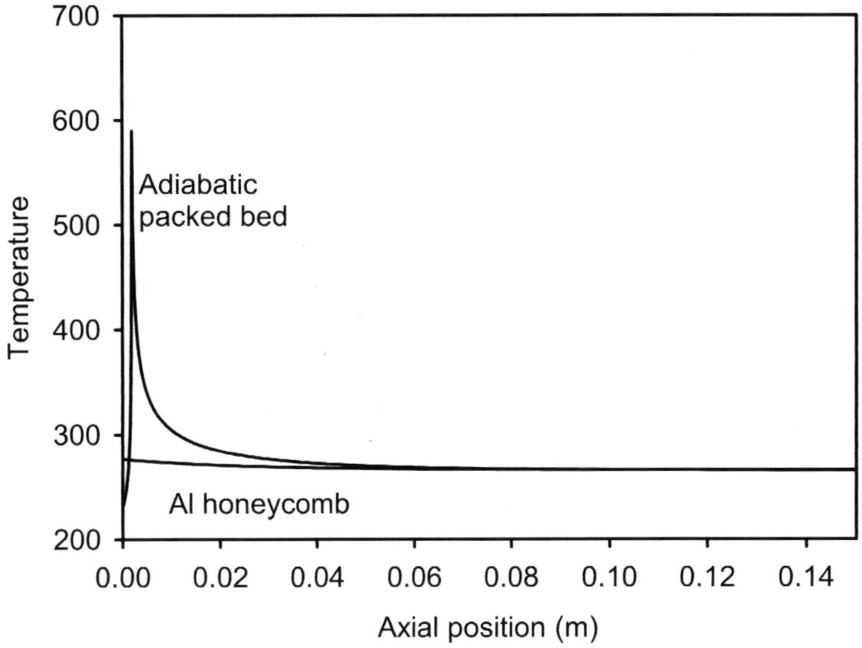

Fig. 2. Temperature profile along the reactor

Fig. 3 shows the effect of the feed flow rate on the conversion, the maximum and the exit temperature of the honeycomb reactor. It is seen that with increasing the flow rate, the conversion decreases, while the maximum temperature as well as the width of the temperature variation along the reactor increases.

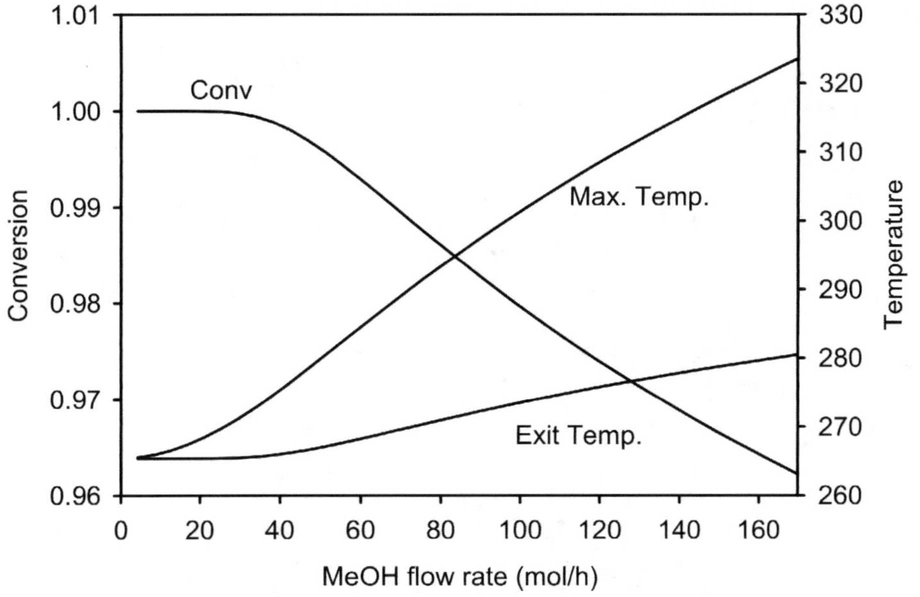

Fig. 3. Effect of feed flow rate on the methanol conversion, the maximum temperature and the exit temperature

The ratio of oxygen to methanol is an important operating variable affecting the performance of the reactor. Its effect is shown in Fig. 4. The methanol conversion increases with increasing oxygen content in the feed. At the same time the maximum and exit reactor temperatures also increase with the oxygen content. It is desirable to keep the oxygen content low while maintaining complete methanol conversion to maximize hydrogen production. Around 13 % in the oxygen/methanol ratio results in the complete methanol conversion for the simulation condition.

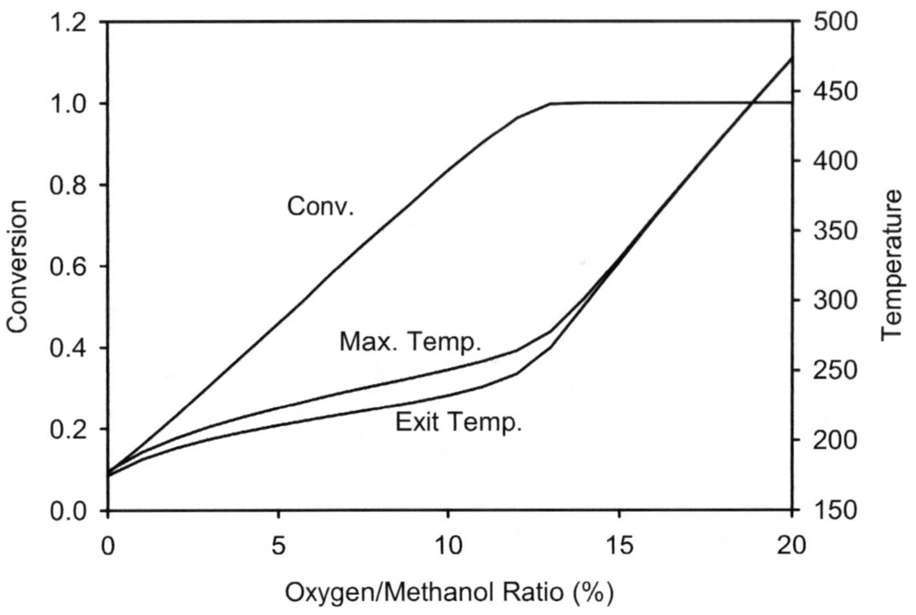

Fig. 4. Effect of oxygen/methanol ratio on the conversion and the reactor temperatures.

The 1500 cm^3 honeycomb reactor can produce 2.6 m^3H_2(STP)/h, which can be used to generate around 3 kwh in a fuel cell.

4.2. Honeycomb experiments

In all the experiments, the measured reactor temperatures at the front, the mid point and the exit of the reactor were within 15 °C along the honeycomb reactor without a hot-spot. This demonstrates that the heat conduction through the structure of the reactor works to eliminate the hot spot and to make the temperature profile smooth. Fig. 5 shows the experimental data (filled circle) for varying the oxygen feed rate and the simulation results (filled square). With increasing oxygen flow rate, the conversion as well as the reactor temperature increase. The agreement between the experiments and the simulation is not satisfactory, probably due to inaccuracy in the kinetics and the inevitable heat loss from the reactor to its surrounding.

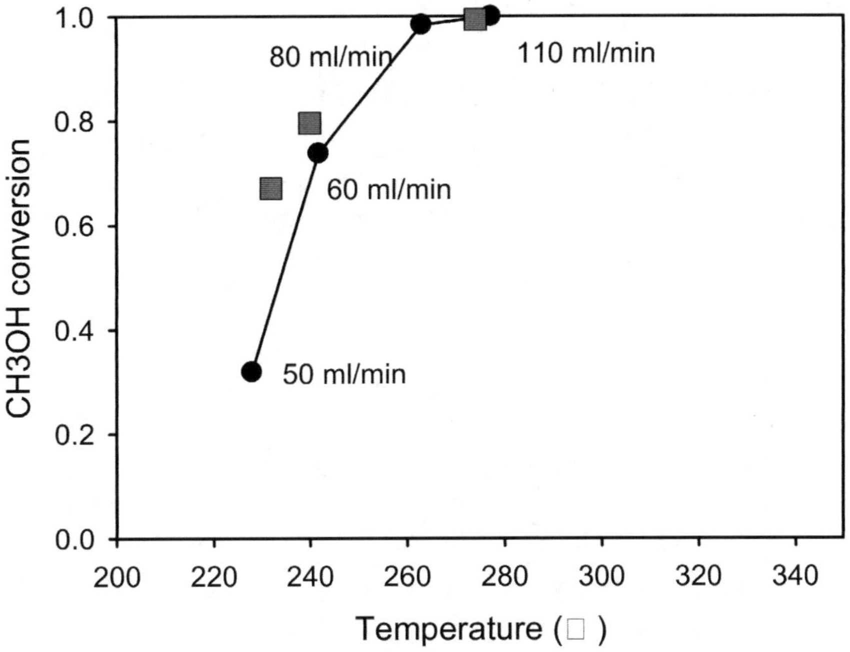

Fig. 5. Effect of oxygen feed flow rate on conversion and average honeycomb reactor temperature.

5. Conclusion

A highly conductive metal honeycomb reactor is developed for methanol autothermal reforming. The reactor facilitates transfer of the heat generated in the front section of the reactor by the partial oxidation of methanol to the latter part of the reactor, where it is used by the steam reforming reaction. Thus a hot spot, which usually forms when the reaction is carried out in a conventional packed-bed reactor and destroys the catalyst, does not occur with the honeycomb reactor. A reactor model for the honeycomb reactor is developed and compared with the experiments. The effects of feed flow rate and oxygen content in the feed on the methanol conversion and reactor temperature are investigated by simulation of the honeycomb reactor with the model.

References

1. J. R. Lattner, M. P. Harold, Catalysis Today 120 (2007) 78-89.

2. J. K. Lee, D. H. Kim, Korean Chem. Eng. Res., 42 (2004) 387-395.

3. L. Alejo, R. Lago, M. A. Pena, J. L. G. Fierro, Appl. Catal. A. 162 (1997) 281-297.

(98j) Methanol Reforming on Structured Catalyst and Microchannel Reactors

Mario Montes[1], Francisco Javier Echave[1], Oihane Sanz[1], Brahim Arhoun[1], Miguel Angel Centeno[2], José Antonio Odriozola[2], Gurutze Arzamendi[3] and Luis M. Gandia[3], (1)Department of Applied Cehmistry, University of the Basque Country (UPV/EHU), 20018 San Sebastian, Spain, (2)Inorganic Chemistry Department and Materials Science Institute, University of Sevilla-CSIC, Sevilla, Spain, (3) Quimica Aplicada, Universidad Publica de Navarra, Pamplona, Spain

Methanol is often considered the best choice of a liquid fuel to generate hydrogen due to its high energy density, low cost and easy transportation. In addition, the reaction of steam reforming of methanol does not require high temperature that results in an easy temperature control. The use of micro-reactors for hydrogen production gains an increasing importance from the point of view of creating a portable and low cost hydrogen source for numerous applications, in particular as feeding for fuel cell. Micro-reactors have the advantages of high surface-to-volume ratio, fast response time, ease integration, and small footprint, which are ideal for portable power systems. In addition, enhanced mass and heat transport properties are also recognized as advantages of micro-reactors.

Catalysts (CuZnAl2O3 or Pd/ZnO) were deposited on the micro-channel walls by the washcoating method. To optimize the coating procedure, preliminary studies were carried out by coating micro-monoliths (presenting the same channels size) of the same alloy, varying different parameters and properties: particle size, solid content, additives, and pH of the slurry. The coating procedure result was evaluated by means of the specific load (mg/cm^2), homogeneity (SEM), adhesion (ultrasound test) and catalytic activity.

The best results were obtained balancing several parameters like the solvent nature, the pH or the additives to improve specific load and adhesion without loosing catalyst activity. Micro-channels blocks showed excellent reforming activity with very low CO production (<1%)

See more of this Session: Fuel Processing for Hydrogen Production (I)
See more of this Group/Topical: Topical 2: Advanced Fossil Energy Utilization

(109a) Prospects for Producing Low Carbon Transportation Fuels From Captured CO_2 In a Climate Constrained World

Thomas G. Kreutz, Princeton Environmental Institute, Princeton University, Princeton, NJ

The climate implications of emerging technologies that capture CO_2 to produce transportation fuels (CCTF) are investigated by studying two examples: biodiesel from microalgae and Sandia National Laboratory's S2P process. Simple performance and economic models for each technology are examined in the context of a bifurcated – "pre-decarbonization" vs. "post-decarbonization" – climate regime in which CCTF uses CO_2 that is captured from, respectively, power plant flue gases vented to the atmosphere or taken from CO_2 in pipelines destined for geologic storage. CCTF promises to improve domestic energy security by converting sunlight and waste CO_2 into transportation fuels; in addition, these fuels are roughly climate neutral when CO_2 is captured from either flue gases or directly from the atmosphere. However, after the power sector becomes largely decarbonized under a stringent climate policy, large point sources of concentrated CO2 are likely to be relatively rare, and unfortunately, fuels made from pipeline CO_2 do *not* have markedly reduced net GHG emissions. Thus, absent the development of economical CO_2 capture from air, it is difficult to see how CCTF can play a significant long term role in decarbonizing the US transportation sector (and thus reaching US climate goals).

See more of this Session: Beneficial Utilization of Carbon Dioxide Separated From Flue Gas
See more of this Group/Topical: Topical 2: Advanced Fossil Energy Utilization

Scoping study of Methanol Production from CO_2 with Zero GHG Electrolysis H_2

Richard D. Doctor
rdoctor@anl.gov; (630) 252-5913
Argonne National Laboratory
9700 S. Cass Avenue
Argonne, Illinois 60439

January 31, 2011

Scoping study of Methanol Production from CO_2 with Zero GHG Electrolysis H_2

Richard D. Doctor, *Argonne National Laboratory*

A scoping study of system costs for producing methanol from CO_2 and H_2 using low-carbon electricity to supply the hydrogen from commercial electrolysis produced methanol via:

$$CO_2 + 3H_2 \rightarrow CH_3OH + H_2O$$

Assuming low-carbon electricity at $75/MW, electrolysis delivers hydrogen at $5/kg and the production of methanol in an $98 MM facility (or $128K per tonne of daily production for an 850,000 tonnes/yr boiler plate — and 800,000 actual tonnes/yr facility) should come to $1,000/tonne or $2.91/gallon of methanol. While process improvements are possible, this base case estimate employs commercially deployed equipment. Current high-CO_2 emission methanol costs about $50/tonne according to Lurgi.

Since methanol is typically considered to have ~60% of the energy value of gasoline in automotive use this comes to $4.64/gallon of gasoline equivalent. Were the methanol then to be converted to either DME or gasoline, the cost per gallon would approximately double, and the attendant CO_2 emissions would also need to be managed.

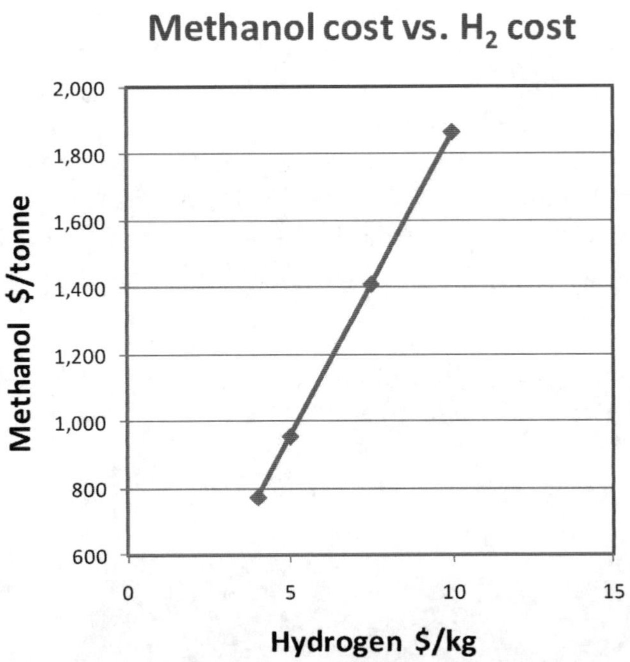

Fig. 1 Hydrogen costs dominates methanol fuel from CO_2 conversion ($1,000/tonne Methanol = $2.91/gallon or approximately $4.85/gallon gasoline equivalent)

It is clear that hydrogen costs dominate the production costs for CO_2 conversion not only for methanol, but for any fuel. A doubling of hydrogen costs to $10/kg increases methanol costs to $1,867/tonne (Fig.1).

As innovation takes place, there are sometimes significant reductions in capital costs. Conversely, periods of high energy costs and high market demands such as the 2002-2007 can lead to marked increases in capital equipment costs. As Fig. 2 shows, methanol production is relatively insensitive to major shifts in capital costs.

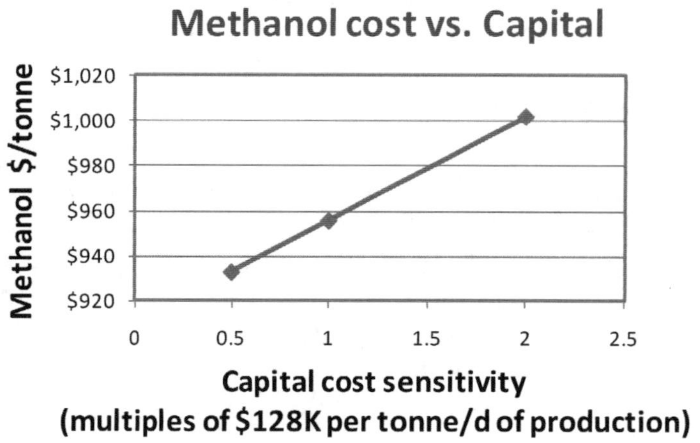

Fig. 2 Methanol fuel from CO_2 conversion is relatively insensitive to capital costs

HYDROGEN PRODUCTION FROM CONVENTIONAL ELECTROLYSIS

Currently, alkaline electrolyzers that convert water to H_2 perform with an efficiency of 70–80% both atmospheric and high pressure systems are available. Costs are comparable. Practical considerations frequently limit performance to the lower end of the efficiency range, with further modest losses from balance-of-plant issues for large-scale systems. One well-established commercial electrolysis system is NorskHydro whose system performance is 4.2 kWh/Nm^3 H_2 = 46.6 kWh/kg H_2 [the theoretical minimum is 39.44 kWh/kg H_2.] Adding in compression, etc. move you up to the range of 50 kWh/kg H_2. A low-pressure cell is shown in Fig. 3.

Fig. 3 Atmospheric pressure electrolyzer (Source: NorskHydro)

NorskHydro electrolyzers were the subject of an earlier design study that used company-supplied capital costs, maintenance, and availabilities to estimate hydrogen costs as a function of electric power costs. Hydrogen costs ranged from \$25/MW electricity = \$2.59/kg H_2 to \$100/MW = \$6.44 kg/ H_2. [Doctor, Richard D., *Flowsheet Optimization of an Integrated High-Temperature Steam Electrolysis System*, Nuclear Hydrogen Initiative: Electrolysis (HiTE) system; CH16EL11 (August 2006)] Further refinement of the cost recovery schedule since the original study would reduce these costs by about 8%. If the hydrogen comes from conventional electrolysis using a NorskHydro system, the anticipated cost of low-carbon electricity becomes the controlling variable for methanol costs (Fig. 4).

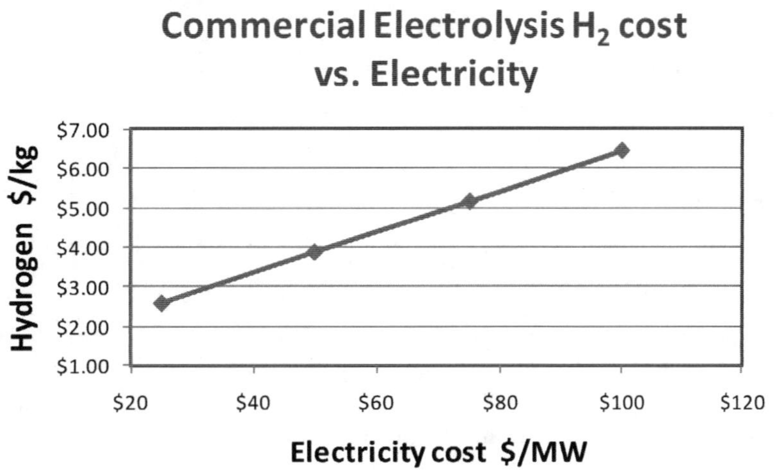

Fig. 4 Costs of Hydrogen from commercial electrolysis

Electrolysis electricity feeding a "merchant plant" is already included in providing for the required H_2 18.2 tonne/h. This is a total of 847.8 $MW_{electric}$

METHANOL PRODUCTION FROM CARBON DIOXIDE

Methanol may be produced in an exothermic reaction from CO_2 and H_2 in a catalyzed reaction:

$$CO_2 + 3H_2 \rightarrow CH_3OH + H_2O \qquad \Delta H_f = -47.16 \text{ kJ/mol}$$

Great Plains reported Cu-Zn on alumina for their system from the 1980s, most of the methanol catalyst literature continues to use this mix. More recently, Zhang et al. [Journal of Natural Gas Chemistry, (16:1) March 2007] investigated the effect of vanadium addition to $Cu/\gamma\text{-}Al_2O_3$ catalyst. They found that the catalytic performance of the Cu-based catalyst improved after V addition. The influence of reaction temperature, space velocity and the molar ratio of H_2 to CO_2 on the performance of $12\%Cu\text{-}6\%V/\gamma\text{-}Al_2O_3$ catalyst were also studied. The results indicated that the best conditions for reaction were as follows: 240 °C, 3600 h^{-1} and a molar ratio of H_2 to CO_2 of 3:1 – although not having some excess of H_2 needs to be studied further.

Typically, commercial systems employ partial oxidation to supply CO as the process feed for methanol production. Catalyst sulfur tolerance is an issue, and this needs to be considered since sulfur may be present in CO_2 recovered from fossil fuel systems. From conventional PC-boilers this will be in the form SO_2, while in gasification systems it will be as H_2S. With proper care, Methanol-synthesis catalysts should be able to last for decades.

METHANOL PRODUCTION CAPITAL COSTS

Methanol production from natural gas Lurgi claims production costs in the range of US$ 50 per tonne and is a well-established technology. A green-field plant located in the methanol complex at Al Jubail in Saudi Arabia was completed in 2003 at a cost of $227 MM. The capacity of the new methanol plant is approximately 850,000t/yr, and it produces 140,000t/yr of carbon monoxide (Fig. 5). The exothermic heat typically is used to raise intermediate pressure steam.

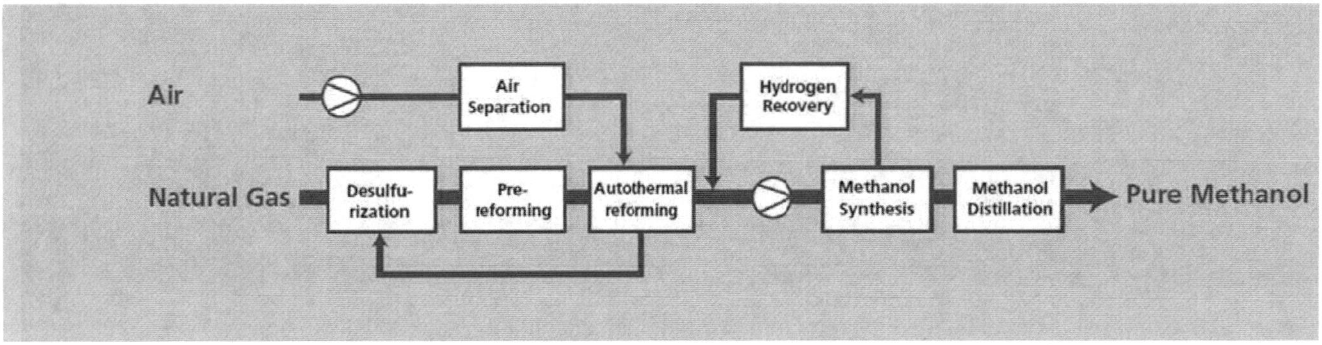

Fig. 5 Commercial methanol production (Source: Lurgi)

If hydrogen is considered as a commodity supplied by a merchant plant, the air-separation and auto-thermal reforming portion of the facility can be eliminated. These are very significant process simplifications that result in deep cost reductions. Eliminating these components would make this system the same as the methanol production unit installed at the Great Plains Coal-gasification plant which scales to $98.4 MM in 2003 dollars for an 850,000t/yr facility (Fig. 6).

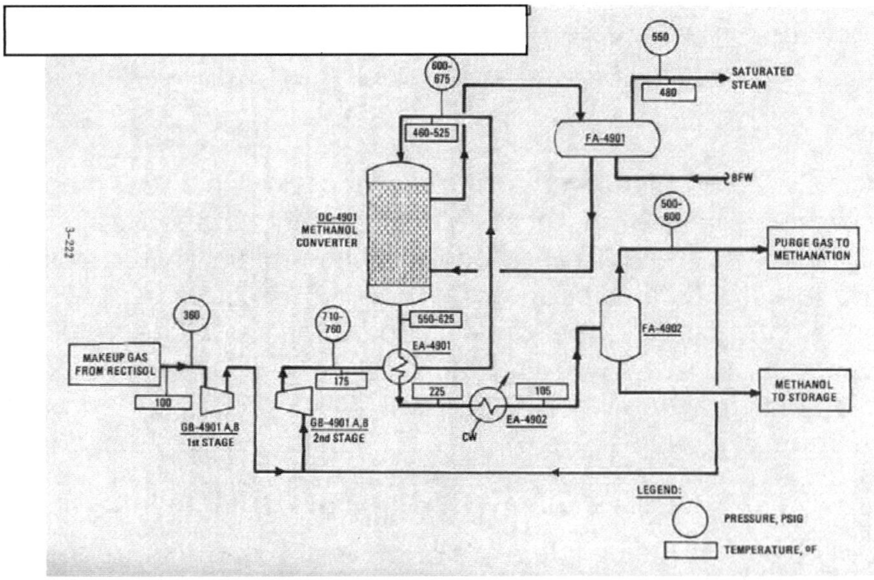

Fig. 6 Commercial methanol production Great Plains (Source: DOE/CH/10088-1874)

These lower capital cost projections were employed in this scoping study. It should be observed that the capital cost sensitivities (Fig. 2) shows that methanol production is relatively insensitive to major shifts in capital costs.

COST ENGINEERING

A cost-engineering linked spreadsheet for this effort has been developed consistent with process design and screening [from Stermole, F.J. and J.M. Stermole, Economic Evaluation and Investment Decision Methods, 9th Ed., Investment Evaluations Corp., Golden, CO (1996) and Gradassi, Michael J., Amoco, Naperville, IL; "Manufacturing Economics of Gas to Liquids," AIChE 2nd International Conference on Refining (1998)]. This approach has been modified by incorporating the Modified Accelerated Cost Recovery Schedule to give a more accurate assessment project returns and as this graphic shows, it follows the yearly revenue stream. The cost sheet below shows an example of this.

The massive increase in capital costs since 2002 markedly impacts project economics. The 20% rate of return assumed here finds that year 6 is revenue neutral (Fig. 7).

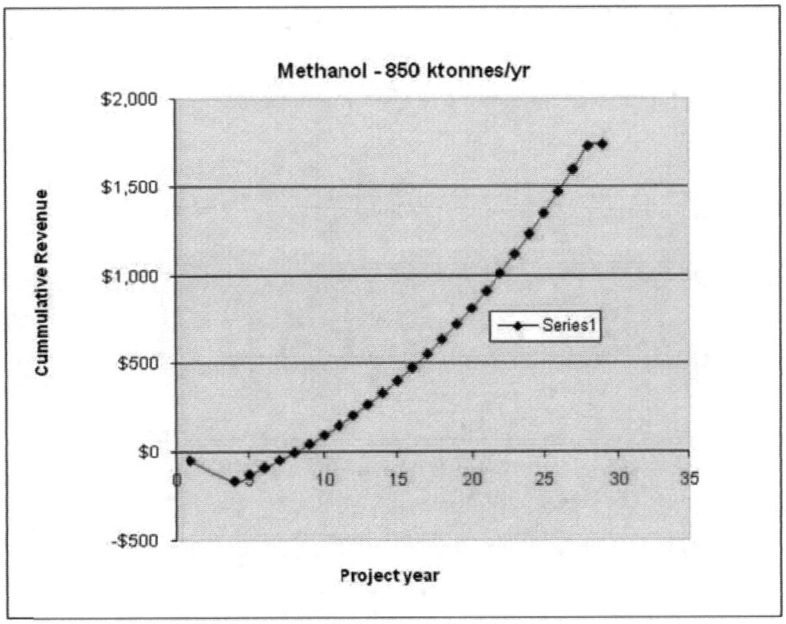

Fig. 7 Facility lifetime revenue for producing methanol from CO_2 and H_2 using low-carbon electricity

Here hydrogen is considered as a commodity supplied by a merchant plant at $5/kg ($5,000/tonne). Taking some credit for CO_2 after the fence at $5/tonne is lost in the noise. No credit was taken for raising steam, however this would not be considered to significantly impact the economics, nor would neglecting the cooling water which typically is used for ~50°C cooling of the condensed methanol water mixture coming off intermediate steam generation.

For this effort, the linked spreadsheet considers:

1.0 PROJECT COST ENGINEERING - SUMMARY OF ASSUMPTIONS
CO_2 to Methanol - 850,000 tonnes/yr

1.1 Project Life - Cash Flow Summary
1.1 Project Life - Cumulative Revenue Graph

2.0 CAPITAL
2.1 Capital Investment — Installed Costs \$98,403,793
Capital Cost Rate \$127,711 \$/thousand tonnes per year capacity

2.2 Capital Basis
Construction 3 year -- Plant construction spending profile 25%; 50%; 25%;
Plant Operations Life 25 year; Salvage 1 year – *Project start 2009*

2.3 Depreciation - MACRS Replacement Equipment (none)
2.3 Depreciation - MACRS Balance of Plant

3.0 CONSUMABLES

3.1	CO_2	133.4 tonne/h	-\$5 \$/tonne
3.2	H_2	18.2 tonne/h	\$5,000 \$/tonne *Electrolysis*

4.0 UTILITIES
4.1 Electric power requirements — *not including electricity for Electrolysis*
7.9 $MW_{electric}$ \$75.00 \$/$MW_{electric}$
Electrolysis already included in H_2 *18.2 tonne/h = 847.8* $MW_{electric}$

4.2 Canal Water (including conditioning) — *not linked into scoping study*

5.0 FIXED AND VARIABLE OPERATING COSTS
5.1 Fixed Operating Costs — Personnel 35; Operator hours per annum 72,327; Burdened labor \$28.00/hr; Overhead and (G&A) 30%

5.2 Variable Operating Costs

CO_2	-\$16,010	\$ 2008-4th Qtr/stream day
H_2	\$2,183,219	\$ 2008-4th Qtr/stream day
Electric power	\$14,252	\$ 2008-4th Qtr/stream day
TOTAL	*\$2,181,461*	*\$ 2008-4th Qtr/stream day*

For Plant On-Stream Factor 100%; VOC – Annual \$796,233,250

5.3 Maintenance Costs — 7.00% ; *Annual \$13,777,000*

6.0 REVENUE STREAM

Year 2013 — *1st year of full operation (as an example)*

Gross Revenue	$851,356,168
Fixed Operating Costs	($3,310,987)
Variable Operating Costs	($823,611,057)
Maintenance	($7,027,270)
Depreciation	($7,836,568)
Salvage	$0
Taxable Income	$9,570,286
-Tax Due	($3,349,600)
+Tax Credit	$0
Net Income	$6,220,686
+Equip Depreciation	$7,836,568
After Tax Cash Flow	*$14,057,254*

7.0 PROJECT LIFE REVENUE Owner's Equity 100% ; General Inflation 2.50%; Working Capital 4.00%; (Federal + State) Effective Tax Rates 35.00%; Plant On-Stream Factor (year 1) 90%; Plant On-Stream Factor (year 2+) 95%

The input set of assumptions for this screening analysis are shown in the summary sheet below.

1.0	**Project Cost Engineering - Summary of Assumptions**		**PROJECT COST ENGINEERING - CO_2 to Methanol - 850,000 tonnes/yr** ***Process Scoping of March 31, 2009***					
	OUTPUT							
	Methanol (99.5% purity)	**850,000**	**tonnes/yr**	**$1,001**	**$/tonne**		2008-4th Qtr	
		97.0	tonne/h	$2.91	$/gallon MeOH			
				$4.85	*~$/gallon gasoline equivalent @ 60% MeOH*			
	INPUT							
3.1	CO_2	133.4	tonne/h	**-$5**	$/tonne		2008-4th Qtr	
3.2	H_2	18.2	tonne/h	**$5,000**	$/tonne	*Electrolysis*	2008-4th Qtr	
4.1	Electric power demand	7.9	$MW_{electric}$	**$69.40**	$\$/MW_{electric}$		2008-4th Qtr	
4.2	Canal Water (including conditioning)	0.0	m^3/h	$0.1035	$/cubic meter		2008-1st Qtr	
	UNIT CAPITAL COST							
2.1	**Installed Costs**			**525.4**	2007-4th Qtr		CEPCI	Baseline
				606.8	2008-3rd Qtr		CEPCI	Current
2.1	Capital Cost Rate	$255,423	$/thousand tonnes per year capacity					
2.1	Plant Construction Period	3	year -- *Plant construction spending profile on reference sheet*					
	Plant Operations Life	25	year					
	Salvage	1	year					
1.1	Project Life (Construction + Operation)	29	year					
7.0	Land surface rights	**$500**	$K					
7.1	Interst during construction (IDC) rate	**5.00%**						
7.1	Salvage value of capital	**3.60%**	percentage of total capital investment					
7.1	Decommissioning costs	**3.50%**	percentage of total capital investment					
5.1	**FIXED OPERATING COSTS**							
5.1	Personnel	35						
5.1	Operator hours per annum	68,466	Burdened labor	**$28.00**	$ 2008-1st Qtr		Overhead and (G&A)	**30%**
	Nelson Farrar - Labor Cost (1980-2007.Nov)	0.501%	0.000001108447					
5.3	Levelized Annual Maintenance	**7.00%**						
5.3	Levelized Annual Maintenance	$13,777	$K					
5.1	Property Tax & Insurance (% fixed capital)	**2%**						
5.2	**VARIABLE OPERATING COSTS**							
	Nelson Farrar - Chemical Cost (1980-2007.Nov)	2.150%	-0.000043246473					
7.1	Plant On-Stream Factor (year 1)	**80%**						
7.1	Plant On-Stream Factor (year 2+)	**94%**						
2.2	**ACCOUNTING SCHEDULE**	25	year - *Operating life*					
2.3	Depreciation - MACRS	21	year - *Balance of plant*					
2.2	Depreciation - MACRS	8	year - *Equipment*					
7.0	Owner's Equity	**100%**						
7.1	General Inflation	**2.50%**						
7.0	Working Capital	**4.00%**						
7.0	(Federal + State) Effective Tax Rates	**35.00%**						
1.1	Internal Rate of Return	**20.0%**						
1.1	Project start year	**2009**						

PROCESS SIMULATION

A process simulation typically should be used as the basis for scoping the overall process economics. Unit level performance should be evaluated by looking at the various system configurations using the latest release of the Aspen Plus® process design software linked when necessary to simulation of complex chemical and electrochemical phenomena in aqueous and mixed solvent environments. Novel configurations may require interfacing with Multi-component species modeling, Computational Fluid Dynamics (CFD) modeling and complex Partial Differential Equation solvers such as COMSOL Multiphysics® at critical stages.

Process Design keeps the focus of the effort on:

- Identification of process rates, reaction kinetics, detailed stream compositions, and equipment sizing so that process optimization may be addressed.
- Identification of novel equipment and consideration of its cost and interface challenges for optimizing performance while assuring that regulatory limits are met.
- Integrating these "state-of-the-art" novel system components with the bulk of "off-the-shelf" components.
- Estimating costs for the system.
- Evaluating opportunities for overall process optimization.

Integrating process design aids the development of more economical production systems by asking the questions that keep the focus of the project on those areas of uncertainty directly impeding commercialization.

(109c) CO_2 to Methanol

Dolly Chitta and Chett J. Boxley, Ceramatec, Salt Lake City, UT

Methanol is a feedstock for scores of chemicals such as formaldehyde, acetic acid and biodiesel. The global demand for methanol in 2008 was 39 MMTPA of which North America consumed approximately 8 million Tons. The proposed research will result into development a novel catalyst to convert CO_2 to methanol. Preliminary economic calculations indicate commercial feasibility of such a schematic and hence a significant market impact is possible. Methanol itself can be converted to gasoline via commercially known processes (e.g. Mobil Methanol-to-Gasoline) process.

Current commercial manufacturing process for methanol utilizes natural gas as feedstock to produce syngas (steam reforming) which is then converted to methanol over a catalyst (Cu-Zn-Alumina). The commercial catalyst is active only in the presence of carbon monoxide.CO_2 activation using existing catalysts is a significant challenge. To convert CO_2 to methanol, new class of catalysts needs to be discovered. The research will focus upon development of novel hierarchical zeolite based promoted catalysts. The primary metal will be selected from Cu, Fe, Co and Ni. Promoters will be chosen from ZrO_2, CeO_2 and other rate earth. The performance of zeolite based catalysts will be compared to that of commercial Cu-Zn-Al_2O_3 based catalyst. The catalyst will advance the state-of-the art in CO_2 conversion.

See more of this Session: Beneficial Utilization of Carbon Dioxide Separated From Flue Gas
See more of this Group/Topical: Topical 2: Advanced Fossil Energy Utilization

(109d) Synthesis of Novel Catalyst for the Trireforming of CO_2-Rich Flue Gases

Tracy J. Benson[1], Jackie W Seaman[1], Juan Cruz[1] and Austin W Prince[2], (1)Chemical Engineering, Lamar University, Beaumont, TX, (2)Electrical Engineering, Lamar University, Beaumont, TX

Today's chemical industry is very complex with most commodity chemicals being produced from some variation of petroleum feedstocks. From transportation fuels to polymers, petroleum is truly the largest source of basic hydrocarbons for which many of the common household products consumers use. Along with this, stems many industrial and environmental issues which must be addressed for long-term stability of the industry. The upgrading of petroleum feedstocks, which are bulky, hydrogen deficient molecules, generally requires vast amounts of hydrogen gas during petroleum upgrading. Hydrogen is typically produced via methane steam reforming reactions. Carbon dioxide, noted for its greenhouse gas potential, is produced as a byproduct of methane steam reforming. CO_2 sequestration is a growing concern and is at the forefront of many political and environmental policies that are being introduced worldwide. One avenue to combat the growing release of CO_2 into the atmosphere is to convert it to useful chemicals via trireforming.

Trireforming is the conversion of carbon dioxide, methane, and water to produce synthesis gas (syngas). The uniqueness of the trireforming technology is that bulk flue gases can be utilized without the separation of CO2 and steam. Only the SOx's should be scrubbed before processing, greatly reducing separation costs. The problems associated with the conversion of CO_2 are overcoming thermodynamic limitations, as carbon dioxide tends to be a very stable molecule. However, through careful catalyst design, a novel catalyst has been synthesized that operates autothermically, neither exothermic nor endothermic, to decrease trireforming processing costs. This dual functioning catalyst is composed of nickel nano-sized particles that promote CO_2 conversion in a semi-oxidative environment. The catalyst, synthesized using a reverse micelle technique, is composed of an MgO – TiO support, which provides for adequate coverage of the active metal and catalyst robustness.

Characterizations of the catalyst have been performed using Scanning Electron Microscopy, X-ray Fluorescence, X-ray Diffraction, and Temperature Programmed Reaction (and Desorption). Initial testing of the trireforming catalyst has indicated that conversions of 90 and 100% of CO_2 and methane, respectively, are possible at reaction temperatures of 750°C. These findings and a postulated reaction site mechanism will be presented.

See more of this Session: Beneficial Utilization of Carbon Dioxide Separated From Flue Gas
See more of this Group/Topical: Topical 2: Advanced Fossil Energy Utilization

(109e) Influence of Carbon Dioxide on the Ignition Temperature of Coal

Ali S. Rangwala, Fire Protection Engineering, Worcester Polytechnic Institute, Worcester, MA and Evan J. Granite, National Energy Technology Laboratory, Pittsburgh, PA

The purpose of this study is to analyze the influence of carbon dioxide on the auto- ignition temperature of coal. It is shown that the minimum ignition energy necessary to ignite a given thickness of coal increases with addition of carbon dioxide locally near the ignition zone. The results are used to illustrate the feasibility of using carbon dioxide or a mixture of carbon dioxide and water vapor towards control and suppression of underground coal fires. Underground coal fires represent a substantial nuisance in terms of property damage, public safety, greenhouse gas generation, and coal resource destruction. Some of the available techniques employed to combat these fires are compared with the current approach.

See more of this Session: Beneficial Utilization of Carbon Dioxide Separated From Flue Gas
See more of this Group/Topical: Topical 2: Advanced Fossil Energy Utilization

(109f) The $10 Million - Carbon Capture and Recycling (CCR) Prize

Matt Peak, Prize Capital, LLC, Pasadena, CA

Prize Capital is working with the X PRIZE Foundation, St. Louis Science Center and two major utilities to create a high profile international competition focused on developing and deploying viable technologies that can cost-effectively recycle carbon emissions from coal-based power plants into brilliant new products such as fuel and building materials.

Innovators will compete for a $10 million prize and an advanced market commitment to deploy at full scale a 400MW coal-based power plant. Through the course of the competition, these innovators will have access to land and resources adjacent to a real world power plant where they can deploy, plug-in, and prove their technologies. They also have the potential to receive capital through the patent-pending Prize Capital funding mechanism, which provides financing to entrants for IP protection as well as tag-along investments.

The CCR Prize will offer unparalleled leverage to create radical breakthroughs as it inspires innovators to develop deployable approaches to solving a global problem.

See more of this Session: Beneficial Utilization of Carbon Dioxide Separated From Flue Gas
See more of this Group/Topical: Topical 2: Advanced Fossil Energy Utilization

Dimethyl ether and methanol steam reforming to hydrogen-rich gas over $CuO-CeO_2/\gamma-Al_2O_3$ catalysts

S.D. Badmaev[1,2], G.G. Volkova[1], V.D. Belyaev[1,2], V.A. Sobyanin [1,2]

[1] *Boreskov Institute of Catalysis, 630090, Novosibirsk, Russia*

[2] *Novosibirsk State University, 630090, Novosibirsk, Russia*

sukhe@catalysis.ru

Abstract

Methanol and dimethyl ether steam reforming to hydrogen-rich gas were studied over $CuO-CeO_2/\gamma-Al_2O_3$ catalysts. The catalysts were shown to provide complete conversion of DME and methanol to the hydrogen-rich gas containing <2 vol.% CO at 300-350°C. The hydrogen production rate in these reactions reached 600 mmol $g^{-1}h^{-1}$. XRD, HRTEM, EDX and FTIR were used for catalysts characterization. High activity of catalysts may be explained by assuming that DME dehydration occurs on the acid sites of $\gamma-Al_2O_3$ and methanol steam reforming takes place on mixed oxide phase $CuO-CeO_2$, solid solution of copper ions in cerium dioxide.

1. Introduction

Currently, fuel cells are considered as an alternative environmentally sound source of electric power. The most of fuel cells are fed by pure hydrogen or hydrogen-rich gas mixtures produced by catalytic conversion of hydrocarbons. Dimethyl ether (DME) and methanol are among the most promising primary sources of hydrogen. DME and methanol are synthetic materials. Comparative analysis of the reactions of DME and methanol steam reforming (temperature range, product distribution) seems highly reasonable in order to determine their promises for fuel cell applications and selection of appropriate sources for hydrogen-rich gas production. Both DME and methanol can be easily and selectively converted to hydrogen-rich gas at relatively low temperature (250-350°C) compared to other fuels such as natural gas, gasoline and LPG [1-3].

Overall DME and methanol SR are expressed by equations:

$CH_3OH + H_2O = 3H_2 + CO_2$ (1)

$(CH_3)_2O + 3H_2O = 6H_2 + 2CO_2$ (2)

It is generally assumed that DME SR proceeds via consecutive two-step reaction mechanism [4].

$CH_3OCH_3 + H_2O = 2CH_3OH$ (3)

$CH_3OH + H_2O = 3H_2 + CO_2$ (4)

The first step of DME SR is hydration of DME into methanol; the second step is methanol steam reforming into hydrogen-rich gas. Besides, during methanol and DME SR, a reverse water gas shift reaction may occur to produce carbon monoxide:

$CO_2 + H_2 = CO + H_2O$ (5)

A design of active and selective catalysts for the DME SR requires knowledge on the properties of both Cu-containing catalysts for methanol SR and solid acids for the DME hydration to methanol. The present work reports the studies of methanol and DME SR to hydrogen rich gas over bifunctional $CuO-CeO_2/\gamma-Al_2O_3$ catalysts [5-7].

2. Experimental

2.1. Catalyst preparation

The Cu-Ce/γ-Al_2O_3 catalysts were synthesized by treating γ-alumina in solutions of copper and cerium salts taken at the given ratio. Samples were dried at 100°C and calcined at 450°C for 3 h.

2.2. Reaction experiments

The reactions of methanol and DME SR were carried out in a U-shaped fixed-bed continuous-flow quartz reactor (i.d. 4 mm) at atmospheric pressure. Reagents and products in all experiments were analyzed by two-on-line gas chromatographs (GC) "Tsvet-500" equipped with TCD. DME, water, methanol, CO_2 and HCHO were separated using a Porapack T column operated at 115°C; and H_2, N_2, CO and CH_4 were separated on a molecular sieve column operated at ambience temperature. Argon was used as a carrier gas.

Experiments on DME and methanol SR were carried out over Cu-Ce/γ-Al_2O_3 catalyst. The catalyst was pre-reduced *in situ* at 300°C for 1 h using 5% H_2 in N_2 before beginning the reaction. Then the temperature was lowered to 473K in flowing H_2 and the catalyst was exposed to a feeding gas mixture composed of 20% DME, 60% H_2O and 20% N_2 (40% methanol, 40% H_2O and 20% N_2) at GHSV = 10000 h^{-1}.

Methanol and DME conversion were calculated as:

$$X_{methanol}(\%) = \frac{C^{in}_{methanol} - C^{out}_{methanol} \times \frac{C^{in}_{N2}}{C^{out}_{N2}}}{C^{in}_{methanol}} \times 100$$

$$X_{DME}(\%) = \frac{C^{in}_{DME} - C^{out}_{DME} \times \frac{C^{in}_{N2}}{C^{out}_{N2}}}{C^{in}_{DME}} \times 100$$

where C^{in}, C^{out} are inlet and outlet concentrations, respectively.

2.3. X-ray analysis

HRTEM images were obtained on a JEM-2010 electron microscope (JEOL, Japan) with a lattice-fringe resolution of 0.14 nm at an accelerating voltage of 200 kV. The high-resolution images of periodic structures were analyzed by the Fourier method. Local energy-dispersive X-ray analysis (EDXA) was carried out on an EDAX spectrometer (EDAX Co.) fitted with a Si (Li) detector with a resolution of 130 eV.

2.4. FTIR spectroscopy

FTIR spectroscopy (Shimadzu FTIR-8300 spectrometer) was applied to determine the acidity of the solid-acid catalysts by monitoring the CO adsorption (a.b. 2150-2175 cm^{-1} for Brønsted acid sites (BAS), a.b. 2175-2240 cm^{-1} for Lewis acid sites (LAS)) according to procedure described in [8]. The procedure allows determining both amount and strength of BAS and LAS. The BAS strength is characterized by proton affinity (PA, kJ/mol), i.e. by the energy of proton releasing; strength of LAS is characterized by the heat of CO adsorption (Q_{co}, kJ/mol).

2.5. XRD

XRD studies were carried out on a diffractometer URD-63 with CuK_α radiation using graphite monochromator with reflected beam. The lattice parameters and dispersion of CeO_2 were determined from line (111), $\Delta\alpha \pm 0.003$ Å.

3. Results and discussion

3.1. Catalytic activity measurements

Fig. 1 demonstrates the temperature dependencies of the methanol (a) and DME (b) conversions and of the outlet H_2, CO_2 and CO concentrations over $CuO–CeO_2/\gamma-Al_2O_3$ catalysts. Fig. 1 presents also the temperature dependencies of respective equilibrium values. Detail thermodynamic calculations are described in [1, 6]. The main reaction products at temperatures 200–370°C were H_2, CO_2 and CO. The CO concentration in the hydrogen-rich gas was considerably lower than the calculated equilibrium value.

As is seen from Fig. 1a, the methanol conversion increases with the temperature and approaches the equilibrium value at 275-300°C. The outlet H_2, CO_2 and CO concentrations increased with increasing temperature. The concentrations of H_2 and CO_2 are close to the equilibrium, whereas the CO concentration in the whole temperature interval was below equilibrium and equaled insignificant value of ~2 vol.% at 300°C. This occurred most likely due to reaction (5) which didn't attain equilibrium under the experimental conditions.

As Fig. 1b shows, the DME conversion and the outlet H_2, CO_2 and CO concentrations also increased with increasing temperature. The DME conversion attained the equilibrium value of 100% at 350 °C. The H_2 and CO_2 concentrations exceeded respective equilibrium values, while the CO concentration was much below the equilibrium value. As the temperature increases up to 370 °C, the DME conversion and the CO_2 and CO concentrations remain the same; the H_2 concentration decreases below the equilibrium value due to methane formation by the reaction of carbon oxides hydrogenation.

Note that the presence of both copper and ceria is of key importance for getting the stable catalysts for methanol and DME SR. Special experiments showed that no DME SR to hydrogen-rich gas proceeds on $\gamma-Al_2O_3$ and $CeO_2/\gamma-Al_2O_3$ at temperature up to 400 °C. The $CuO/\gamma-Al_2O_3$ catalysts are unstable under the DME SR conditions; the hydrogen productivity of these catalysts decreases with time and becomes half as much just in 3h on-stream (Fig.2).

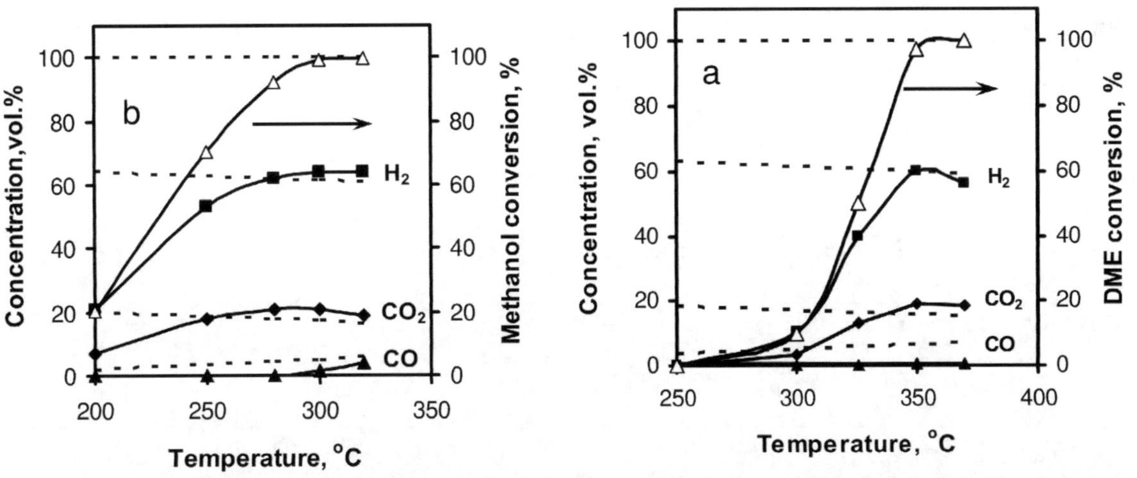

Fig. 1. Effect of temperature on DME/methanol conversion and H_2, CO_2, CO concentrations in methanol (a) and DME (b) steam reforming over $CuO-CeO_2/\gamma-Al_2O_3$ catalyst. Solid lines – experiments, dotted lines – thermodynamic equilibrium values. Reaction conditions: P = 1 atm, GHSV = 10000 h^{-1}; $CH_3OH:H_2O:N_2$ = 2:2:1 ($CH_3OCH_3:H_2O:N_2$ = 1:3:1).

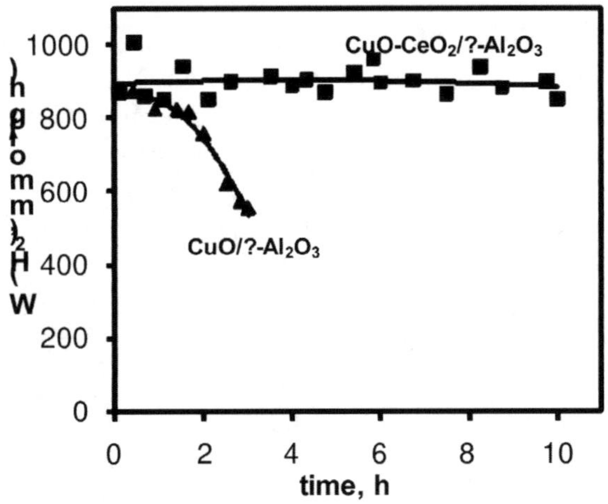

Fig. 2. Effect of CeO_2 on the stability of $CuO\text{-}CeO_2/\gamma\text{-}Al_2O_3$ catalyst for DME SR.

Reaction conditions: T = 350°C, P = 1 atm, GHSV = 30000 h^{-1}; $CH_3OCH_3{:}H_2O{:}N_2$ = 1:3:1.

The results obtained prove $CuO\text{-}CeO_2/\gamma\text{-}Al_2O_3$ to be efficient catalyst for methanol and DME SR to hydrogen-rich gas with low CO content. As shown in Table 1, the $CuO\text{-}CeO_2/\gamma\text{-}Al_2O_3$ catalyst is more efficient for DME SR than the other samples [9, 10]. With this catalyst, the hydrogen production rate reached 600 mmol $H_2/(g_{cat}\cdot h)$ at 350°C, GHSV = 10000 h^{-1}. Besides, as Fig. 2 shows, the catalyst provided good stability at high temperature 350°C – no catalyst decay was observed during 10 h on stream.

Essential advantage is that DME and methanol SR over $CuO\text{-}CeO_2/\gamma\text{-}Al_2O_3$ catalysts produces hydrogen-rich gas containing ~2 vol.% CO, that can be used directly for fuelling, for example, high-temperature polymer electrolyte membrane fuel cells (HT PEM FC) which tolerate up to 5 vol.% of CO in the feed gas. For fuelling low-temperature PEM FC, it is needed only to clean up the hydrogen-rich gas from CO to the level of 10 ppm. Since no individual WGS stage, that is necessary in most other hydrogen production processes, is involved here, the miniaturization of the fuel processors for hydrogen production by methanol or DME SR is quite feasible.

Table 1. Comparison of catalyst activities in DME SR

Catalyst	GHSV, h^{-1}	DME conv., %	W(H2), mmol/(g_{cat} h)	Refs.
$Cu\text{-}Ce/\gamma\text{-}Al_2O_3$	10000	70-100	470-610	Present work
$Cu\text{-}Zn/\gamma\text{-}Al_2O_3$	3600	58	50	[9]
$Cu\text{-}Zn/\gamma\text{-}Al_2O_3$	1700	100	60-80	[10]

3.2. Catalyst characterization

As supposed earlier, the activity and stability of bifunctional $CuO–CeO_2/\gamma\text{-}Al_2O_3$ catalyst for DME SR are most likely associated with the $\gamma\text{-}Al_2O_3$ acidic centers and with the properties of copper, which are responsible for, respectively, DME hydration reaction and methanol SR reaction. XRD, HRTEM, EDX and FTIR were used to specify the kind of active site on the $CuO–CeO_2/\gamma\text{-}Al_2O_3$ catalysts.

Results of FTIR and XRD for $CuO–CeO_2/\gamma\text{-}Al_2O_3$ catalysts are presented in Table 2. Only Lewis acid sites (with Q_{co} = 30 kJ/mol) typical for γ-alumina were registered on the surface of all $CuO–CeO_2/\gamma\text{-}Al_2O_3$ catalysts.

It can be seen from Table 2 that the unit cell parameter of CeO_2 decreased from 5.417 Å to 5.402-5.378 Å for $CuO–CeO_2/\gamma\text{-}Al_2O_3$. It indicates the formation of the solid solution of copper-cerium oxides. TEM observations presented in Fig. 3 show the formation of ~50 nm aggregates of mixed $CuO\text{-}CeO_2$ oxide. It means that $CuO\text{-}CeO_2$ solid solution is composed of elementary particles (35-42 Å) agglomerated in large particles of ~50 nm in size [6]. These particles of mixed $CuO\text{-}CeO_2$ oxide are located within the large pores of γ-alumina.

Our aim was to provide the formation of the $CuO\text{-}CeO_2$ solid solution on the surface of alumina and at the same time to prevent the blocking of alumina acid sites. We have succeeded for the first time in design of bifunctional catalysts for DME SR with two types of active sites on the surface of alumina: solid solution of copper-cerium oxides and acid sites.

Table 2. Characterization of $CuO–CeO_2/\gamma\text{-}Al_2O_3$ by FTIR and XRD

Catalyst	Lewis acid sites		$a(CeO_2)$, Å
	Q_{co}, kJ/mol	N, μmol/g	
5% CuO – 5% $CeO_2/\gamma\text{-}Al_2O_3$	30	110	5.381
10%CuO – 2.5% $CeO_2/\gamma\text{-}Al_2O_3$	30	70	5.378
15%CuO – 5% $CeO_2/\gamma\text{-}Al_2O_3$	30	50	5.402
12.5 $CeO_2/\gamma\text{-}Al_2O_3$	n.d.	n.d.	5.417
CeO_2	-	-	5.416

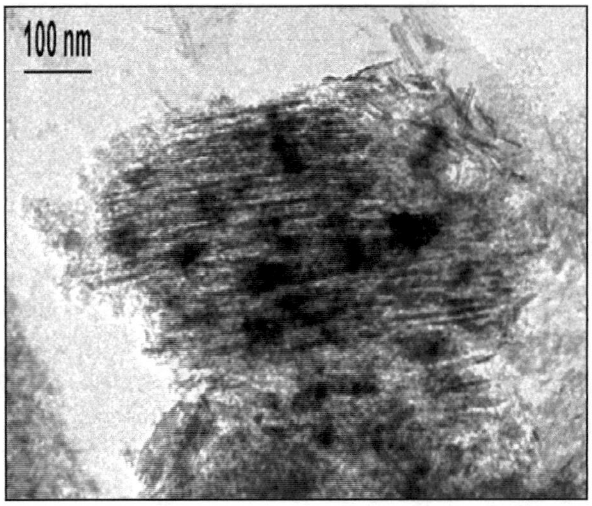

Fig. 3. TEM micrograph of $CuO–CeO_2/\gamma-Al_2O_3$ catalyst.

4. Conclusions

Highly active and selective bifunctional $CuO–CeO_2/\gamma-Al_2O_3$ catalysts for methanol and dimethyl ether steam reforming were developed. Two types of active sites were registered on the surface of these catalysts: Lewis acid sites for DME hydration into methanol and $CuO–CeO_2$ solid solutions for steam reforming of methanol to hydrogen-rich gas.

Acknowledgements

The work supported by Government Contract № P960 in the framework of Federal Targeted Program «Scientific and Research-Educational Staff for Innovative Russia 2009-2013».

References

1. V.A. Sobyanin, S. Cavallaro, S. Freni, Energy Fuels 14 (2000) 1139.
2. T.H. Fleisch, R.A. Sills, M.D. Briscoe, J. Natural Gas Chem., 11 (2002) 1.
3. International DME Association Website: www.aboutdme.org
4. V.V. Galvita, G.L. Semin, V.D. Belyaev, T.M. Yurieva, V.A. Sobyanin, Appl. Catal. A: Gen. 216 (2001) 85.
5. S.D. Badmaev, G.G. Volkova, V.D. Belyaev, L.M. Plyasova, V.A. Sobyanin, Rassian Patent 2 286 210 (2006). (in Russian).
6. G.Volkova, S. Badmaev, V.Belyaev, L. Plyasova, A. Budneva, E. Paukshtis, V. Zaikovsky, V. Sobyanin, Stud. Sur. Sci. Catal. 167 (2007) 445.
7. S.D. Badmaev, P.V. Snytnikov, Int. J Hydrogen energy 33 (2008) 3026.
8. E.A. Paukshtis, IR spectroscopy for heterogeneous acid-base catalysis. Nauka, Novosibirsk, 1992, 253 p. (in Russian).
9. T.A. Semelsberger et.al, Appl. Catal. A: Gen.309 (2006) 210
10. N. Kenji et.al,, Japan Patent 2003-38957 (2003)

(118b) Role of Support Vacancies on the Reactivity of Gold Catalysts In the CO-PROX Reaction

Oscar H. Laguna, Willinton Y. Hernández, María Isabel Domínguez, Francisca Romero-Sarria, Svetlana Ivanova, Miguel Angel Centeno and **José Antonio Odriozola**, Inorganic Chemistry Department and Materials Science Institute, University of Sevilla-CSIC, Sevilla, Spain

The search for renewable energy sources that can replace fossil fuels has driven the interest in CO abatement reactions. Hydrogen production through either reforming fossil or renewable fuels results in mixtures of H_2 and carbon oxides as main products. For fueling PEM fuel cells hydrogen must contain ultralow levels of carbon monoxide (typically bellow 10 ppm) to prevent poisoning of the Pt-based PEMFC catalyst. The preferential oxidation with air of the pre-cleaned reformate (PROX) is a cheap and effective solution, since the range of working temperature matches the one at which PEM fuel cells operate [1].

Although there is no agreement in the literature on which is the active gold species in the CO oxidation reaction, it is well established that small gold NPs (2-3 nm) deposited on reducible supports like TiO2 or CeO2 are more active than when supported on non-reducible supports [2]. We have previously shown that the creation of oxygen vacancies in CeO2, results in a decreasing of the average size of the gold particles deposed on it. [3] We have also shown that the support structural defects play a key role on the activity of gold catalysts in the oxidation of CO, being the catalytic activity a function of defect concentration [3-5]. Therefore, the modification of the reducible oxides network with different cations will result in modified defect structures that will depend on the nature of the modifier. The study of the effect of the interaction of these modified supports with metallic gold particles and their influence in the catalytic activity in the PROX reaction will allow understanding the synergy in these systems.

The incorporation of yttrium into the titania lattice, that is accompanied by creation of oxygen vacancies, favors the conversion of CO at low temperatures. Our results show that the amount of dopant cations introduced into the titania controls the number of surface oxygen vacancies created and, as a result, the gold particle size, directly influencing the catalytic activity; and the band gap of the resulting solid, existing a linear relationship between the catalytic activity and the band gap values found for the studied solids. These results agrees fairly well with theoretical data on these Systems [6].

On the other hand, the modification of CeO_2 with three doping metallic elements Zr, Zn and Fe, resulted in materials whose structure depends on the nature of the modifier. Solid solutions are formed on doping with Zr and Fe cations, whereas ZnO segregation is observed on doping with Zn. Whatever the case an enhancement of the reducibility or the oxygen exchange ability is observed. However, oxygen vacancies depends on the nature of the modifier, being enhanced or decreased the Lumber of vacancias on the nature of the dopant cation. Nevertheless, a high dispersion of gold nanoparticles is achieved in all cases. These results allow stating the formation of oxygen vacancies is not the only way to increase the dispersion of gold nanoparticles. In general, surfaces sites with increased electronic density act in a similar way that oxygen vacancies, being sites for the preferential deposition of the gold nanoparticles.

References

[1] F. Mariño, G. Baronetti, M. Laborde, N. Bion, A. Le Valant, F. Epron, D. Duprez, Int. J. Hydrogen Energy 33, 2008, 1345.

[2] M. Schubert, S. Hackenberg, A.C. van Veen, M. Muhler, V. Plzak, R.J. Behm, J. Catal. **197**, 2001, 113.

[3] F. Romero-Sarria, L.M. Martìnez, M.A. Centeno, J.A. Odriozola, J. Phys. Chem. C **111**, 2007, 14469.

[4] W.Y. Hernández, F. Romero-Sarria, M.A. Centeno, J.A. Odriozola, J. Phys. Chem. C 114 (2010) 10857–10865. 870

[5] M.I. Domínguez, F. Romero-Sarria, M.A. Centeno, J.A. Odriozola, Appl. Catal. B 87, 2009, 245.

[6] J.J. Plata, A.M. Márquez, J. Fdez. Sanz, R. Sánchez Avellaneda , F. Romero-Sarria, M.I. Domínguez, M.A. Centeno, J.A. Odriozola, Topics Catal. In the press

See more of this Session: Fuel Processing for Hydrogen Production (II)
See more of this Group/Topical: Topical 2: Advanced Fossil Energy Utilization

(118c) Development of Membranes for Hydrogen Production From Coal

T. H. Lee, Y Lu, C. Y. Park, S. E. Dorris and U (Balu) Balachandran, Energy Systems Division, Argonne National Laboratory, Argonne, IL

We are developing dense hydrogen transport membranes (HTMs) for separating hydrogen from product streams generated during coal gasification. Among the various HTMs we developed, membranes made of 50 vol.% Pd with Y_2O_3-stabilized ZrO_2 have been extensively studied. Hydrogen separation with these membranes is nongalvanic (i.e., requires no electrical power), and the hydrogen separated from the feed stream is of high purity, making post-separation purification steps unnecessary. Using various feed gas mixtures, we measured the hydrogen permeation rate for the membranes in the temperature range of 400-900°C. This rate varied linearly with the inverse of membrane thickness and reached ≈26 and ≈50 cm^3 [STP]/min-cm^2 at 400 and 900°C, respectively, for a ≈20-μ m-thick membrane supported on porous substrate when 90% H_2 at ambient pressure was used as the feed gas. The long-term chemical stability of these membranes has been measured under various simulated coal gases and temperatures. We are also developing non-precious metallic membranes that have the potential to meet DOE flux and cost targets at 350°C. The present status of membrane development will be presented in this talk.

See more of this Session: Fuel Processing for Hydrogen Production (II)
See more of this Group/Topical: Topical 2: Advanced Fossil Energy Utilization

(118d) Coal Gas Assisted Hydrogen Production Using Mixed-Conducting Oxygen Transport Membranes

C. Y. Park, Tae H. Lee, Stephen E. Dorris, Yunxiang Lu and U.(Balu) Balachandran, Energy Systems Division, Argonne National Laboratory, Argonne, IL

COAL GAS ASSISTED HYDROGEN PRODUCTION USING MIXED-CONDUCTING OXYGEN TRANSPORT MEMBRANES

C.Y. Park, T.H. Lee, S.E. Dorris, Y. Lu, and U. (Balu) Balachandran

Energy Systems Division

Argonne National Laboratory

Argonne, IL 60439-4838

ABSTRACT

Due to their ability to separate oxygen without a need for electrodes, or electrical circuitry, oxygen transport membranes (OTMs) attract considerable interest in applications such as oxygen separation from air, upgrading of natural gas for production of liquid fuels, and oxygen-enriched combustion. The objective of this study is to develop dense OTMs that can produce hydrogen via coal/coal gas-assisted water dissociation.

Tubular-type OTMs were used to produce hydrogen via water splitting. Two different OTM materials, $La_{0.7}Sr_{0.3}Cu_{0.2}Fe_{0.8}O_{3-d}$ (LSCF) and $BaFe_{0.9}Zr_{0.1}O_{3-d}$ (BFZ), were prepared by a conventional solid-state technique. In tests with an LSCF thin-film tube (thickness of ª30 mm, active area of 15.3 cm^2) and a BFZ self-supporting tube (thickness of 1.05 mm, active area of 13.5 cm^2) as the OTMs, hydrogen was produced by flowing CO/CO_2 gas mixtures on the oxygen-permeation side of the OTM and steam on the other side, the so-called hydrogen-generation side. In this method, the simulated coal gas on the oxygen-permeate side drives the removal of oxygen from the hydrogen-generation side of the OTM, where hydrogen and oxygen are produced by water splitting. With CO (99.5% purity) flowing on the oxygen-permeate side, the hydrogen production rates of LSCF and BFZ at 900 °C were measured to be ª19.6 and ª22.0 cm^3/min, respectively, indicating that hydrogen can be produced at a significant rate by using product streams from coal gasification.

To be practical for hydrogen production, OTMs must remain chemically and mechanically stable in corrosive environments at elevated temperatures. The OTMs also should be available in a shape with a large active area, like a tubular film. Details on the preparation of thin films, microstructural behavior, flow rate effects, and stability tests will be discussed. Also, the oxygen permeation flux values of new promising membrane materials will be presented in this talk.

Work supported by the U.S. Department of Energy (DOE), Office of Fossil Energy, National Energy Technology Laboratory's Advanced Fuels Program, and Energy Efficiency and Renewable Energy, Office of Fuel Cell Technologies Program, under Contract DE-AC02-06CH11357.

See more of this Session: Fuel Processing for Hydrogen Production (II)
See more of this Group/Topical: Topical 2: Advanced Fossil Energy Utilization

(118e) Comparison Between CO-PROX and CO-SMET Catalytic Processes as Final Step for Syngas CO-Clean-up

Stefania Specchia, Samir Bensaid, Camilla Galletti and Vito Specchia, Dept. of Materials Science and Chemical Engineering, Politecnico di Torino, Torino, Italy

Thanks to their characteristics of high energy efficiency and low emissions, Polymer Electrolyte Fuel Cells (PEM-FCs) are the most promising candidates for electric power generator in transportation applications. At present, the most probable fuel is H_2-rich gas coming from reformed or partially oxidized hydrocarbon fuels, which downstream the Water Gas Shift step contains at least 0.5-1% of CO. Unfortunately, the performance of the Pt, the most effective electro-catalyst for H_2 oxidation in PEM-FCs, is seriously depressed from the presence of only 5 to 10 ppm of CO. There are several methods suitable to remove CO from syngas. For PEM-FCs applications however, the feasible choice may be only between the preferential CO oxidation (CO-PROX) and the selective CO methanation (CO-SMET) because of the limited available spaces and low operating pressures in such application systems. The CO-PROX process has so far been extensively tested since it is somehow reliable to remove CO down to 10 ppmv by raising the O_2 availability unlimitedly. Nonetheless, the technology requires a closely controlled low O_2 supply to keep the possibly lowest H_2 parallel oxidation and a working temperature window suitably wide for control purposes. This, while making the method costly and complex, obviously hampers its application to low-power PEM-FCs where very small oxidant rates have to be provided by use of well-refined expensive mass flow meters. Without any additional reactant required, the CO-SMET can avoid the above-mentioned shortcomings of the CO-PROX application. Furthermore, the CO methanation is less exothermic than the CO and H_2 oxidations. Thus, a CO-SMET reactor is inherently more easily controllable than a CO-PROX one. The present work deals with a comparison of catalysts performance for CO abatement via CO-PROX or CO-SMET.

Considering the CO methanation reaction ($\Delta°_{298} = -206.2$ kJ mol^{-1}), the potential chemical energy of reactants is practically conserved considering the heat developed by the reaction itself and the combustion energy of the produced CH_4. Moreover, the removal of 1 CO mole via methanation requires 3 mol of H_2. Of these, 2 mol are restored as CH4 product; the latter in the case where the FPU is integrated with the PEM-FC, as in APU systems, can be reused downstream the stack by recirculating the anode off-gas to the steam reformer burner where it is burnt for heating purposes with the main fuel rate to the burner. In a PEM-FC system, in general, the unreacted H_2 in the anode off-gas may reach 10–20% of its original feed rate. Considering H_2 consumption, for CO-SMET the actual H_2 loss should be of 3 mol per mole of removed CO, if the methanation selectivity toward CO is equal to 1 (i.e., no CO_2 methanation occurs).

For the CO-PROX there is theoretically no H_2 loss if the O_2 selectivity towards CO is equal to 1. This theoretical reaction efficiency, however, is hardly possible to be reached in practice over commercial catalysts, but also none is willing to risk the stack safety by operating at λ = O/CO = 1.0 v/v, even if the catalyst could allow, also taking into account that an actual PEM-FC system fluctuates intrinsically. Therefore, some O_2 excess allows a safer reactor management but, as a consequence, some H_2 loss, via H_2 oxidation, surely exists in practical CO-PROX reactors. This loss, based on literature reports, should be at least from 1 to 3 mol of H_2 per mole of removed CO because actual COPROX reactors run generally at λ = O/CO ratios of 2.0–4.0 v/v.

Hence, if one considers the possible heat recovery (to maintain as highest as possible the transformation efficiency of primary fuel to final H_2, FPU requires thermal energy, especially the "noble" one at high temperature), the CO-PROX reactor (assuming average λ of 3 and heat transfer efficiency of 1) allows a reaction heat recovery about 3.7 times higher compared to the CO-SMET reactor. Though, the heat from CO-PROX is at low temperature, so thermodynamically of not high significance. On the contrary, CO-SMET process, through the combustion of produced CH4, allows also obtaining, at high temperature and hence thermodynamically very useful for the FPU necessities, an amount of heat per mole of removed CO a little bit higher than the heat released from the CO-PROX reaction.

So, considering the global heat recovery (directly from the reactor and from the produced CH4 combustion), that from CO-SMET is about 1.3 times the one obtainable from CO-PROX and about 80% of this thermal energy (the one coming from CH_4 combustion) is available at a higher temperature level compared to the CO-PROX case. The availability of high temperature heat is very appreciated and useful since it positively affects the APU efficiency: APU, in fact, needs high temperature heating in the FPUs, otherwise satisfied by primary fuel burning.

On the other hand, considering the power losses from the two clean-up process final steps, using the above assumptions, CO-SMET consumes 1 H_2 mole per mole of removed CO more than CO-PROX (3 mol compared to 2 mol, respectively); assuming an FC average efficiency of 0.5 and anode stoichiometry of 1.2, the power loss for CO-SMET is independent from the stack power, but dependent on the residual CO concentration. Hence, the power loss for CO-SMET is higher compared to CO-PROX and ranges from 1.2% to 2.5% of the stack power for a CO residual inlet concentration in CO-SMET reactor of 0.5% or 1% b.v., respectively.

Thus, considering pros and cons, the CO-SMET process can be proposed as a promising route to remove CO in the reformate stream and as an alternative process to CO-PROX in FPUs for mobile and residential PEM FCs applications. Obviously, the development of catalysts able to remove the CO from percent levels (generally about 0.5–1%) down to ppmv level (10–50ppmv and lower) and characterized by high selectivity values towards the CO methanation in the presence of high CO_2 concentrations is a mandatory issue.

The present paper deals with the comparison of CO-PROX and CO-SMET catalysts for the final step of CO clean-up in FPs. The comparison has been carried out both experimentally, by developing and testing suitable catalysts for both the clean-up processes, and by modeling a FPU for stationary or mobile applications, carrying on as clean-up reatcot the CO-PROX or CO-SMET units. Comparisons between the energetic aspects of the modelled FPS will be presented in the full paper.

On the experimental pointof view, concerning the CO-PROX, Rh-based catalysts carried on Al_2O_3 and 3A zeolite were studied, whereas Ru-based catalysts carried on ZrO_2 and CeO_2 were employed for the CO-SMET. All the catalysts were prepared by a conventional impregnation method and CO removal was determined by varying the operating conditions (inlet $CO/CO_2/H_2/H_2O$ concentrations, temperature, weight space velocity WSV). Among the prepared catalytic materials, the 1% Rh-zeolite 3A catalyst, tested at a WSV of 0.66 $NL\times min^{-1}\times g_{cat}^{-1}$, was found to be the most suitable one for the CO-PROX at low temperature: it reduced the inlet CO concentration below 10 ppmv within a temperature range of at least 80-120 °C without the appearance of undesirable side reactions. On the contrary, for the CO-SMET the best results were obtained with the 5% Ru/CeO_2 catalyst: the CO inlet rate was completely transformed into methane in a temperature range of 250-280 °C at WSV of 0.45 $NL\times min^{-1}\times g_{cat}^{-1}$, and simultaneously the parasite reactions (CO_2–MET and R-WGS) rate remained at an acceptable level.

See more of this Session: Fuel Processing for Hydrogen Production (II)
See more of this Group/Topical: Topical 2: Advanced Fossil Energy Utilization

(118f) The Oxidative Steam Reforming (OSR) Performance of Pt-Ni/Al_2O_3 System with Mixed Propane-Methane Feed

Feyza Gökaliler, Z. Ilsen Önsan and A.Erhan Aksoylu, Chemical Engineering, BogaziÁi University, Istanbul, Turkey

Combined fuel processor-fuel cell (FP-PEMFC) systems are estimated to be well proliferated as energy production units in decentralized applications. Hydrogen production from various hydrocarbons via fuel processing needs a complicated system having up to five serially connected reactor units in which three catalytic reactions occur. The reactions, namely reforming, water gas shift and preferential oxidation, have their own difficulties arising from either thermodynamic limitations, reaction conditions and/or the FP product specifications imposed by the PEMFC. The energy efficiency of the overall FP-PEMFC system is one of the key parameters to be maximized; energy efficiency becomes even more important considering the fact that there may be economical/availability considerations that may impose the use of various hydrocarbon fuels as the FP feed. Our previous experimental studies with propane and LPG feeds [1, 2, 3] showed that OSR should be preferred as the reforming reaction; when the Pt-Ni bimetallic system is used as the catalyst, the high activity/selectivity characteristics are obtained at relatively low reaction temperatures due to the energy efficiency of the catalyst. The energy efficiency/exergy analysis of the FP-FC system has been also performed on a model by our group where experimental results obtained from the OSR, WGS and PROX catalysts under realistic conditions were used. In the system configuration, the rejected stream from the PEMFC is fully fed to a burner and then is used in vaporization of the water stream fed to the OSR unit before the burner effluent is emitted to the atmosphere. The results of the analysis showed that high temperature level and unused CH_4 content of the exhaust stream of the FP-FC combined system decreases the energy efficiency drastically when there is no partial-recycle to the OSR unit.

The aim of the current work is to analyze the performance of the 0.2 wt% Pt- 10 wt% Ni/Al_2O_3 system in propane OSR reaction when the hydrocarbon feed has various levels, between 5-20%, of methane. The effects of temperature, steam to carbon and carbon to oxygen ratios of the feed as well as the effect of residence time on the catalyst performance have been studied in an experimentally designed fashion. The experimental results showed that adding methane to the propane feed led to an enhancement in the hydrogen yield without significantly changing the H_2/CH_4 ratio at the OSR exit.

References:

[1] B. S. Ça•layan, A. K. Avcı, Z. I. Önsan and A. E. Aksoylu; *Appl. Catal. A* 280 (2005) pg. 181

[2] B. S. Ça•layan, Z. I. Önsan and A. E. Aksoylu; *Catal. Lett.* 102 (2005) pg. 63

[3] F. Gökaliler, B. A. Göçmen and A. E. Aksoylu, *Int. J. Hydr. Ener.* 33 (2008) pg 4358

See more of this Session: Fuel Processing for Hydrogen Production (II)
See more of this Group/Topical: Topical 2: Advanced Fossil Energy Utilization

SYNTHESIS AND CHARACTERIZATION OF $CuO/ZnO-Al_2O_3$ WASHCOAT CATALYSTS WITH CeO_2 SOLS ONTO A MICRO-CHANNEL REACTOR FOR STEAM REFORMING OF METHANOL

Kuen-Song Lin*, Abhijit Krishna Adhikari, Hsiu-Ping Yeh, and Chuin-Tih Yeh

Department of Chemical Engineering and Materials Science/Fuel Cell Center, Yuan Ze University, Chung-Li 320, Taiwan, R.O.C.

Porous alumina support CuO/ZnO mixed with CeO_2 sol washcoat catalysts have been characterized for autothermal reforming of methanol in the present study. Experimentally, XRD patterns of as-synthesized washcoat catalysts coated with increasing content of CeO_2 sols onto the surface of micro-reactor indicated that the Cu and Zn species were oxidized to CuO and ZnO, respectively. The intensities of Cu content from XRD data indicate that Al_2O_3 assign with Cu(0) to from $CuAl_2O_4$. The EXAFS data reveals that the Cu species have a Cu-O bonding with a bond distance of 2.13 Å and a coordination number of 4.82. The copper oxides may be interacted with the Cu(II) species in ceria identified using EPR spectroscopy. Moreover, the SRM performance tests were carried out at 483–573 K with the same weight of coated catalysts packed in the micro-reactor. A hydrogen production rate of 2.16 L h^{-1} is obtained and the corresponding methanol conversion is 100% at 543 K using the CuO/ZnO with CeO_2 sols washcoat catalysts. In addition, this work also showed that the feasibility of as-synthesized catalysts and the effects of CeO_2 binders for hydrogen production. With this approach, the feasible potential for production of high purity hydrogen can be achieved and a final processing to make it suitable for an integrated proton exchange membrane fuel cell (PEMFC) system.
Keywords: washcoat, microchannel, ceria sol, reforming, XANES/EXAFS, Fuel Cell

Introduction

Recently, primary fuel (such as methanol) oxidation through the catalysts are receiving considerable attention because of their potential role in the environmental friendly fuel cell technologies.[1–3] In the reported literatures, steam reforming of methanol (SRM) (equation 1) and partial oxidation reaction of methanol (POM) (equation 2) are consisted with production of hydrogen and are belonging with endothermic and exothermic reaction, respectively.[3] Combination of SRM and POM is effectively known as autothermal reforming (ATR) process and is attenuated with decreasing the reactor volume and design.[3,4] A SRM is the main reforming reaction provides the stoichometric conversion of methanol to hydrogen. The SRM produces a relatively small amount of carbon monoxide at low temperatures; where as carbon monoxide is known to be very poisonous for Pt catalysts in a proton exchange membrane fuel cell (PEMFC) system.[5] In addition, autothermal reforming of methanol (ATRM) process leads to produce higher amount of hydrogen and consists with CO as a by-product which has negative impact on PEMFC operation.

The catalytic reaction for SRM and ATRM are generally based on copper or palladium. Where, Pd has consisted with higher thermal stability and the production of large content of CO with lower selectivity from methanol decomposition. On the other hand, Cu-based catalyst has consisted with low cost, higher reactivity, and selectivity.[6,7] In addition, copper catalyst is effectively dispersed in ZnO or ZnO/Al_2O_3 to generate mesoporosity, with high surface area and good thermal stability, exhibited interesting catalytic activity in the CO preferential oxidation reaction. Thus, $CuO/ZnO–Al_2O_3$ is one of the most used catalysts for hydrogen generation by ATRM and is coated on a microchannel reactor.[1,2] As well, cerium promoted copper catalyst is existed for the recovery of hydrogen from methanol with the optimization of carbonmonoxide concentration.[8] As an important component in catalysts, ceria promotes high oxygen storage capacity (OSC), oxygen vacancy, high oxygen ion conductivity and affective binder on microchannel reactor. Reduction of Ce(IV) to Ce(III) by oxygen ion leads to the generation of surface oxygen vacancy. These oxygen vacancies can act as promoting sites for the optimization of carbon monoxide concentration. Furthermore, ceria hindered Cu sintering in Cu/Al_2O_3 catalysts and increased thermal stability and catalytical reactivity.[9] According with the Germani et al.[2] binder play a major role because of metal complexes formation is influenced the catalytic activities on the methanol decomposition reaction in the microchannel reactor. However, it is difficult to incorporate catalyst onto the microchannel reactor.[2,10,11] Several coating techniques have been applied for microchannels such as chemical vapor deposition, anodic oxidation, sol-gel process, washcoating, and electro-deposition to investigate the ATRM reaction.[12–14] The activities of catalysts are changed by the concentration differences for pre-sols and the treatment procedures of the drying and calcination temperatures.[13] Stutz and Poulikakos.[14] indicate that a thinner catalyst layer is not sufficient to obtain higher catalyst activity in the ATRM reaction. In this present work, we were prepared $CuO/ZnO-Al_2O_3$ catalyst washcoats with CeO_2 sols and then to investigate the autothermal reforming of methanol in a microchannel reactor for a fuel cell system. In addition, the fine structures of catalyst washcoats were also identified and characterized using Brunauer-Emmer-Teller (BET) adsorption isotherms, field-emission scanning electron microscopy / energy dispersive X-ray spectroscopy (FE-SEM), X-ray powder diffraction (XRD), X-ray photoelectron spectroscopy (XPS), electron paramagnetic resonance (EPR), X-ray absorption near edge structure (XANES), and extended X-ray absorption fine structure (EXAFS) techniques.

Experimental Section

Microchannels in this microreactor were made of stainless steel SUS 304 by an etching process and included with a flow route of reactants and products. The size of the plate with microchannel was 60 (L) × 35 (W) × 0.6 (T) mm. The plate was designed with thirty-four microchannels with the plate of 35 (L) × 0.4 (W) × 0.2 (D) mm. CH_3OH and O_2 or water mixtures were passed through the right or left upper inlet and were distributed over the plates sequentially to the parallel microchannels. Finally, H_2/CO_2 mixtures were collected through the outlet at the bottom side. A graphite gasket was used between microchannel plates for gastight and the whole reactor was contained in stainless-steel housing.

As-synthesized homogeneous sol was prepared by adding 20 g as-synthesized catalyst (CuO 40%, ZnO 50%, and Al_2O_3 10%) and reosmosis water (ROM) (Catalyst:ROM = 1:2) with a magnetic stirrer for continuous stirring at 500 rpm. Diluted homogeneous solution was prepared by dropwise addition of HNO_3 and NaOH to maintain the pH value at 9. The homogeneous solution was continuously stirred for 30 min to 6 h to obtain the catalyst slurry. After, stainless steel substrate surfaces were cleaned with acetone/ $HNO_{3(aq)}$ and then the catalyst slurry (S20) was washcoated on the surface of microchannels with CeO_2 sol (as a binder (B) 1 to 10 wt.% of solid content, respectively), followed by drying at 25 °C for 5h and then calcined at 350 °C for 2 h. The CeO_2 sol was used as a binder to enhance the adhesion between catalyst slurry and the SUS 304 or substrates. After calcinations CeO_2 film was formed between catalyst and substrate to obtain catalyst washcoat and denoted as S20-B1, S20-B2, and S20-B10, respectively. The specific surface areas of as-synthesized $CuO/ZnO–Al_2O_3$ catalyst and the CeO_2 sols were measured according to the BET equation by nitrogen adsorption isotherms at 77 K (-196 °C) with a Micromeritics ASAP 2010 Instrument. For the BET surface area measurement, 0.2 g catalyst samples were ground, sieved, and filled into sample tubes. Prior to measurement, all catalyst samples were degassed at 423 K (150 °C) for 6 h. For the calculation of the BET surface areas, the relative pressure range P/P_0 of 0.05 to 0.2 was used. The pore radius distribution was determined using the method of Barrett, Joyner, and Halenda (BJH method). The morphologies, microstructures, and the elemental composition of the washcoat catalysts were determined by FE-SEM/EDS (JEOL 5600). Identification of the solid phases and crystalline structure of as-prepared catalyst sample was performed using XRD (Shimadzu Labx XRD–6000). All samples were therefore scanned from 20 to 80°(2θ) with a scan rate of 4°(2θ) min^{-1} and monochromatic CuK_α radiation. The recorded specific peak intensities and 2θ values can be further identified by a computer database system (JCPDS).

The XANES and EXAFS spectroscopies were collected at the Wiggler beamline 17C1 in the National Synchrotron Radiation Research Center (NSRRC) of Taiwan. The electron storage ring was operated with the energy of 1.5 GeV and a current of 100–200 mA. A Si(111) DCM was used for providing highly monochromatized photon beams with an energy of 4 to 15 keV (BL17C1) and resolving power ($E/\Delta E$) of up to 7000. Data were collected in fluorescence or transmission mode with a Lytle ionization detector.[15] for Cu (8979 eV) K edges experiments at room temperature. Local structural parameters such as the bond distance (R), coordination number (CN), and Debye-Waller factor (σ) for different coordination shells surrounding the absorbing atoms were obtained through non-linear least-square fitting route. Moreover, the raw absorption data in the region of 50 to 200 eV below the edge position were also fit to a straight line using the least-square algorithms.[15–17] The XANES was extended to energy of the order of 50 eV above the edge. The k^2-weighted and EXAFS spectra were Fourier transformed to R space over the range between 2.5 and 12.5 $Å^{-1}$. The EXAFS data were analyzed by using the UWXAFS 3.0 program and FEFF 8.2 codes.[16,17]

The performance of methanol conversion tests were conducted in a lab-scale five-plate microchannel reactor using porous alumina support CuO/ZnO with CeO_2 sol catalysts washcoat. After the catalyst was applied in each plate then five plates were staked together with well sealed and integrated in a house. The total weight of as-synthesized catalyst coated on the five plates was 100 mg. A pre-mixture of methanol and water (1:1.3 molar ratio) solution was introduced to the microreactor using an automotive syringe pump with He carrier gas and then vaporized through a feed line heated at 182 °C. The feed rate of methanol liquid reactant mixtures was 0.05–0.2 mL min^{-1} with 16.2 $g_{MeOH}\ h^{-1}\ g_{cat.}^{-1}$ weight-hourly space velocity in mass methanol per time and mass of catalyst (WHSV). The ATRM reaction was performed in the temperatures ranged of 210–300 °C. The microreactor was enclosed and heated by an electrically heated jacket and operated under isothermal with isobaric conditions including a temperature controller. Noncondensible gases stream of H_2, CO, and CO_2 generated in the microreactor were directly passed through an ice water cooler at 0 °C and were sampled or analyzed using a gas chromatography/thermal conductivity detector (GC/TCD) (Aligent 6890N).

Results and Discussion

According with the FE-SEM microphotos, the significant morphological differences between the as-prepared catalysts and the slurries are observed in Figure 1. The irregular shapes of crystals in catalysts and notably aggregated particles in $CuO/ZnO\text{-}Al_2O_3$ catalyst washcoats with CeO_2 sols were also found. In addition, catalyst has been gradually changed the color from deep brown into greenish blue after the addition of cerium sol solution and isopropyl alcohol (IPA). In fact, absorbance of water molecules shrink the electron density of Cu(II) ions. Therefore, it reveals that the copper species are associated with *d-d* transition state and Cu(II) ions coordination state is reflected the color changing behavior for washcoat slurry materials. Furthermore, BET N_2 adsorption for as-synthesized catalysts and catalytic mixtures with a CeO_2 sol washcoat were 26, 31, 43, and 62 $m^2\ g^{-1}$, respectively as shown in Table 1. It indicates that the addition of CeO_2 sols may promote to increase the surface area of the catalyst.[18] Moreover, porosity of the catalyst is decreased significantly after certain amount of higher loading of the binder as confirmed with the investigation of FE-SEM micrographs and BET measurements.

Figure 1. FE-SEM microphotos of (a) $CuO/ZnO\text{-}Al_2O_3$ catalyst, catalyst washcoats of (b) S20-B1, (c) S20-B2, and (d) S20-B10, respectively.

The X-Ray diffraction patterns for the $CuO/ZnO\text{-}Al_2O_3$ catalyst and catalyst with CeO_2 binder washcoats are shown in Figure 2 From Figure 2(a), the weak peaks at Bragg angles 2θ of 43.3, 50.14 and 74.1° for metallic copper (JCPDS 89-2838); 35.5° was attributed to CuO (JCPDS 89-5899) and also 36.52 and 73.7° was represented as Cu_2O (JCPDS 65-3288), respectively.[8,9,18–20] The characteristic peaks of ZnO were at 2θ of 31.80, 34.51, 36.30, 62.82 and 69.10° (JCPDS 89-1397), respectively.[18–20] A peak of Al_2O_3 at 35.15, 37.77, 43.34, 57.50, 66.59°(JCPDS 82-1468) and spinal $CuAl_2O_4$ was also observed in the XRD spectra at 31.28, 38.85 and 44.82°(JCPDS 78-0556), respectively.[9] While the characteristic weak peak of stainless steel at 2θ of 62.79 and 76.80° were attributed into Figure 2. As shown in Figure. 2(b) to 2(d), XRD patterns of as-synthesized washcoat catalysts coated with increasing content of CeO_2 sols at 28.55, 47.50 and 56.40° (JCPDS 81-0792) onto the surface of microreactor indicated that the Cu species were oxidized mainly into CuO[8.9] Closer inspection of the XRD patterns revealed that the as synthesized washcoat catalyst has been consisted with the particle size around 18 nm and gradually decrease the size (as 13 ± 1 nm) with the increases of the loading of CeO_2 sols.[9,18,19] The peak width of copper for the increasing of CeO_2 loading is gradually affected to broaden than as synthesized sols and that represented the smaller copper particle sizes, as shown in

Figure 2 (b) to 2(d). This suggested that CuO was finely dispersed on the surface of ceria or a Ce-O-Cu solid solution was formed.[8,19] Therefore, it was improved to facilitate the copper surface area and that is verified with the BET nitrogen adsorption isotherms analyses.

Table 1. Specific Surface Area and Average Pore Diameters of Catalyst Slurries

Sample	Specific surface area (m^2 g^{-1})	Average pore diameter[a] (nm)
$CuO/ZnO-Al_2O_3$	26.0	18.6
Slurry S20-B1[b]	31.0	30.0
Slurry S20-B2[c]	43.0	30.0
Slurry S20-B10[d]	62.0	10.0

[a]Average pore diameters were detemined by BJH equation;
[b]Slurry of 20 wt.% solid content with 1 wt.% CeO_2 binder solid at pH =7;
[c]Slurry of 20 wt.% solid content with 2 wt.% CeO_2 binder solid at pH =7;
[d]Slurry of 20 wt.% solid content with 10 wt.% CeO_2 binder solid at pH =7.

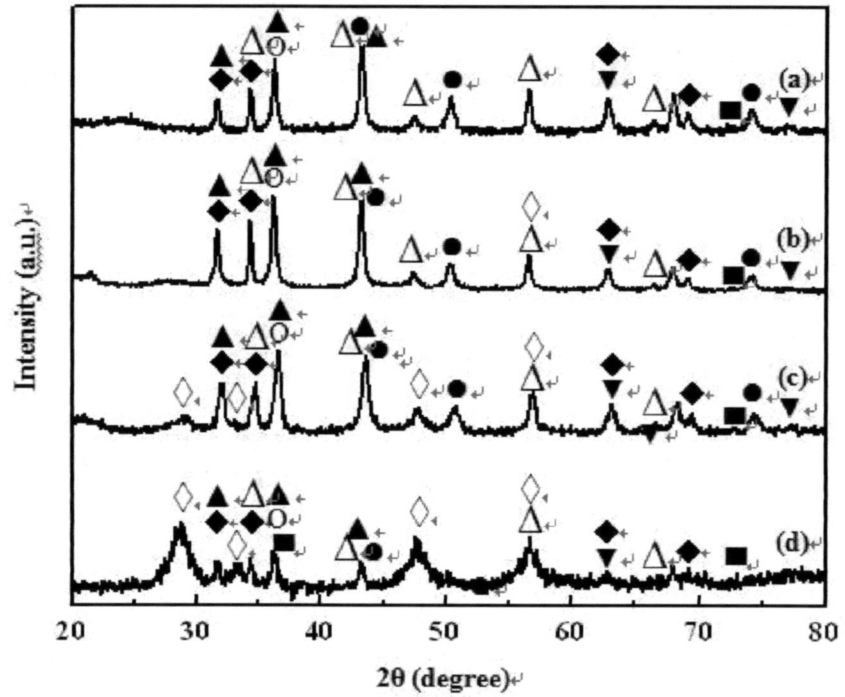

Figure 2. XRD patterns of reduced (a) $Cu/ZnO-Al_2O_3$ catalyst, catalyst with CeO_2 binder washcoats of (b) S20-B1, (c) S20-B2, and (d) S20-B10, respectively (●: Cu; O: CuO; ■ : Cu_2O; ◆ : ZnO; ▲: $CuAl_2O_4$; ◊: CeO_2; Δ: Al_2O_3; ▼SUS 304).

The ATRM reaction was performed in order to investigate the as-synthesis catalyst activity in the stacked plate-typed reactor as revealed in Figure 3. An ATRM performance was carried out at 210–300 °C with the same weight of coated catalysts packed in the microreactor. After the activation of the catalyst (S20-B2); methanol was started to decompose and the conversion was above 95% at 240 °C. In addition, a higher temperature was required for higher methanol conversion but decreasing with the increasing of feed flow rate and was related with the steam to carbon (methanol) ratio.[20–26] According with the Seo et al.[26] methanol conversion was achieved 90% at 260 °C and the feed flow rate was 0.2 cm^3 min^{-1}. A hydrogen production rate of 2.16 L h^{-1} was obtained at a feed rate of methanol liquid mixtures of 0.05–0.2 mL min^{-1} with a WHSV of 16.2 g_{MeOH} h^{-1} $g_{cat.}^{-1}$. Moreover, the selectivity of hydrogen and carbon dioxide was high enough and a typical dry gas composition of the present system was 73-74% H_2, 24-25% CO_2, and <0.1% CO gas products, respectively. Nevertheless, a T_r> 260± 10 °C was also required in the cases to achieve high methanol conversion.[9–13,27] The addition of ceria caused the synergistic effect to give the activity enhancement. However, methanol conversation was decreased to 95% at 270 °C for S20-B10. Due to the morphology, porosity, BET surface area significantly affects the catalytic activity, the XRD patterns were also confirmed.[25–28] This confirms that the as-synthesized catalyst are coated on to the microchannel reactor participates in an ATRM reaction and that is suitable for the integrated PEMFC application on energy generation.

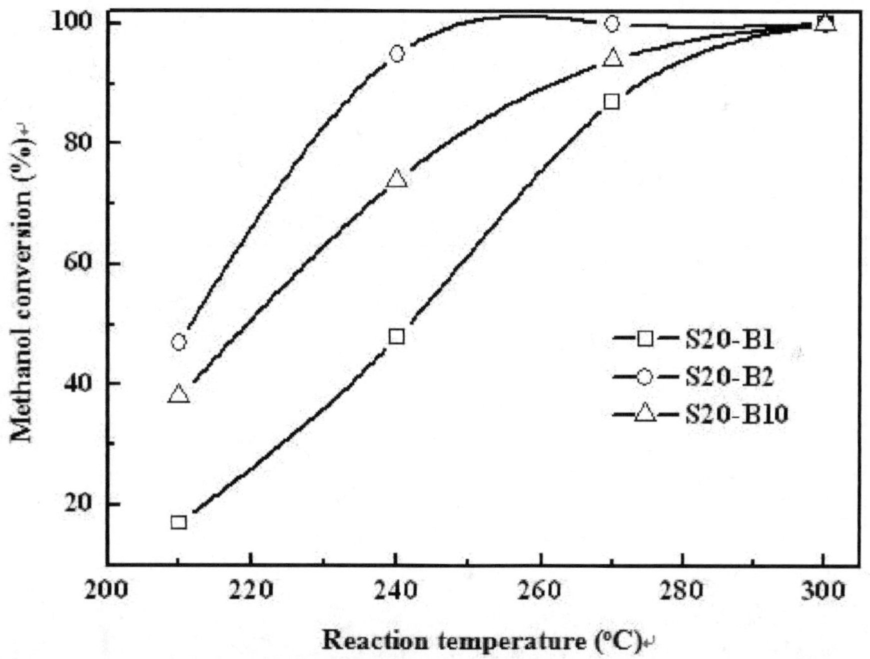

Figure 3. Methanol conversion of Cu/ZnO-Al_2O_3 catalyst with CeO_2 binder washcoats of (a) S20-B1, (b) S20-B2, and (c) S20-B10, respectively for the steam reforming of methanol in a microreactor (steam/ MeOH = 1.3; WHSV = 16.2 g_{MeOH} h^{-1} $g_{cat.}^{-1}$; loading weight of washcoat catalyst = 20 mg/plate).

Generally, Cu K-edge EXAFS spectroscopy can provide the information on the atomic arrangement of sorbents in terms of bond distance, coordination number, and kind of neighbors. A high reliability of the EXAFS data fitting for Cu species in as-synthesized and after reaction were obtained. Fourier transformation was performed on k^3-weighted oscillations over the range of 2.2 and 10 $Å^{-1}$. The XANES spectra of several different standards (Cu-foil, Cu_2O, and CuO) and the different samples of as-synthesized $CuO/ZnO-Al_2O_3$ catalysts with CeO_2 sols slurry and after reaction are shown in Figure 4. The K-edge positions of standard Cu foil, Cu_2O, and CuO are 8978.2, 8979.36, and 8982.23 eV, respectively (Figure 4(b)-(d)) and catalyst washcoat before and after the SRM reaction have the K-edge points at 8982.4, and 8982.15 eV, respectively (Figure 4(a),(e)). It is found that the K-edge points of all the samples are very close to the Cu(II). Compared with the standard Cu(II) the offset is 0.25 eV due to *1s* to *3d* transition.[8] Since Cu ions are not entirely surrounded by oxygen atoms and form oxides (OH^-) with the covalent bond, it gives slight shift. However, the well-defined shoulders at 8985.3 eV are attributed to the *1s* to $4p_{xy}$ transition that indicates the existence of Cu(II) species in the washcoat slurries.[8,28,29] As prepared and after reaction, it has been evident from the XANES spectra that the observed samples can be ascribed the slight deflection of Cu(II) during the methanol reduction process in Figure 4(e) and consisted with XRD analyses. To obtain the information on local structure around Cu(II) species in catalyst washcoat of the $Cu/ZnO–Al_2O_3$ catalysts with CeO_2 sols were analyzed with EXAFS spectra as shown in Table 2. Interestingly, the as-synthesized $Cu/ZnO–Al_2O_3$ catalysts with CeO_2 sols (S20-B2 sample before ATRM reaction) have consisted with the oxygen atom directly bonded to central Cu atom. It has the Cu-O bond length of 1.99 Å with the coordination number of 3.24. After the ATRM reaction, structures of the washcoat catalyst samples were transformed into copper oxide and the coordination number changed to 10.2 with a Cu–O bond distance of 1.95 Å. Two approaches were taken according with the fitting of the crystallographic data. In the first case, the coordination number was held at the crystallographic values with the radial distance and Debye–Waller factor were varied. Generally, according with the entire EXAFS data analyses of the Debye-Waller factors ($\Delta\sigma$) were less than 0.01 Å. The apparent shortening of the Cu–O bond length probably was caused by the random motion of surface atom on Cu particles instead of a true shortening of the bond distance. Thus is responsible for the low-temperature ATRM reaction and consistent with XRD data analyses.

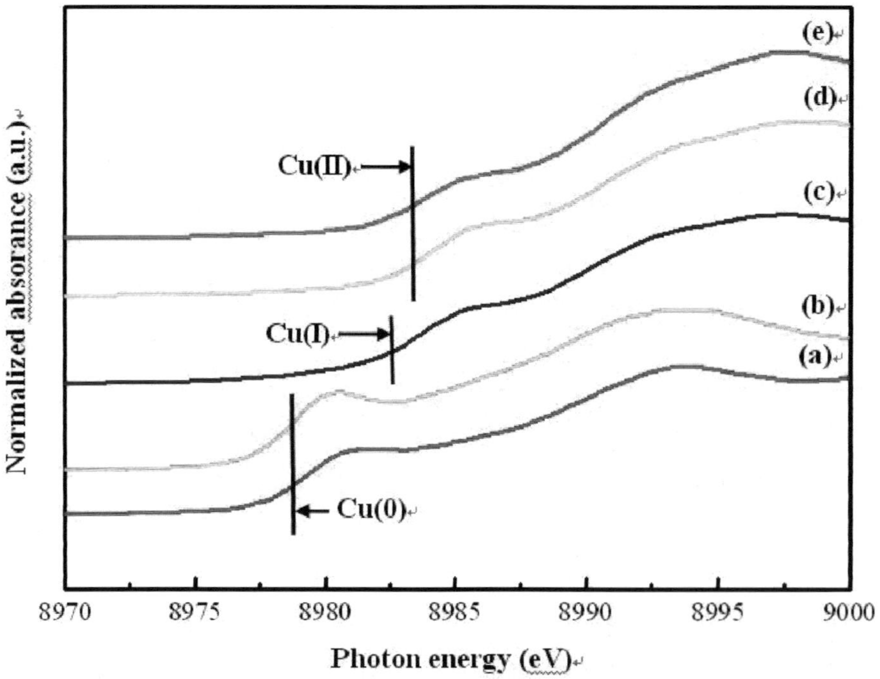

Figure 4. XANES spectra of copper standards and copper species in the $Cu/ZnO-Al_2O_3$ catalyst with CeO_2 binder washcoats (S20-B2) (a) before reaction, (b) Cu foil, (c) Cu_2O, (d) CuO, and (e) after reaction for the steam reforming of methanol in a microreactor.

Table 2. Structural Parameters of Slurry (20 wt% Solid Content with 2% CeO_2 Binder Solid at pH =7) Before and After Reaction Analyzed by EXAFS

Bonding pair	CN^a (±0.05)	R^b (Å) (±0.02 Å)	$\Delta\sigma^{2c}$ ($Å^2$)
Fresh S20-B2			
Cu-O	3.24	1.99	0.00398
Used S20-B2			
Cu-O	10.22	1.95	0.00431

^{a}CN: Coordination number; ^{b}R: Bond distance;
^{c} σ : Debye-Waller factor.

In addition, coordination and valance states of the paramagnetic Cu(II) species in the as-synthesized and after SRM reaction were also studied by EPR spectroscopy at room temperature as shown in Figure 5. Since the minimum *g*-factor of the catalyst is 2.11 (g_{DPPH} factor = 2.0037 at 25 °C), it is likely that the Cu(II) species has a square-plane structure in as-synthesized $CuO/ZnO\text{-}Al_2O_3$ catalyst washcoats with CeO_2 sols slurry.[30,31] According with Figure 5(a), the EPR spectrum of the fresh Cu catalyst at 25 °C is consisted with as-synthesized slurry. On the contrary, the spectrum (Figure 5(c)) exhibits a dislocated broad signal. The spectrum is unlike that of any monomeric tetragonal copper (II) species. Furthermore, signal K at 3000 and 3568 G in the spectra corresponds to Cu(II)-Cu(II) ion pairs arising from the coupling between unpaired electrons of two Cu(II) ions.[30,31] This may suggests that the copper species may be interacted with ceria and consisted with the EXAFS/XANES and XRD analyses.

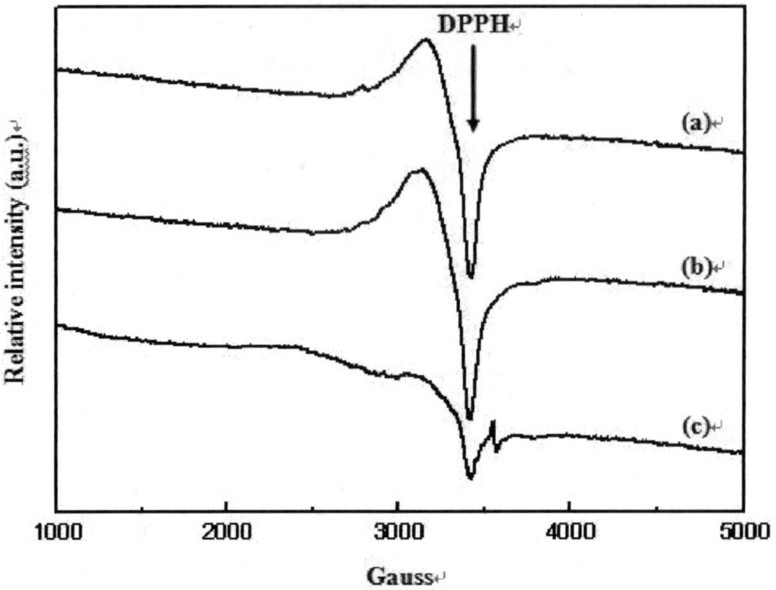

Figure 5. EPR spectra of (a) fresh Cu/ZnO-Al_2O_3 catalyst, catalyst washcoat (S20-B2) (b) freshly prepared, and (c) after reaction measured at 300 K (ν_{EPR} = 9.784 GHz; Field (Gauss) = 3475 or g_{DPPH} = 2.0037; S20-B2 denotes "slurry of 20 wt.% solid content with 2 wt.% CeO_2 binder solid at pH =7").

Conclusions

A significantly high active and stable $CuO/ZnO-Al_2O_3$ catalyst with CeO_2 sols washcoat catalyst was successfully investigated in a microchannel system for hydrogen generation integrated to a PEMFC system. The as-synthesized catalysts with the addition of CeO_2 sol were adversely affected the surface area and indicated high methanol conversations. According with the XRD patterns, catalyst sizes decrease with increasing CeO_2 sols. In addition, the XANES or EXAFS spectra identified the oxidative states and fine structures of the Cu species. A hydrogen production rate of 2.16 L h^{-1} was obtained at a feed rate of methanol liquid mixtures of 0.05–0.2 mL min^{-1} with a WHSV of 16.2 g_{MeOH} h^{-1} $g_{cat.}^{-1}$. Moreover, the methanol conversion was about 95% at 240 ^{o}C. This work showed that the feasibility of as-synthesized catalysts and the effects of CeO_2 binders for hydrogen production. With this approach, the feasible potential for production of high purity hydrogen can be achieved and final processing to make it suitable for an integrated PEMFC system.

References

[1] Horny C., Renken A. *Catal. Today*. 2007;120:45–53.
[2] Germani G., Stefanescu A., Schuurman Y., Veen A.C.V. *Chem. Eng. Sci.* 2007;62:5084–5091.
[3] Patel S., Pant K.K. *Appl. Catal. A: Gen.* 2009;356:189–200.
[4] Geissler K., Newson E., Vogel F., Truong T.B., Hottinger P., Wokaun A., *Phys. Chem. Chem. Phys.* 2001;3:289–293.
[5] Wild P.J.D., Verhaak M.J.F.M. *Catal. Today.* 2000;60:3–10.
[6] Yu X., Tu S.T., Wang Z., Qi Y. *J. Power Sources.* 2005;150:57–66.
[7] Kim T., Lee D.H., Park D.E., Kwon S. *ASME J. Fuel Cell Sci. Technol.* 2008;5:011008 (1–6).
[8] Maciel C.G., Profeti L.P.R., Assaf E.M., Assaf J.M. *J. Power Sources.* 2011;196:747–753.
[9] Yu X., Tu S.T., Wang Z., Qi Y. *Chem. Eng. J.* 2006;116:123–132.
[10] Park D.E., Kim T., Kwon S., Kim C.K., Yoon E. *Sens. Actuators. A.* 2007;135:58–66.
[11] Park G.G., Yim S.D., Yoon Y., Lee W.Y., Kim C.S., Seo D.J. *J. Power Sources.* 2005;145:702–706.
[12] Qi A., Peppley B., Karan K. *Fuel Process. Technol.* 2007;88:3–22.
[13] Lim M.S., Kim M.R., Noh J., Woo S.I. *J. Power Sources.* 2005;140:66–71.
[14] Stutz M.J., Poulikakos D. *Chem. Eng. Sci.* 2008;63:1761–1770.
[15] Lytle F.W. *J. Synchrotron Rad.* 1999;6:123–134.
[16] Resslert T. *J. Synchrotron Rad.* 1998;5:118–122.
[17] Nesvizhskii A.I., Rehr J.J. *J. Synchrotron Rad.* 1999;6:315–316.
[18] Patel S., Pant K.K. *Chem. Eng. Sci.* 2007;62:5436–5443.
[19] Polster C.S., Nair H., Baertsch C.D. *J. Catal.* 2009;266:308–319.
[20] Liu Y., Hayakawa T., Suzuki K., Hamakawa S. *Catal. Comm.* 2000;2:195–200.
[21] Moretti E., Lenarda M., Storaro L., Talon A., Montanari T., Busca G., Castellón E.R., López A.J., Turco M., Bagnasco G., Frattini R. *Applied Catalysis A: General.* 2008;335:46–55.
[22] Moretti E., Lenarda M., Storaro L., Talon A., Frattini R., Polizzi S., Castellón E.R., López A. *J. Appl. Catal. B: Environ.* 2007;72:149–156.
[23] Avgoropoulos G., Ioannides T. *Appl. Catal A: Gen.* 2003;244:155–167.
[24] Yi N., Si R., Saltsburg H., Stephanopoulos M.F. *Energy Environ. Sci.* 2010;3:831–837.
[25] Avgouropoulos G., Papavasiliou J., Ioannides T. *Chem. Eng. J.* 2009;154:274–280.
[26] Seo D.J., Yoon W.L., Yoon Y.G., Park S.H., Park G.G., Kim C.S. *Electrochimica Acta.* 2004;50:719–723.
[27] Huang G., Liaw B.J., Jhang C.J., Chen Y.Z. *Appl. Catal. A: General.* 2009;358:7–12.

[28] Patel S., Pant K.K. *J. Power Sources.* 2006;159:139–143.
[29] Lamberti C., Bordiga S., Bonino F., Prestipino C., Berlier G., Capello L., D'Acapito F., Xamena F.X.L.I., Zecchina A. *Phys. Chem. Chem. Phys.* 2003;5:4502–4509.
[30] Bera P., Priolkar K.R., Sarode P.R., Hegde M.S., Emura S., Kumashiro R., Lalla N.P. *Chem. Mater.* 2002;14:3591–3601.
[31] Aad E.A., Bechara R., Grimblot J., Aboukais A. *Chem. Mater.* 1993;5:793–797.

Improved preparation and high catalytic performance of core shell Ni-based catalysts for H_2-rich gas production from ethanol

Guangming Zeng, Ruixue Gu, Yongdan Li[*]
Tianjin Key Laboratory of Applied Catalysis Science and Technology and State Key Laboratory for Chemical Engineering (Tianjin University), School of Chemical Engineering, Tianjin University, Tianjin 300072, China

1. Introduction

Hydrogen-rich gas production through steam reforming of ethanol is an attractive technology since ethanol is a renewable raw material obtained from fermentation of various agricultural products (e.g., energy crops, forestry residue, organic waste). Although steam reforming has been extensively studied in the literature and industry, the development of suitable catalysts is still a significant challenge from the standpoint of catalyst deactivation.

In the last two decades there has been an increased interest in using hydrotalcite (HT)-like compounds as precursors for mixed oxide catalysts in various reactions [1-3]. Preparation of heterogeneous catalyst from HT-like precursors may afford the oxide by calcination that showed interesting properties, such as high surface area, "memory effect", basic character and thermal stability. Many researchers [4-7] have reported that Ni/Mg-Al catalysts derived from HT-like precursor presented high and stable activities for reforming reactions due to stable and highly dispersed Ni metal particles. However, one of the shortcomings of HT-derived Ni/Mg-Al catalyst is: the mixed oxide uniformly contains Ni species from the surface skin to the core of the catalyst particles by forming Ni-Mg-O solid solution. As a result, a large amount of Ni included in the catalyst bulk becomes irreducible and therefore ineffective for catalysis [8].

In heterogeneous catalysis, the active chemical composition of a catalyst can work more efficiently only in combination with a well-organized structure. A catalyst having a finely designed structure usually exhibits improved physical and chemical properties and hence is potentially useful in a broader range of applications. Core shell structure materials have attracted great attention over the last decade because of the tailored, optical, and surface properties. It is desirable to prepare Ni/Mg-Al catalysts in core shell structure, i.e., active Ni species are concentrated in the surface layer of the catalyst particles where mutil-phase reactions preferentially proceed in the micro- or mesoporous space [9].

Recently, Takehira et al. [10] have systematically studied the reconstitution of HT-derived mixed oxide leading to core shell type Ni loading catalyst. In the present study, the supported Ni catalysts with core shell structure were prepared using an optimized synthesis method and their physicochemical properties and catalytic behaviors on ethanol steam reforming were investigated in comparison with the conventional HT-derived Ni-based catalyst.

2. Experimental

2.1. Catalyst preparation

The core shell Ni@Mg-Al catalysts were prepared by adopting "memory effect"

[*] Corresponding author. Tel.: +86 22 27405613; Fax: +86 22 27405243. E-mail address: ydli@tju.edu.cn.

through dipping Mg-Al mixed oxide in an aqueous solution of Ni^{2+} nitrate, followed by the drying and calcinations. The detailed preparation procedure was described in previous publications [8,9] with important improvement based on our group's abundant experience derived from a long-term study of HT-derived catalysts. Mg-Al HT-like precursor was prepared by co-precipitation method. An aqueous solution containing the nitrates of Mg^{2+} and Al^{3+} ($M^{2+}/M^{3+}=3$) was added slowly with vigorous stirring into an aqueous solution of sodium carbonate (1.0 mol L^{-1}). By adjusting pH of the solution to 9.0 with the same solution of sodium carbonate, heavy slurry precipitated, followed by the crystal growth by aging the solution at 60 °C for 12 h. After the solution was cooled to room temperature, the precipitate was filtered, washed with de-ionized water, and dried overnight at 120 °C. The HT-like precursors were calcined in a muffle furnace from ambient temperature to 600 °C at a rate of 2 °C min^{-1} and kept at 600 °C for 6 h. The mixed oxides were pressed, crushed and sieved to particles of 40-80 mesh. The conventional Ni/Mg-Al (15 wt.% Ni) catalyst was prepared through the same procedure except being calcined at 700 °C for 6 h at a rate of 5 °C min^{-1}.

Core shell loaded Ni catalysts were prepared by accelerating the reconstitution of Mg-Al HT structure in the surface layer of the Mg-Al mixed oxide particles. The particles were dipped in the aqueous solution of Ni^{2+} nitrate (0.6 mol L^{-1}) at pH=4.0, 60 °C for 15-60 min, dried in air at 120 °C and calcined at 700 °C for 6h with a heating rate of 5 °C min^{-1}. Thus, reconstitution was happened just on the surface layer of Mg-Al mixed oxide particle, a part of the Mg^{2+} sites were replaced by Ni^{2+} upon immersing the particles in aqueous solution of Ni^{2+} nitrate. The particles with various dipping time were name Mg-Al-15, Mg-Al-30, Mg-Al-45 and Mg-Al-60, where the number after elements indicates the dipping time.

2.2. Characterization of catalyst

ICP measurement was carried out with a Varian Vista MPX inductively coupled plasma mass spectrometer (ICP-AES). X-ray diffraction patterns were obtained at room temperature using a Rigaku D/max 2500 v/pc instrument (Rigaku Corp. Japan) with CuKα radiation, 40 kV and 200 mA power. The step-scans were taken over the range of 2θ from 5 to 90° in steps of 0.02° and the intensity data for each one were collected for 0.15 s. JADE5 software was utilized for data analysis. Scanning electron micrograph (SEM) image of cross section of the particle after dipping was obtained with a Philips XL30ESEM intrument equipped with EDX. Surface morphologies of the catalyst particles were examined using a FEI NanoSEM 430 intrument. N_2 adsorption/desorption (77 K) was used to examine both BET surface area and the porous property of the particles. The measurement was carried out on a Quantachrome Autosorb-1 intrument.

2.3. Catalyst testing

Steam reforming of ethanol was carried out in a fixed bed stainless-steel tubular reactor, operating at atmospheric pressure, loaded with 200 mg of catalyst mixed with 1.0 g of quartz particles (both of them 40~80 mesh sieved). The reaction temperature was measured and controlled by two Type-k thermocouples. One of the thermocouples was inserted into a thermowell and centred with the catalyst bed, while the other one was placed into the annular space between the reactor and tubular furnace. The catalyst was pre-reduced at 700 °C for 2 h with 30 ml min^{-1} H_2 prior to the introduction of vapor phase ethanol-water feed mixture. A 3:1 molar ratio of H_2O:C_2H_5OH solution was fed by a micro-pump at a rate of 4 ml h^{-1} to the vaporizer, which was maintained at 200 °C. 20 ml min^{-1} N_2 was added to the vaporized reactant feed as an internal standard to aid in analysis of the product gas composition and determination of the product flow. The feed and reaction products were analyzed on-line using a PerkinElmer Clarus 500 GC equipped with TCD and FID detectors.

3. Results and discussion

3.1. Material characterization

The preparation of core shell Ni@Mg-Al catalyst were characterized using the

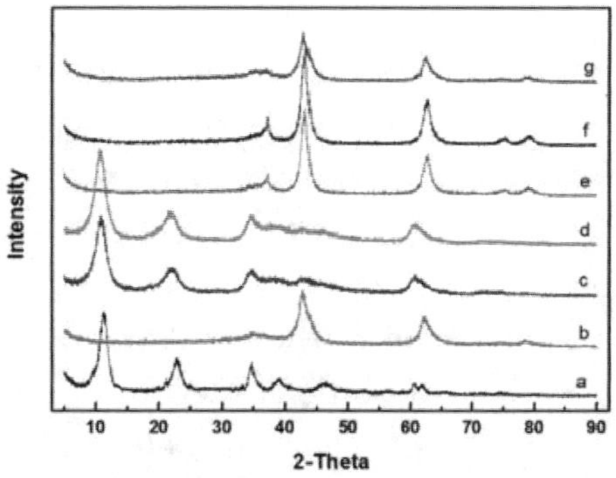

Fig. 1. XRD patterns of (a) Mg-Al HT-like precursor; (b) Mg-Al mixed oxide calcined at 600 °C; (c) after dipping the sample (b) in Ni^{2+} nitrate for 15 min; (d) after dipping the sample (b) in Ni^{2+} nitrate for 60 min; (e) after calcining the sample (c) at 700 °C; (f) after calcining the sample (d) at 700 °C; (g) Ni/Mg-Al catalyst calcined at 700 °C.

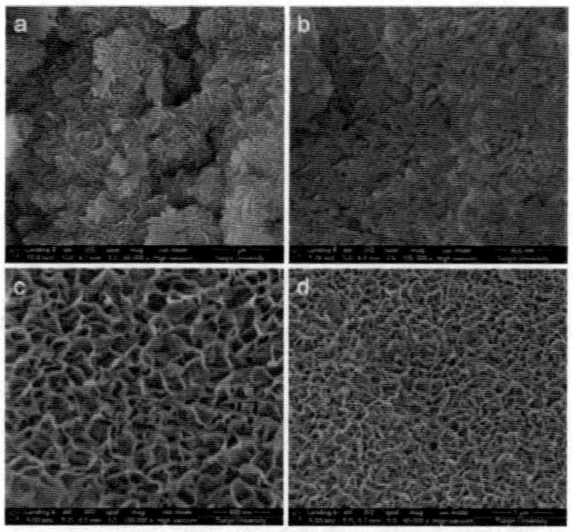

Fig. 2. Morphologies of the surface of particles during the preparation of Mg-Al-30 catalyst: (a) Ma-Al HT-like precursor; (b) after calcined at 600 °C, pressed and sieved; (c) and (d) after dipping in Ni^{2+} nitrate for 30 min, followed by drying at 120 °C.

XRD patterns in Fig. 1. Mg-Al HT-like precursor was formed by co-precipitation, and converted to Mg-Al mixed oxide by the calcination at 600 °C. The curve of Fig. 1(b) only showed the diffraction lines of MgO, no other oxide has been detected. When the Mg-Al mixed oxide was pressed, crushed and sieved to the particles and dipped in an aqueous solution of Ni^{2+} nitrate (0.6 mol L^{-1}) at 60 °C for 15 min, the reconstitution of Mg-Al HT-like structure was clearly observed (Fig. 1(c)). Increasing the dipping time to 60 min (Fig. 1(d)) resulted in no significant change in the XRD pattern, suggesting that the reconstitution may proceed rapidly and almost complete in a time as short as 15 min. By ICP analyses (Table 1), the content of Ni in the sample dipping for 15 min is 8.1 wt.%, while the conresponding Mg showed an obvious decrease of 6.0 wt.% compared with Mg-Al mixed oxide. Increasing the dipping time, the content of Ni increased slowly, whereas that of Mg showed a conformable small decrease, suggesting that some of the Mg^{2+} sites in the HT-like phase were replaced by Ni^{2+} ions during the reconstruction of HT-like structure.

Table **1.** Chemical composition, BET surface area and pore volume of the HT-derived catalysts.

Sample	Chemical composition (wt.%)			BET surface area ($m^2 g^{-1}$)		Pore volume ($cm^3 g^{-1}$)	
	Ni	Mg	Al	Uncalcined	Calcined	Uncalcined	Calcined
Ni/Mg-Al	15.0	31.6	14.0	---	154.1	---	0.445
Mg-Al	---	42.2	15.9	---	194.4	---	0.468
Mg-Al-15	8.1	36.0	15.9	10.8	159.5	0.017	0.324
Mg-Al-30	9.1	35.5	15.6	11.6	153.8	0.02	0.306
Mg-Al-45	9.0	35.6	15.6	8.29	160.4	0.013	0.317
Mg-Al-60	10.8	34.7	15.2	6.46	159.0	0.014	0.314

The morphology observation of the surface of particles during the preparation of Mg-Al-30 sample is shown in Fig. 2. The HT-like precursor just after the drying showed lamellar structure, each lamella has a size around 200 nm in diameter and 20 nm in thickness by the nanoSEM. After calcined at 600 °C for 6 h, pressed and sieved to particles, the Mg-Al mixed oxide still retain the lamellar structure (Fig. 2(b)), but a smaller size around 150 nm. This is in good accordance with the report that the morphology of brucite sheets was kept after the calcination at 650 °C and still remained in the resulting periclase-like Mg-Al-O solid solution, and even in the structure after the calcination at 950 °C [11]. When the particles were dipped in 0.6 mol L^{-1} Ni^{2+} nitrate for 30 min, follow by drying at 120 °C, the surface of each particle was covered by net-like structures formed through the connection of the abundant lamellar precursor (Fig. 2(c) and (d)).

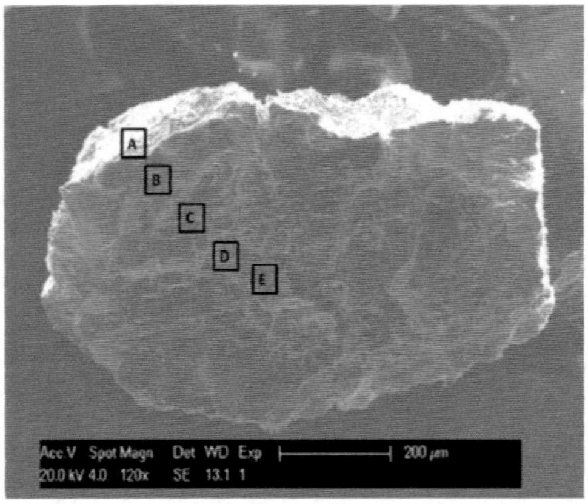

Fig. 3. SEM image of cross section of Mg-Al-30 and the locations where EDX spectra were taken.

Table **2.** EDX results of cross section of Mg-Al-30 sample at locations marked in Fig. 3.

Area	Chemical composition (wt.%)		
	Ni	Mg	Al
A	78.68	13.31	8.01
B	27.74	47.27	24.99
C	5.58	60.09	34.32
D	3.86	60.68	35.45
E	0.64	60.06	39.30

After dipping, core shell Ni loading occurred, as shown in Fig. 3. It is clearly shown in Table 2 that Ni was concentrated, while Mg was diluted, on going to the surface layer from the centre of the particle.The amount of Al showed a drastically varivation, especially between area A and C, which could attribute to the strongly decreased Ni content, Al^{3+} partly dissolve in acidic solution and area A is almost all on the particle surface. These again suggest that a part of Mg^{2+} (and/or Al^{3+}) in HT-like phase, formed by reconstitution, were selectivity replaced by Ni^{2+} during the dipping.

Interestingly, BET surface area and pore volumes of the Mg-Al mixed oxide particles drastically decreased (from 194 to 6.5 m^2 g^{-1}, 0.47 to 0.01 cm^3 g^{-1}) after the dipping treatment and drying at 120 °C, indicating that the surface was covered by thick dense layer. This thick dense layer is probably composed of Ni-Mg-Al HT-like compounds which covered the particle surface, possessed porous structure and hindered the further penetration of Ni^{2+} nitrate, as evidenced by Fig 2(c) and (d). After the calcination at 700 °C, all samples recovered the surface area to a comparable value with conventional Ni/Mg-Al catalyst. It can be concluded that the preparation conditions significantly affected the core shell Ni concentrated layer, and a balance between the reconstitution rate of Ni-Mg-Al HT-like structure and the penetration rate of Ni^{2+} nitrate may decide the Ni content [10].

3.2. Catalytic performance

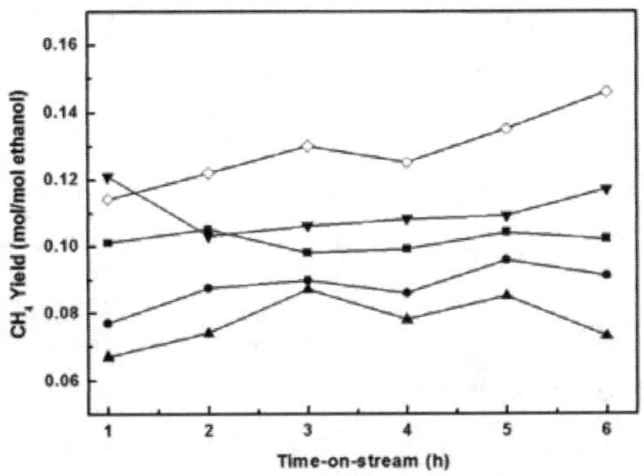

Fig. 4. The comparison of CH_4 yields from SCR over core shell and conventional Ni-loaded catalysts during 6 h time-on-stream (mol/mol ethanol). (■) Mg-Al-15; (●) Mg-Al-30; (▲) Mg-Al-45; (▼) Mg-Al-60; (◇) Ni/Mg-Al.

Table 3. The average product distribution of core shell and conventional catalysts during 6 h time-on-stream (mol/mol ethanol).

Sample	H_2	CO_2	CH_4	CO
Ni/Mg-Al	4.62	0.54	0.129	1.33
Mg-Al-15	4.75	0.67	0.102	1.22
Mg-Al-30	4.89	0.62	0.088	1.30
Mg-Al-45	4.81	0.63	0.077	1.29
Mg-Al-60	4.75	0.68	0.111	1.21

[illegible]talysts were carried out in ethanol steam reforming at 700 °C and the data during 6 h were collected, averaged and compared with the conventional Ni/Mg-Al catalyst, as illustrated in Table 3. Additionally, CH_4 by-product is one of the key factors that may lead to high H_2 and CO_x product because the generation of 1 mol of the CH_4 will result in loss of more than 2 mol of H_2 in steam reforming. Therefore, the time dependence of CH_4 product yield during 6 h time-on-stream is given in Fig. 4. The Activity results showed that all catalysts are highly active and stable for steam reforming of ethanol even with a stoichiometric feed composition. Ethanol and the reaction intermediates were entirely reformed into H_2 and C1 products. Compared with the conventional Ni/Mg-Al catalyst, the core shell Ni-loaded catalysts with lower Ni content presented higher H_2 and CO_2, lower CH_4 (also shown in Fig. 4) and CO production. This may due to the surface enrichment of active Ni species by loading in core shell structure, which promoted the application efficiency of Ni metal and weakened the effect of intra-particle mass transfer. Among the core shell catalysts, according to results shown in Table 3 and Fig. 4, the activity changed slightly during the dipping time, and the highest H_2 production was observed at the dipping time of 30 min. However, in the case of 45 min, the lowest CH_4 production was achieved. These results indicated that the best dipping time of core shell Ni@Mg-Al catalyst for high H_2 and low CH_4 production may exist between 30 and 45 min aiming to approach a Ni content of 9.0%.

4. Conclusion

Core shell Ni@Mg-Al catalyst was achieved by applying "memory effect" through dipping HT-derived Mg-Al mixed oxide in an aqueous solution of Ni^{2+} nitrate. A low heating temperature, a low calcination temperature, an appropriate pH of the dipping solution (pH=4.0), a gentle temperature of water bath, and a dipping time of 30-45 min were preferable for the core shell Ni loading by balancing the reconstitution and the penetration to form the thick dense layer of Ni-Mg-Al HT-like precursor.

Compared to conventional HT-derived Ni/Mg-Al catalyst, core shell Ni@Mg-Al catalysts with lower Ni content showed a higher activity and better stability in ethanol steam reforming process. It is concluded that the better performance may due to the surface enrichment of active Ni species, which promote the reactant achievable active Ni metal and weakened the effect of intra-particle mass transfer.

Reference

1. Debecker DP, Gaigneaux EM, Busca G. Exploring, tuning, and exploiting the basicity of hydrotalcites for applications in heterogeneous catalysis. Chem Eur J. 2009;15:3920-3935.
2. Takehira K, Shishido T. Preparation of supported metal catalysts starting from hydrotalcites as the precursors and their improvements by adopting "memory effect". Catal Surv Asia. 2007;11:1-30.
3. Li YD, Chen JL, Qin YN, Chang L. Simultaneous production of hydrogen and nanocarbon from decomposition of methane on a nickel-based catalyst. Energy Fuels. 2004;14:1188-1194.
4. Tsyganok AI, Tsunoda T, Hamakawa S, Suzuki K, Takehira K, Hayakawa T. Dry reforming of methane over catalysts derived from nickel-containing Mg-Al layered double hydroxides. J Catal. 2003;213:191-203.
5. Takehira K, Shishido T, Wang P, Kosaka T, Takaki K. Autothermal reforming of CH4 over supported Ni catalysts prepared from Mg-Al hydrotalcite-like anionic clay. J Catal. 2004;221:43-54.
6. Olafsen A, Slagtern A, Dahl IM, Olsbye U, Schuurman Y, Mirodatos C. Mechanistic features for propane reforming by carbon dioxide over a Ni/Mg(Al)O hydrotalcite-derived catalyst. J Catal. 2005;229:163-175.
7. Li MS, Wang XD, Li SR, Wang SP, Ma XB. Hydrogen production from ethanol steam reforming over nickel based catalyst derived from Ni/Mg/Al hydrotalcite-like compounds. Int J Hydrogen Energy. 2010;33:1008-1013.
8. Takehira K, Shishido T, Shoro D, Murakami K, Honda M, Kawabata T, Takaki K. Preparation of egg-shell type Ni-loaded catalyst by adopting "memory effect" of Mg-Al hydrotalcite and it's application for CH4 reforming. Catal Commun. 2004;5:209-213.

9. Takehira K, Shishido T, Shouro D, Murakami K, Honda M, Kawabata T, Takaki K. Novel and effective surface enrichment of active species in Ni-loaded catalyst prepared from Mg-Al hydrotalcite-type anionic clay. Appl Catal A. 2005;279:41-51.
10. Takehira K, Kawabata T, Shishido T, Murakami K, Ohi T, Shoro D, Honda M, Takaki K. Mechanism of reconstitution of hydrotalcite leading to eggshell-type Ni loading on Mg-Al mixed oxide. J Catal. 2005;231:92-104.
11. Millange F, Walton RI, O'Hare D. Time-resolved in situ X-ray diffraction study of the liquid-phase reconstruction of Mg-Al-carbonate hydrotalcite-like compounds. J Mater Chem. 2000;10:1713-1720.

Decomposition of Carbon Dioxide over Zinc Ferrite Nanoparticles with Methane Recycling

Abhijit Krishna Adhikari, Kuen-Song Lin, Yu-Pei Chen, and Hung-Bin Tsai*
Chung-Li City Taiwan, R.O.C.

Abstract

The decomposition of CO_2 over zinc ferrite nanoparticles (ZFNs) at 573 K was studied and the oxidation states of Fe and Zn species were investigated using XANES/EXAFS of fresh and reduced ZFNs with fine structure analyses. Oxygen deficiency in ZFNs was obtained by reducing the ZFNs catalyst in hydrogen. Oxygen deficient ZFNs were found highly active to decompose CO_2 within few minutes. Decomposition of CO_2 was occurred by the incorporation of oxygen in the vacancies of ZFNs and deposition of carbon occurred on the surface. The deposited carbon was converted into methane upon treatment with hydrogen while regenerating used ZFNs to oxygen deficient ZFNs. Oxygen and carbon rather than CO were produced in the decomposition process. Furthermore, the deposited carbon on the surface converts into methane (methanation) upon treatment with H_2 (hydrogenation) while regenerating used ferrite to oxygen deficient ferrite. The complete decomposition of CO_2 was possible because of higher surface-to-volume ratio of ZFNs than bulk zinc ferrite. The pre-edge XANES spectra of Fe in ZFNs exhibits an absorbance feature at 7115 eV for the *1s* to *3d* transition which is forbidden by the selection rule in case of perfect octahedral symmetry. The EXAFS data showed that the fresh ZFNs had two central Fe atoms coordinated by primarily Fe–O and Fe–Fe with bond distances of 1.89 Å and 3.05 Å, respectively. It also revealed that the ZFNs maintained its spinel structure after reduction at 573 K, even though a minor amount of zinc metal was formed. Recovery of valuable methane by decomposition of CO_2 over ZFNs with recycling heat energy of offgas produced from power generation plant or steel industry is also an appealing resource recovery alternative.

Keywords: CO_2 decomposition, Zinc ferrite nanoparticle, Greenhouse effect gas, Methanation, XANES/EXAFS.

Introduction

Carbon dioxide is a highly oxidized and very thermodynamically stable compound, and CO_2 reduction scarcely occurs at room temperature under ambient pressure [1–4]. Carbon dioxide is also of the greatest interest as C_1 feedstock because of the vast amounts and low cost of bulk CO_2. Hydrogenation of CO_2 to produce petrochemical feed stocks, such as oxygenated or hydrocarbons has been extensively explored [2–8], but the high cost of H_2 has limited most applications. Photoreduction of CO_2 has been widely developed into two conceptually different categories that utilize semiconductor electrode/colloid materials [4,5]. Unfortunately, photo-reduction of CO_2 is notoriously inefficient, so that it has not been recognized as an effective CO_2 utilization alternative [6–10]. The mitigation of CO_2 emission has also been a major environmental concern. One of the most effective solutions may be the *in-situ* decomposition of CO_2 to carbon by catalytic route at lower temperature in each stationary source.

Recently, decomposition of CO_2 to carbon and oxygen over nanophase ferrite catalysts may be an innocent and innovatory technique to reduce greenhouse gas effects [11–14]. Ferrite, a stable and nontoxic material has been widely used in various technical applications including in catalysis. Nanophase ferrites show novel properties that are often significantly different from the bulk due to fundamental changes in coordination, symmetry, and confinement [11–13]. Therefore, the use of nanophase ferrites may lead to not only CO_2 decomposition but also production of CH_4 (methanation) which can be used as an energy source. The dynamic alternation between stoichiometry and non-stoichiometry in spinel ferrites ($M(II)Fe(III)_2O_4$), where M denotes e.g. Zn, Ni, Co, Cu etc. metal is an important feature that can be exploited in gas reactions. These reactions involve lattice oxygen in two different ways, either through the transfer of oxygen atoms from the ferrite to the gas (oxidation reaction) or through oxygen vacancies being replenished by oxygen from the gas (reduction reaction). The reduction process requires the non-stoichiometric form which can be obtained by the use of a reducing agent such as hydrogen. These non-stoichiometric or oxygen-deficient ferrites can be used to decompose water to hydrogen and carbon dioxide to carbon [15]. Oxygen-deficient ferrites are interesting because of their high decomposition efficiencies for both water and carbon dioxide at low temperatures. Tamaura et al. [15] reported that oxygen-deficient magnetite (Fe_3O_4) decomposes carbon dioxide to carbon with efficiencies up to 100% at low temperature as below 290°C with the formation of trace amount of carbon monoxide. As the operating temperature is very low, the spinel structure of the ferrite remains stable and able to accommodate interchanges between stoichiometry and non-stoichiometry. Oxygen-deficient ferrites are expressed as $M(II)Fe(III)_2O_{4-\delta}$, where δ is the degree of oxygen deficiency. A high δ in magnetite leads to increased Fe^{2+} reduction potentials and enhanced decomposition efficiencies. The M metal influences the reactivities of reduced ferrites toward decomposing carbon dioxide by steering the maximum degree of oxygen-deficiency and thereby the reduction potential of Fe^{2+} [16–19]. Significantly, Kato et al. [16] showed that both the rate of reduction and decomposition are greatly improved when the A metal is Zn^{2+} instead of Fe^{2+}, whereas Tabata et al. [17,18] reported that the opposite is true when the A metal is Mn^{2+}. The valency and fine structure of Zn or Fe atom have not been well studied, though the decomposition of CO_2 has been studied by many groups. The major metal components of the nanophase ferrites are nickel, zinc and iron. X-ray absorption spectroscopy (XAS) is an excellent technique for characterizing the valency and local structure of metals in catalysts with short-range order [20,21]. X-ray absorption near edge structure (XANES) spectroscopy provides information of electronic configuration, stereochemistry and the oxidation states of the metallic atoms in nanophase catalysts being investigated. In addition, extended X-ray absorption fine structure (EXAFS) spectroscopy can investigate the information on the atomic arrangement of nanophase catalysts in terms of bond distance, number and kind of near neighbors, thermal and static disorder [20,21]. The XANES and EXAFS also offer the basic knowledge of understanding the oxidation states and fine structures of Zn or Fe in nanophase ferrites to a further study on the mechanism of surface reaction of nanophase ferrites. Thus, the main objectives of the present work were to envisage the conversion of CO_2 and the fine structures of Zn and Fe species in the nanophase ferrite catalysts.

Experimental

Preparation of nanophase $ZnFe_2O_4$

$ZnFe_2O_4$ ($Zn_xFe_{3-x}O_{4-\delta}$, $0<\delta<1$) were synthesized by oxidation in air with aqueous suspensions of Fe(II) and Zn(II) hydroxide mixture based on a hydrothermal method. The mixture of zinc sulfate and iron nitrates were neutralized with aqueous ammonia solution until the pH value reached to 8.5. The solution were transferred into a 250 mL PTFE-lined stainless-steel autoclave and heated at 450 K for 4 h. After the hydrothermal reaction, the solid residues of $ZnFe_2O_4$ crystals were separated from the solution by centrifugation, washed repeatedly and thoroughly with deionized water, and finally dried at 378 K overnight.

Decomposition of CO_2

Decomposition of CO_2 was catalyzed by prepared nanophase $ZnFe_2O_4$ and the reactions were performed in a continuous flow method using a fixed bed reactor connected to a gas-phase chromatography (GC/TCD, Hewlett-Packard 5890) equipped with a mass sensitive detector. About 1.5-2.0 g $ZnFe_2O_4$ was placed in a quartz cell and heated from ambient temperature to 573 K in a hydrogen flow (80 mL min^{-1}) and kept this temperature for 3 h. After flowing with helium (80 mL min^{-1}, 10 min), CO_2 gas (5 mL min^{-1}) was decomposed at 573 K over the reduced ZFNs catalyst. The gaseous products, were detected by GC/TCD. A Porapak R stainless steel column (80/100, 9.15 mL × 3.175 mm (O.D.)) was programmably heated up at 308 K and held for 11 min and then heated up to 448 K at 30 K min^{-1} to obtain a resolvable separation of CO, CO_2, and CH_4 gases. Helium was used to elute the waste gases for clearing as a carrier gas. After the CO_2 decomposition, the reactor was flushed again with hydrogen gas and monitoring the outlet gases for hydrocarbons.

Characterization

X-ray diffraction (XRD) measurements were conducted using standard powder diffraction procedures. The sample was smear-mounted onto a glass slide and then analyzed by Cu Kα radiation (RIGAKU Model D/MAX III-V) at 30 kV and 20 mA with a scan rate of 4°(2θ) min^{-1}. The specific peak intensities and 2θ values recorded were further identified by a computer database system (JCPDS). FE-SEM (S-4700 Type II) and TEM (Zeiss 10C) were performed to identify the morphologies, crystallinities, and particle size distribution. Nitrogen adsorption isotherms of the samples were measured at 77 K using an ASAP 2010 (Micromeritics).

X-ray absorption (EXAFS and XANES) spectra were collected at the Wiggler beam line 17C1 at NSRRC of Taiwan. The electron storage ring were operated with energy of 1.5 GeV and a current of 100-200 mA. A Si (111) double-crystal monochromator (DCM) was used for selection of energy with an energy resolution of 1.9×10^{-4}. Data were collected in fluorescence or transmission mode with a Lytle detector [22] in the regions of the Zn (8333 eV) and Fe (7112 eV) K edges at room temperature. The photon energy was calibrated by characteristic pre-edge peaks in the absorption spectra of Zn and Fe standards. The raw absorption data below the edge position were fit to a straight line using the least-square algorithms. The fitted pre-edge background curves were extrapolated throughout all data range, and subtracted and normalized to minimize the effect of sample thickness. The near-edge structure in an absorption spectrum covers the range between the threshold and the point at which the EXAFS begins [11,23]. The k^2 and k^3-weighted and EXAFS spectra were Fourier transformed over the range of interatomic distance (R). The EXAFS data were further analyzed by using the UWXAFS 3.0 program and FEFF 8.0 codes [11,23].

Results and discussion

Results from the decomposition of carbon dioxide at 573 K over reduced nanophase and bulk zinc ferrites are shown in Fig.1. For both materials, the highest activity was observed within first 6 min of reaction and thereafter gradually stabilizes at constant levels. After stabilized the decomposition activity, the CO_2 decomposition rate for ZFNs was 97~99% while it was about 44% for bulk zinc ferrite and there was no further change was occurred during 25 min.

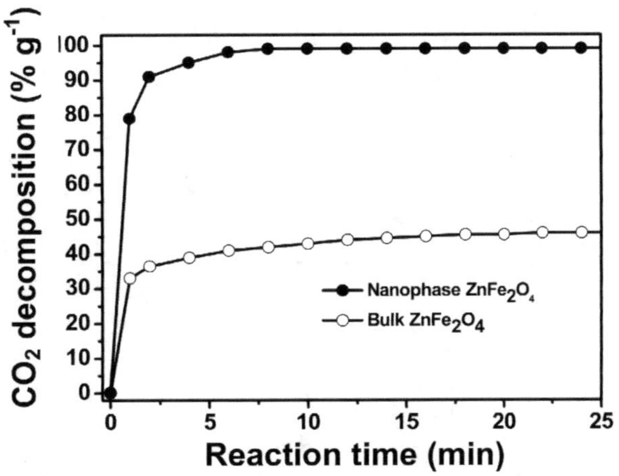

Fig.1. Decomposition of CO_2 over nanophase $ZnFe_2O_4$ and bulk $ZnFe_2O_4$ catalysts at 573 K and 1 atm.

Prior to the CO_2 decomposition both samples were reduced by hydrogen at same temperature and time that makes the materials more active to adsorb oxygen from CO_2. Due to the slight changes in particle size, the ZFNs were highly effective in decomposing CO_2 to C and O_2 within few minutes. As expected, due to more easy access of oxygen vacancies created by reduction, ZFNs with a higher surface area was more reactive to CO_2 decomposition. There are two reaction routes for decomposing CO_2 by oxygen deficient ferrites. One is the decomposition to carbon and oxygen while the other one is decomposition into carbon monoxide and oxygen which can be shown in following equations where δ represents the degree of oxygen deficiency [24–28].

$$\delta/2\ CO_2 + ZnFe_2O_{4-\delta} \rightarrow \delta/2\ C\ (s) + ZnFe_2O_4 \quad \text{(i)}$$
$$\delta\ CO_2 + ZnFe_2O_{4-\delta} \rightarrow \delta\ CO + ZnFe_2O_4 \quad \text{(ii)}$$

Generally, the reactions take place simultaneously which is reported by different scientists for similar oxygen deficient ferrites [19,24]. In our experiment, ZFNs produced little or no CO while bulk zinc ferrite resulted in a small amount of undesired CO. The reactivity toward CO_2 decomposition is based on the redox processes of ZFNs. If the redox reactions of the ferrite catalysts with H_2 and CO_2 are accelerated in the presence of catalyst, the CO_2 decomposition reactivity of nanophase ferrite can be improved by impregnation of catalyzing metals such as zinc, copper, cobalt, zinc, manganese etc. on them [25–28]. During the redox reaction, the oxygen atoms of adsorbed CO_2 molecules are incorporated into the oxygen-deficient site of the activated ZFNs, and the carbon is deposited on the solid surface to form elementary and/or polymerized carbon.

Table 1. The physicochemical properties of bulk and nanophase zinc ferrite catalysts

Properties	Bulk $ZnFe_2O_4$	Nanophase $ZnFe_2O_4$
Partical size (nm)	160-380	15-35
Surface area (m^2g^{-1})	12-18	155-156
Shape uniformity	Medium	Good
Reaction activity	Medium	Exellent
Crystalliity of ferrite	Medium	Poor
Selectivity of CO_2 decompositon	0.38-0.46	0.87-0.99

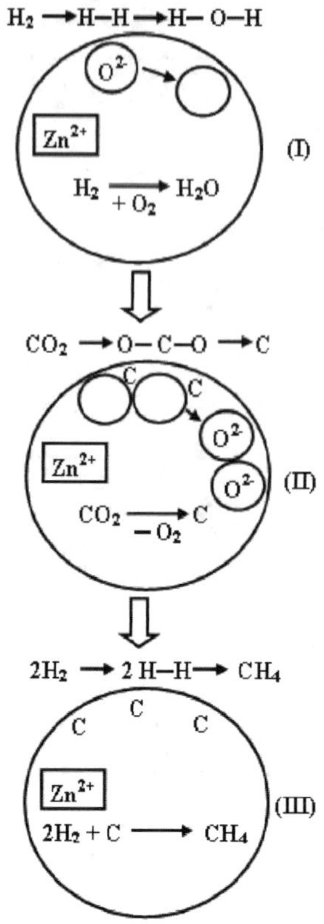

Fig.2. Symmetric flow diagram of CO_2 decomposition over $ZnFe_2O_4$ nanoparticles. (I) H_2 gas decomposes and produces the H_2O molecule which makes oxygen deficient $ZnFe_2O_4$, (II) oxygen deficient $ZnFe_2O_4$ adsorbs oxygen atoms from CO_2 and produces carbon on the surface, and (III) the produced carbon reacts with H_2 and produces CH_4 by methanation.

Table 1 shows the comparison between nanophase and bulk zinc ferrite materials. It shows that ZFNs have the surface area of 160 m^2g^{-1} while bulk zinc ferrite has an average surface area of 15 m^2g^{-1} only. Therefore, the bulk zinc ferrite was less efficient in CO_2 decomposition than the ZFNs. In addition, the bulk zinc ferrite led to a higher fraction of CO formation compared to nanophase one. It indicated that the complete decomposition of CO_2 may carried out through the use of ZFNs because their surface-to-volume ratio. and it was the key point to make more oxygen deficient active catalyst. The mechanism of CO_2 decomposition by the oxygen deficient ferrites is shown in Fig. 2. In principle, when ZFNs were treated with hydrogen, they led to the reduction of Fe^{3+} and a charge compensated by ejecting oxygen anions (creation of oxygen vacancies) from the spinel structure. When CO_2 contacts with oxygen deficient ferrites, decomposition of CO_2 occurs by the incorporation of oxygen anions in the vacancies in the oxygen deficient ferrite thereby restoring the ferrite to stoichiometry. At the same time, electrons are donated from the oxygen deficient ferrite to produce carbon or carbon monoxide. Next, the deposited carbon on the surface converts into methane (methanation) upon treatment with H_2 (hydrogenation) while regenerating used ferrite to oxygen deficient ferrite. In the second step, carbon monoxide may form instead of carbon when the ZFNs were less active or in the other word, when the deficiency of oxygen will be minimized and the structure reached into stoichiometry. The relative oxygen donating ability determines the activity of the ZFNs. Tabata et al. [17] reported that the divalent iron which is formed by reduction acts as the electron donor. Thus, the higher degree of oxygen deficiency or higher content of Fe^{2+} of the ZFNs lead to higher degree of conversion of CO_2. Moreover, the δ value for a particular ferrite depends on metal A while the structure can be denoted as AFe_2O_4. Different metal have different reduction ability to

Fe^{3+}. For example, the δ values of nickel and zinc reported as 0.19 and 0.06 by Tabata et al. [19] while 0.58 and 0.16 by Nordhei et al. [28] respectively. The δ value is much higher for nickel metal as it has $3d^8$ orbital while zinc has $3d^{10}$. In this work, we get the δ value for zinc metal is about 0.42 which is higher than earlier reported value. From the viewpoint of energy saving, recovery of valuable methane by decomposition of CO_2 over ZFNs with recycling heat energy of offgas produced from plants is also an appealing resource for recovery of energy.

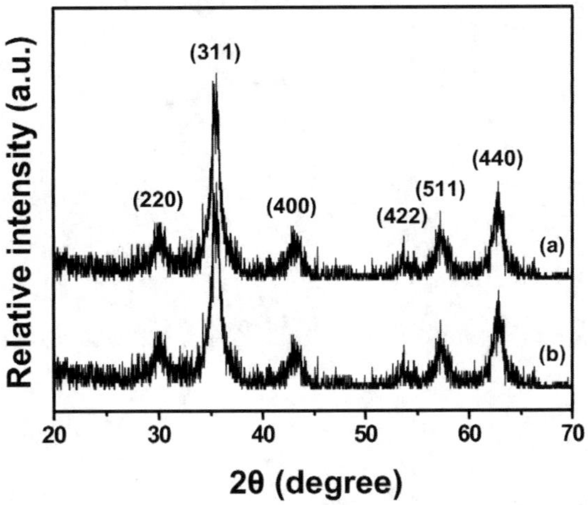

Fig. 3. XRD patterns of (a) fresh and (b) reduced zinc ferrite nanoparticles.

The XRD patterns of the fresh and reduced ZFNs materials are plotted in Fig. 3(a) and 3(b) respectively. As shown in Fig. 3, both fresh and reduced samples are crystalline as they show sharp diffraction peaks, and the position and relative intensity of the diffraction peaks of the sample match well with the standard XRD data for spinel zinc ferrite. Moreover, the diffraction peaks of our sample prepared in this study matched well with the already published XRD patterns on the same ZFNs [25,29,30]. The intensive peaks appearing at 2θ angles in the XRD pattern of ZFNs are characteristics of porous materials which possess numerous pores or cavities. It can be seen that the structure remain spinel after the reduction process. It was observed that the diffraction peaks for both fresh and reduced ZFNs materials corresponding to (220), (311), (400), (422), (511), and (440) diffraction planes of ZFNs were present [31]. A spinel can be represented by the formula AB_2O_4, the A-B magnetic interactions between the magnetic atoms on the A (tetrahedral) site and the B (octahedral) sites are stronger than A-A interactions and B-B interactions. That is, nanophase $ZnFe_2O_4$ crystallizes in the spinel structure which contains two different cation sites: eight tetrahedral A sites and 16 octahedral B sites. From the XRD data, in the lattice of ZFNs with a completely inverse-spinel structure, an equal number of the Fe cations are at the tetrahedral and octahedral sites, respectively. The recoiless fractions of Fe nucleus at the tetrahedral and octahedral sites are nearly equal. The oxygen-deficient ferrite synthesized by activation of spinel ferrite with H_2 at 573 K was represented by $Zn_xFe_{3-x}O_{4-\delta}$ (O≤x≤1, δ>0) where δ denoted the oxygen deficiency. The average crystallite size of ZFNs is calculated to be 17 nm using Scherrer formula from the (311) plane (which corresponds the strongest peak). The particle size of synthesized ZFNs and reduced ZFNs were around 10~30 nm and 15~40 nm identified by FE-SEM micrographs shown in Figs. 4(a) and 4(b), respectively. The fresh ZFNs were well dispersed and spherical in shape with irregular size. However, the reduced ZFNs were also in spherical shape with a little bigger diameter. It may attribute, in higher reduction temperature the samples aggregated and thus gives the bigger diameter than that of fresh ZFNs. This phenomenon also revealed from extensive TEM examinations shown in Fig. 4. TEM images of fresh and reduced ZFNs samples are shown in Figs. 4(c) and 4(d), respectively, where it can be seen that the particle size of the reduced sample is little bit bigger than that of fresh ZFNs. TEM images revealed that the particle size of the fresh ZFNs were about 10~25 nm while the range was larger for reduced ZFNs.

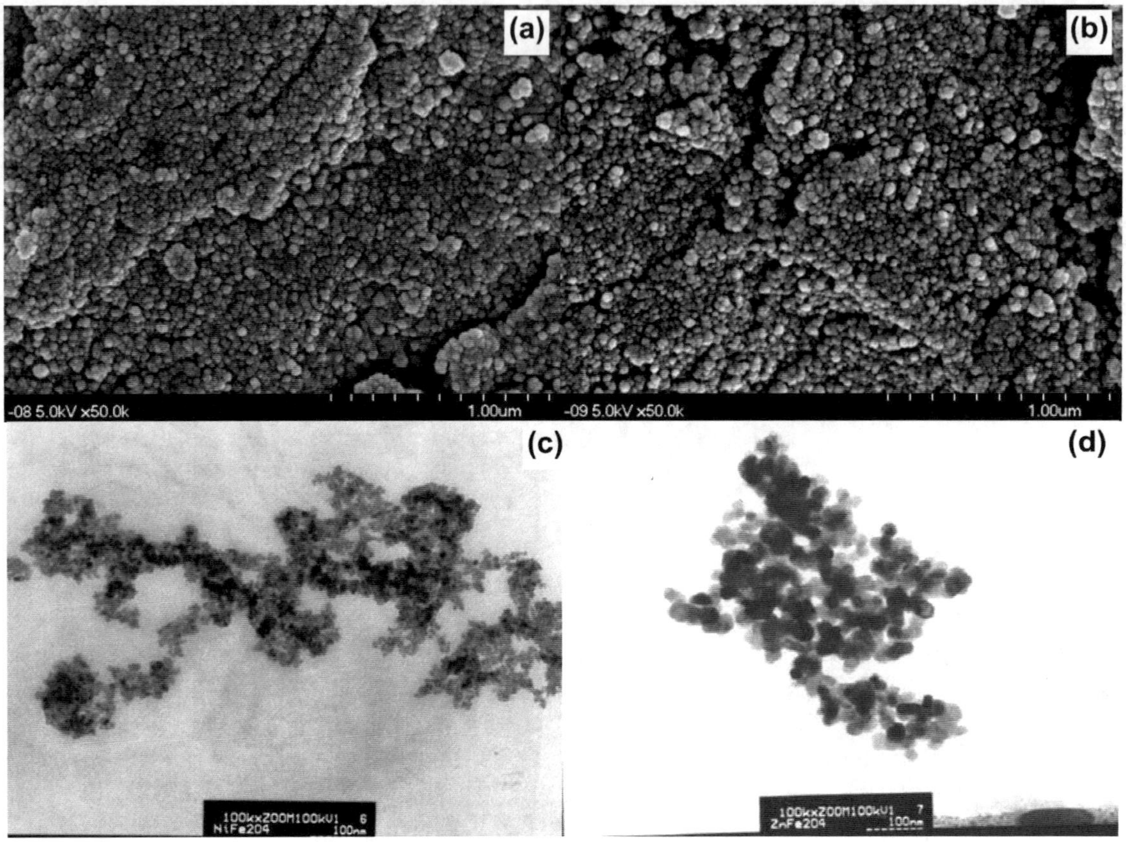

Fig. 4. FE–SEM and TEM micrographs of fresh (a) (c) and reduced (b) (d) zinc ferrite nanoparticles, respectively for CO_2 decomposition at 573 K and 1 atm.

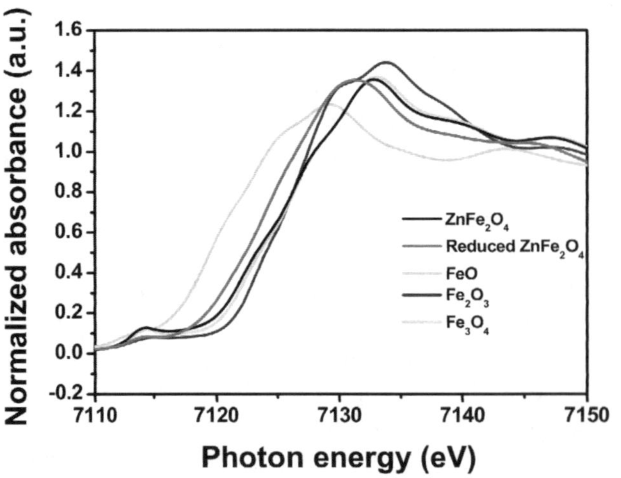

Fig. 5. Normalized Fe K-edge XANES spectra of fresh nanophase $ZnFe_2O_4$, reduced $ZnFe_2O_4$ nanoparticle, FeO, Fe_3O_4, and Fe_2O_3 standards.

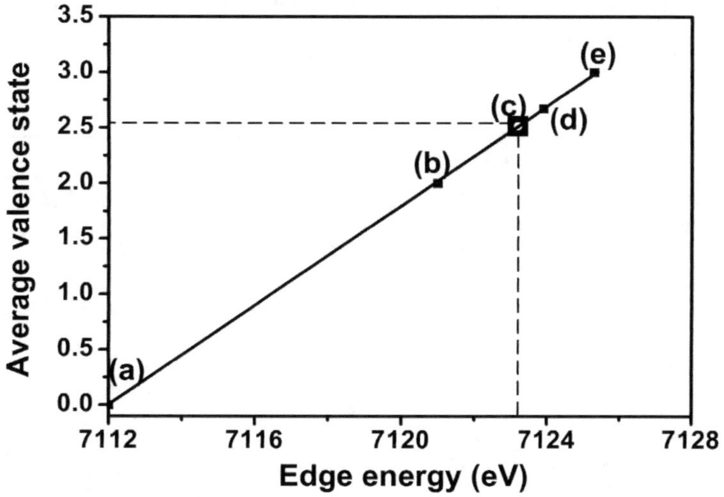

Fig. 6. Edge energies of (a) Fe metal, (b) FeO, (c) Fe_3O_4, (d) reduced $ZnFe_2O_4$, and (e) Fe_2O_3 as a function of their average valence state.

Fig. 5 shows the normalized Fe K-edge XANES of the fresh and reduced zinc ferrites and the references FeO, Fe_2O_3, and Fe_3O_4. The Fe absorption K-edge shifts to lower energies upon reduction in hydrogen which is consistent with a reduction in the average valence state of Fe^{3+}. Removing oxygen from the spinel structure of $ZnFe_2O_4$ mainly influences the valence state of iron with the shift being largest for ZFNs. The pre-edge XANES spectra of Fe in ZFNs exhibit an absorbance feature (7115 eV) for the 1s to 3d transition which was forbidden by the selection rule in the case of perfect octahedral symmetry. The sharp feature at 7134 eV, due to the dipole-allowed of 1s to $4p_{xy}$ electron transition, indicated the existence of Fe^{3+}. The intensity of the 1s to $4p_{xy}$ transition was proportional to the population of Fe^{3+} in ZFNs. Due to the reduction in the flowing hydrogen, a shoulder at 7121 eV and an intense feature at 7126 eV were attributed to the 1s to $4p_{xy}$ transition that indicated the existence of Fe^{2+} species in reduced ZFNs catalysts. Oxygen and zinc are the major atoms coordinated to the two central Fe atoms in fresh nanophase ferrites. The edge energies of the four references FeO, Fe_2O_3, Fe_3O_4, and iron metal as 2, 3, 2.67, and 0, respectively were plotted against their average valence state shown in Fig. 6. This plot was used to estimate the valence states of the reduced ferrite as it displays a linear relationship between the edge energy and valence state as exemplified by the reduced ZFNs (Fig. 6(c)). The edge energies (±0.5 eV) were defined as being half the edge height. The local environments in the oxide references and the ferrites are similar; hence the contribution from the chemical shifts to the absorption edges is also similar. As the iron atoms are in both Fe^{+2} and Fe^{+3} valence states, the spinel Fe_3O_4 is especially appropriate as a reference. Therefore, it is valid to use Fig. 6 to ascertain the valence states of the reduced ferrite and the estimated chemical formula of the reduced ferrite is $ZnFe_2O_{3.58}$.

The ZFNs were further studied by using EXAFS to understand the fine structure and the behavior in the decomposition of carbon dioxide. Fe K-edge EXAFS spectroscopy can provide the information on the Fe atomic arrangement of catalysts in terms of bond distance, number, and kind of near neighbors. Fig. 7 shows the k^3-weighted and least-square fitted Fe K-edge EXAFS and their Fourier transforms of fresh (Figs. 7(a) and 7(b)) and reduced (Figs. 7(c) and 7(d)) ZFNs, respectively. Since, XANES shows significant changes between fresh and reduced ZFNs, it is interesting that the EXAFS data shown in Table 2 are virtually identical. It shows that the ZFNs have two center Fe atoms coordinated by primarily Fe–O and Fe–Fe bonding. The fine structural parameters of ZFNs analyzed by the EXAFS data also suggested that Fe atoms in ZFNs may be mirror-symmetrically (D_d) surrounded by two Zn atoms and four oxygen atoms.

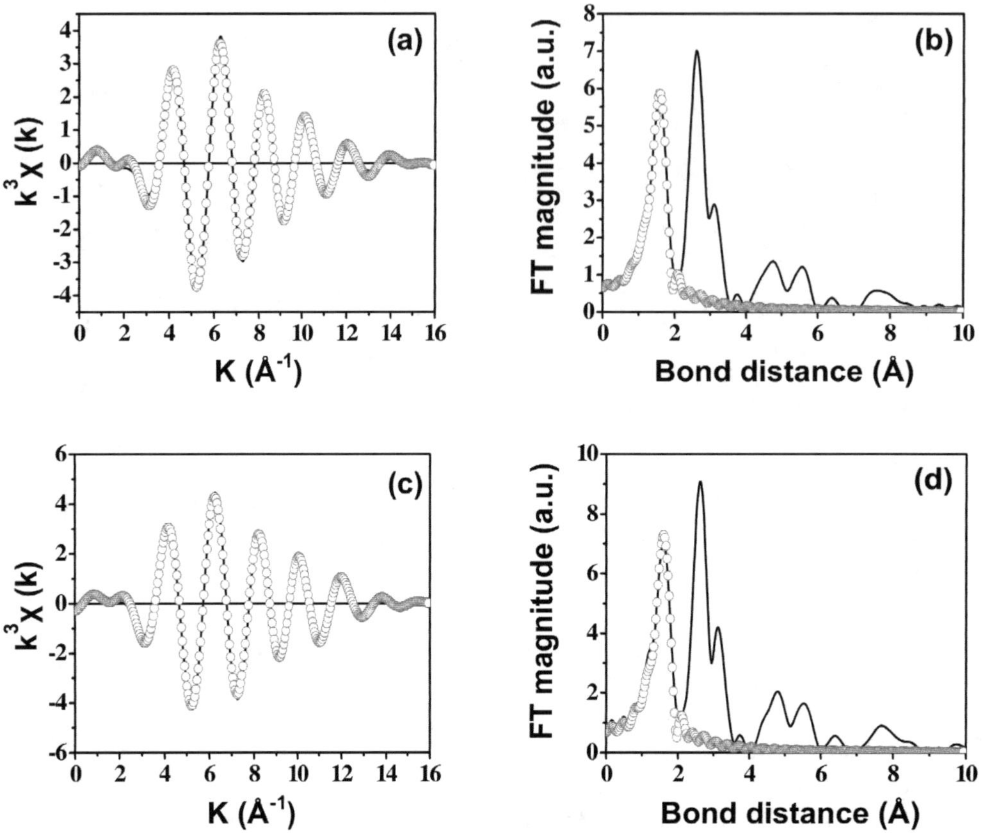

Fig. 7. Fe K-edge EXAFS oscillation $k^3 \chi$ (k) and Fourier transform (FT) spectra of (a) (b) fresh nanophase $ZnFe_2O_4$, and (c) (d) reduced $ZnFe_2O_4$ nanoparticles, respectively. The best fitting of the EXAFS spectra are expressed by the dotted lines.

Table 2. Fine structural parameters of fresh and reduced $ZnFe_2O_4$ nanoparticles analyzed by using EXAFS at Fe and Zn K-edges

Shell		[1]CN (± 0.05)	[2]R (± 0.02 Å)	$^3\Delta\sigma^2(Å^2)^c$
Fresh $ZnFe_2O_4$				
Fe–O	1st	5.72	1.85	0.006
	2nd	2.04	1.99	0.008
Fe–Fe	1st	4.30	3.03	0.012
	2nd	5.71	4.32	0.018
Zn–O	—	4.02	1.96	0.005
Reduced $ZnFe_2O_4$				
Fe–O	1st	5.49	1.88	0.005
	2nd	1.97	2.03	0.011
Fe–Fe	1st	4.07	3.06	0.010
	2nd	5.14	4.34	0.019
Zn–O	—	3.81	2.01	0.006

[1]Coordination number; [2]Bond distance; [3]Debye-Waller factor.

The EXAFS data also showed that the fresh ZFNs had two central Fe atoms coordinated by primarily Fe–O and Fe–Fe with bond distances of 1.85 ± 0.02 Å and 3.03 ± 0.02 Å with the coordination number of 5.7 and 4.3, respectively. After the reduction, the bond distances of Fe–O and Fe–Fe changes into 1.88 ± 0.02 Å and 3.06 ± 0.02 Å, respectively. This is a statistical effect; this means that in spite of removing a small amount of oxygen atoms from the lattice, the overall structure is maintained with little electronic changes. In contrast to the Fe K-edge, the Zn K-edge Fourier transforms of fresh (Figs. 8(a) and 8(b)) and reduced ZFNs (Figs. 8(c) and 8(d)) show significant differences. Minor amounts of zinc metal are formed after reduction. Moreover, the bond length of Zn–O of fresh ZFNs is 1.96 ± 0.02 Å while it is 2.01 ± 0.02 Å for the reduced ZFNs shown in Table 2. The lack of any structural changes confirms that the spinel structure is maintained after reduction, even though a minor amount of zinc metal is formed. Hence, the EXAFS suggests that over reduction does not lead to a complete disintegration of the structure but rather to ejection of metal from the oxygen deficient spinel. These results are similar to those reported by Zhang et al. [32] where the oxygen-deficiency is constant regardless of the quantity of metal formed. Importantly, this means that, in spite of over-reduction, the material is still active in the reaction with carbon dioxide.

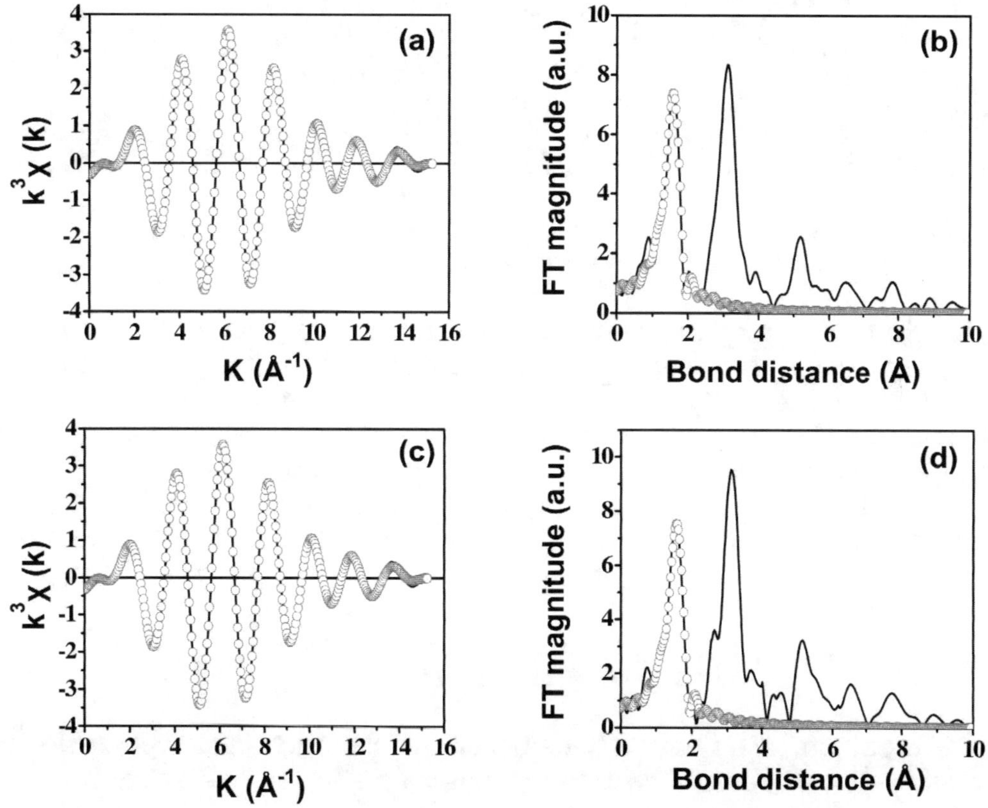

Fig. 8. Zn K-edge EXAFS oscillation $k^3 \chi$ (k) and Fourier transform (FT) spectra of (a) (b) fresh nanophase $ZnFe_2O_4$, and (c) (d) reduced $ZnFe_2O_4$ nanoparticles, respectively. The best fitting of the EXAFS spectra are expressed by the circled lines.

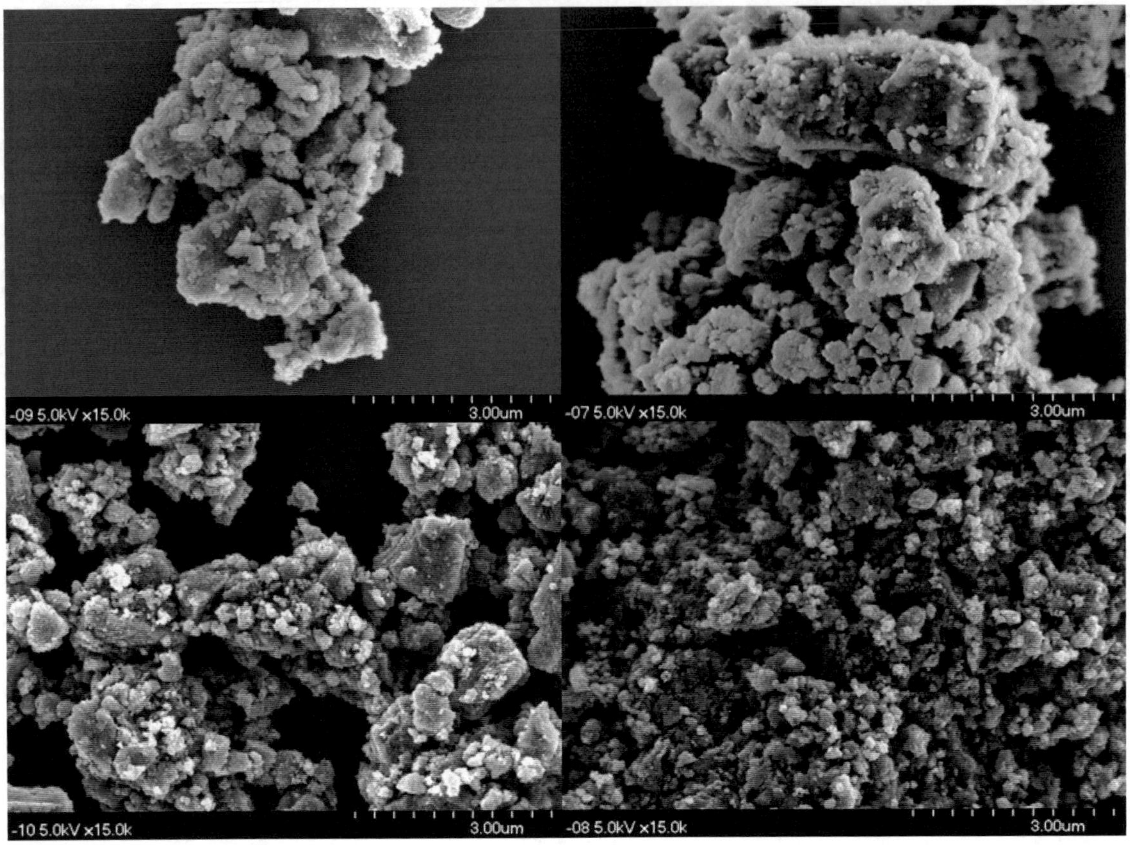

Fig. 9. FE-SEM micrographs of $ZnFe_2O_4$ nanoparticles after (a) 2, (b) 4, (c) 6, and (d) 8 h of methanation.

The FE-SEM images of the reduced ZFNs during reactivated or hydrogenated at different time interval of 2, 4, 6, and 8 h were shown in Fig. 9. It can be seen that at the beginning of reduction process the surface of the ZFNs is covered by whether carbon or the intermetallic compounds formed by produced carbon and the expelled zinc metal from the structure. As the reduction time increased, the amount of the carbon or the intermetallic compounds removed due to the methanation process. The decomposition of CO_2 by the 6 h reactivated ZFNs showed good result with more than 85% decomposition rate. Interestingly, the particle size also increased. It might happen due to the aggregation of the particles at that higher temperature of 573 K.

Conclusions

From XRD pattern, ZFNs having the smaller particle sizes and lower crystal structure showed a relatively broad peak. TEM photographs also showed fine spherical features of the ZFNs with the diameters ranged of 10~25 nm. The ZFNs were all separated and the sizes were narrowly distributed. Due to the slight changes in particle sizes, the ZFNs catalysts are highly effective in decomposing CO_2 to C and O_2 in few minutes. As expected, due to more easy access of oxygen vacancies created by reduction, ZFNs with a higher surface area were more reactive to CO_2 decomposition. In addition, the bulk zinc ferrite led to a higher fraction of CO compared to nanophase one. It indicated that the complete decomposition of CO_2 is possible through the use of ZFNs because of their higher surface-to-volume ratio. The pre-edge XANES spectra of Fe species in ZFNs exhibit an absorbance feature (7115 eV) for the 1s to 3d transition which is forbidden by the selection rule in the case of perfect octahedral symmetry. The EXAFS spectra also showed that the ZFNs had two central Fe atoms coordinated by primarily Fe–O and Fe–Fe with bond distances of 1.85 Å and 3.03 Å, respectively.

References

1. Joos F., Plattner G.K., Stocker T.F. and Schmittner A.K. (1999), "Global Warming and Marine Carbon Cycle Feedbacks on Future Atmospheric CO_2," Science 284 pp. 464–467.
2. Jessop P.G., Noyori R. (1995), "Homogeneous Hydrogenation of Carbon Dioxide," Chem. Rev. 95 pp. 259–272.
3. Borodko Y. and Somorjai G.A. (1999) "Catalytic hydrogenation of carbon oxides – a 10-year perspective," Appl. Catal., A 186 pp. 355–362.
4. Nomura N., Tagawa T. and Goto S. (1998) "In situ FTIR study on hydrogenation of carbon dioxide over titania-supported copper catalysts," Appl. Catal., A 166 pp. 321–326.
5. Li D., Lchikuni N., Shimazu S. and Uematsu T. (1999) "Hydrogenation of CO_2 over sprayed Ru/TiO_2 fine particles and strong metal–support interaction," Appl. Catal., A 180 pp. 227–235.
6. Wang H.Y. and Au C.T. (1997) "Carbon dioxide reforming of methane to syngas over SiO_2-supported rhodium catalysts," Appl. Catal., A 155 pp. 239–252.
7. Wang S. and Lu G.Q. (1998) "Reforming of methane with carbon dioxide over Ni/Al_2O_3 catalysts: Effect of nickel precursor," Appl. Catal., A 169 pp. 271–280.
8. Yamasaki M., Komori M., Akiyama E., Habazaki H., Kawashima A., Asami K., and Hashimoto K. (1999), "CO_2 methanation catalysts prepared from amorphous Ni–Zr–Sm and Ni–Zr–misch metal alloy precursors," Mater. Sci. Eng. A 267 pp. 220–226.
9. Wang J.-Y., Xia G.-G., Huang A., Suib S.L., Hayashi Y.and Matsumoto H. (1999), "CO2 Decomposition Using Glow Discharge Plasmas," J. Catal. 185 pp. 152–159.
10. Furukawa S., Okada M. and Suzuki Y. (1999), "Isolation of Oxygen Formed during Catalytic Reduction of Carbon Dioxide Using a Solid Electrolyte Membrane," Energy Fuels 13 pp. 1074–1081.
11. Tsuji M., Wada Y., Yamamoto T., Sano T. and Tamaura Y. (1996), "CO_2 decomposition by metallic phase on oxygen-deficient Ni(II)-bearing ferrite," J. Mat. Sci. Let. 15 pp. 156–158.
12. Choung J.W., Xu Z. and Finch J.A. (1999), "Role of Complexing Agents in Ferrite Formation under Ambient Conditions," Ind. Eng. Chem. Res. 38 pp. 4689–4693.
13. Rondinone A.J., Samia A. C. S. and Zhang Z. J. (2000), "A Chemometric Approach for Predicting the Size of Magnetic Spinel Ferrite Nanoparticles from the Synthesis Conditions," J. Phys. Chem. B pp. 7919–7922.
14. Luo H. Y., Yue Z. X. and Zhou J. (2000), "Synthesis and high-frequency magnetic properties of sol–gel derived Ni–Zn ferrite–forsterite composites," J. Magn. Magn. Mater. 210 pp. 104–108.
15. Tamaura Y. and Tabata M. (1990), "Complete reduction of carbon dioxide to carbon using cation-excess magnetite," Nature 346 pp. 255–256.
16. Kato H., Kodama T., Tsuji M., Tamaura Y. and Chang S.G. (1994), "Decomposition of carbon dioxide to carbon by hydrogen-reduced Ni(II)-bearing ferrite," J. Mater. Sci. 29 pp. 5689–5692.
17. Tabata M., Nishida Y., Kodama T., Mimori K., Yoshida T. and Tamaura Y. (1993), "CO_2 decomposition with oxygen-deficient Mn(II) ferrite," J. Mater. Sci. 28 pp. 971–974.
18. Tabata M., Akanuma K., Nishizawa K., Mimori K., Yoshida T., Tsuji M. and Tamaura Y. (1993), "Reactivity of oxygen-deficient Mn(II)-bearing ferrites (MnxFe3-xO4-δ, O≤x≤1, δ>0) toward CO_2 decomposition to carbon," J. Mater. Sci. 28 pp. 6753–6760.

19. Kodama T., Tabata M., Sano T., Tsuji M. and Tamaura Y. (1995), "XRD and Mössbauer Studies on Oxygen-Deficient Ni(II)-Bearing Ferrite with a High Reactivity for CO2 Decomposition to Carbon," J. Solid State Chem. 120 pp. 64–69.
20. Meitzner G., Via G.H., Lytle F.W. and Sinfelt J.H. (1992), "Analysis of x-ray absorption edge data on metal catalysts," J. Phys. Chem. 96 pp. 4960–4964.
21. Zabinsky S.I., Rehr J.J., Ankudinov A., Albers R. C. and Eller M.J. (1995), "Multiple-scattering calculations of x-ray-absorption spectra," Phys. Rev. B 52 pp. 2995–3009.
22. Lytle F.W. (1999), "The EXAFS family tree: a personal history of the development of extended X-ray absorption fine structure,"J. Synchrotron Rad. 6 pp. 123–134.
23. A.I. Nesvizhskii, J.J. Rehr, (1999) "L-edge XANES of 3d-transition metals," Synchrotron Rad. 6 pp 315–316.
24. Tabata M., Akanuma K., Togawa T., Tsuji M. and Tamaura Y. (1994), "Mössbauer study of oxygen-deficient ZnII-bearing ferrites (ZnxFe3 –xO4 –δ, 0 x 1) and their reactivity toward CO2 decomposition to carbon,"J. Chem. Soc. Faraday Trans. 90 pp. 1171–1175.
25. Sarangi P. P., Vadera S.R., Patra M.K. and Ghosh N.N. (2010), "Synthesis and characterization of pure single phase Ni–Zn ferrite nanopowders by oxalate based precursor method,"Powder Technol. 203 pp. 348–353.
26. Shin H.-C., Choi S.-C., Jung K.-D. and Han S.-H. (2001), "Mechanism of M Ferrites (M = Cu and Ni) in the CO2 Decomposition Reaction," Chem. Mater. 13 pp. 1238–1242.
27. Nordhei C., Mathisen K., Safonova O., Beek W. van, Nicholson D.G. (2009), "Decomposition of Carbon Dioxide at 500 °C over Reduced Iron, Cobalt, Nickel, and Zinc Ferrites: A Combined XANES-XRD Study," J. Phys. Chem. C 113 pp. 19568–19577.
28. Nordhei C., Mathisen K., Bezverkhyy I. and Nicholson D. (2008), "Decomposition of Carbon Dioxide over the Putative Cubic Spinel Nanophase Cobalt, Nickel, and Zinc Ferrites," J. Phys. Chem. C 112 pp. 6531–6537.
29. Nordhei C., Ramstad A.L. and Nicholson D.G. (2008), "Nanophase cobalt, nickel and zinc ferrites: synchrotron XAS study on the crystallite size dependence of metal distribution," Phys. Chem. Chem. Phys. 10 pp. 1053–1066.
30. Kim J.-S. and Ham C.-W. (2009), "The effect of calcining temperature on the magnetic properties of the ultra-fine NiCuZn-ferrites," Mater. Res. Bull. 44 pp. 633–637.
31. Verma A., Goel T.C., Mendiratta R.G. and Alam M.I. (1999), "Dielectric properties of NiZn ferrites prepared by the citrate precursor method," Mater. Sci. Eng., B. 60 pp. 156–162.
32. Zhang C.L., Li S., Wang L.J., Wu T.H., Peng S.Y. (2000), "Studies on the decomposition of carbon dioxide into carbon with oxygen-deficient magnetite: I. Preparation, characterization of magnetite, and its activity of decomposing carbon dioxide," Mater. Chem. Phys. 62 pp. 44–51.

AUTHOR INDEX

AUTHOR INDEX

AUTHOR INDEX